Lecture Notes in Control and Information Sciences 415

Editors: M. Thoma, F. Allgöwer, M. Morari

Graziano Chesi

Domain of Attraction

Analysis and Control via SOS Programming

Springer

Author

Prof. Graziano Chesi
University of Hong Kong
Dept. Electrical & Electronic Engine
Pokfulam Road
Hong Kong
Chow Yei Ching Bldg
China, People's Republic
Telephone: 85222194362
Fax: 85225598738
Email: chesi@eee.hku.hk

ISBN 978-0-85729-958-1 e-ISBN 978-0-85729-959-8

DOI 10.1007/978-0-85729-959-8

Lecture Notes in Control and Information Sciences ISSN 0170-8643

Library of Congress Control Number: 2011935380

Typeset & Cover Design: Scientific Publishing Services Pvt. Ltd., Chennai, India.

Printed on acid-free paper

9 8 7 6 5 4 3 2 1

springer.com

To my parents, Adele and Luigi

And to my family, Shing Chee, Isabella and Sofia

That sacrificed many weekends to allow me to write this book

Preface

Stability of equilibrium points plays a fundamental role in dynamical systems. For nonlinear dynamical systems, which represent the majority of real plants, investigating stability requires characterizing the domain of attraction (DA) of an equilibrium point, i.e. the set of initial conditions from which the trajectory of the system converges to such a point. Unfortunately, it is well-known that estimating the DA, or even more attempting to control it, are very difficult problems due to the complex relationship of this set with the model of the system. This book addresses the estimation and control of the DA of equilibrium points via the newborn SOS programming, i.e. optimization techniques that have been recently developed based on polynomials that are sum of squares of polynomials (denoted hereafter as SOS polynomials) and that amount to solving convex optimization problems with linear matrix inequality (LMI) constraints, also known as semidefinite programs (SDPs). It is shown for the first time in the literature how these issues can be addressed under a unified framework in various cases depending on the nature of the considered nonlinear systems, including the cases of polynomial systems, uncertain polynomial systems, and nonlinear non-polynomial systems (possibly uncertain). The methodology proposed in this book is illustrated through various examples with fictitious and real systems, such as chemical reactors, electric circuits, mechanical devices, and social models. The book also offers a concise and simple description of the main features of SOS programming which can be used for general purpose in research and teaching, in particular introducing various classes of SOS polynomials and their characterization via LMIs, and addressing typical problems such as establishing positivity or non-positivity of polynomials and matrix polynomials, determining the minimum of rational functions, and solving systems of polynomial equations, in both cases of unconstrained variables and constrained ones.

Organization of the Book

The book is organized in two parts: Part I, which addresses SOS programming, and Part II, which addresses the estimation and control of the DA.

For Part I, Chapter 1 introduces the representation of polynomials via power vectors and via the square matrix representation (SMR), the concepts of SOS polynomials, SOS parameter-dependent polynomials and SOS matrix polynomials, their characterization via the SMR and LMIs, the concept of SOS index, the extraction of power vectors from linear subspaces, and the gap between positive polynomials and SOS polynomials.

Chapter 2 describes the use of SOS polynomials for unconstrained optimization (in particular, establishing positivity or non-positivity of a polynomial and determining the minimum of a rational function through the introduction of the generalized SOS index), constrained optimization (in particular, determining the minimum of a rational function, solving systems of polynomial equations and establishing positivity of a matrix polynomial over semialgebraic sets through the Positivstellensatz), and optimization over special sets (in particular, establishing positivity or non-positivity of a polynomial over ellipsoids and over the simplex). This chapter also addresses the search for a positive semidefinite SMR matrix of a SOS polynomial with the largest rank.

For Part II, Chapter 3 introduces some preliminaries of nonlinear systems, in particular the concepts of equilibria, stability, DA, and their robustness properties in the presence of model uncertainties.

Chapter 4 investigates the DA in polynomial systems, in particular considering the problems of establishing whether a sublevel set of a Lyapunov function (LF) is an inner estimate of the DA, estimating the largest estimate of the DA (LEDA) for such a function and verifying the tightness of the found estimate, presenting an alternative technique based on homogeneous polynomials for the case of quadratic LFs, searching for optimal estimates and establishing global asymptotical stability by considering variable LFs, and designing static output controllers for enlarging the DA.

Chapter 5 addresses the robust DA (RDA) in uncertain polynomial systems, in particular considering the problems of establishing whether a sublevel set of a common LF or a parameter-dependent sublevel set of a parameter-dependent LF is an inner estimate of the RDA, computing the largest estimate of the RDA (LERDA) for such functions according to various criteria, estimating the intersection of parameter-dependent estimates, searching for optimal estimates and establishing robust global asymptotical stability by considering variable LFs, and designing static output controllers for enlarging the RDA.

Chapter 6 addresses the case of nonlinear non-polynomial systems by introducing truncated Taylor expansions of the non-polynomial terms and by taking into account their worst-case remainders over the considered estimate. This is done in both cases of certain dynamics and uncertain ones, by considering the problems of establishing whether a sublevel set of a LF is an inner estimate of the DA or the RDA, estimating the LEDA or the LERDA for such a function, searching for optimal estimates and establishing global asymptotical stability by considering variable LFs, and designing static output controllers for enlarging the DA or the RDA.

Lastly, Chapter 7 describes the use of SOS programming in some miscellaneous problems related to the DA, in particular the estimation of the DA via union of a continuous family of estimates for nonlinear systems, the computation of outer estimates of the set of admissible equilibrium points for uncertain nonlinear systems, and the computation and the minimization via constrained control design of the extremes of the trajectories of nonlinear systems over a set of initial conditions. This chapter also provides some remarks on the estimation of the DA and the construction of global LFs for degenerate polynomial systems.

Acknowledgements

I would like to express my most sincere thanks to my Laurea and PhD supervisors, Alberto Tesi and Antonio Vicino, for having introduced me to the field of control systems and in particular to the topics addressed in this book, and for their wise, fruitful, and constant guidance in many years.

And I wish to thank Andrea Garulli, for his valuable advices since the beginning of my career, and Y. S. Hung, for his kind support and inspiring suggestions since I moved to Hong Kong.

I also want to thank various colleagues for the fruitful discussions that contributed to the composition of this book, in particular A. Aghdam, F. Blanchini, P.-A. Bliman, S. Boyd, A. Caiti, L. Chen, P. Colaneri, E. Crisostomi, Y. Ebihara, Y. Fujisaki, R. Genesio, L. Gruene, D. Henrion, H. Ichihara, J. Lam, Z. Lin, R. Nagamune, D. Nesic, Y. Oishi, A. Packard, A. Papachristodoulou, D. Peaucelle, M. Peet, L. Qiu, B. Reznick, M. Sato, C. Scheiderer, C. W. Scherer, B. Tibken, U. Topcu, L. Zhi.

Lastly, I am grateful to the Department of Electrical and Electronics Engineering of the University of Hong Kong for providing a good environment that allowed me to write this book.

Please send reports of errors to the email address `chesi@eee.hku.hk`. An up-to-date errata list will be available at the homepage `http://www.eee.hku.hk/~chesi`

This book was typeset using LaTeX. All computations were done using MATLAB and SeDuMi. The techniques presented in this book are available in the freely downloadable MATLAB toolbox SMRSOFT [27].

Hong Kong
July 2011

Graziano Chesi

Contents

Notation

Basic Quantities and Sets

$\mathbb{N}$	space of natural numbers (including zero)
$\mathbb{R}$	space of real numbers
$\mathbb{C}$	space of complex numbers
0	null matrix of size specified by the context
0_n	$n \times 1$ null vector
$0_{m \times n}$	$m \times n$ null matrix
I	identity matrix of size specified by the context
I_n	$n \times n$ identity matrix
$\mathbb{R}_0^n$	$\mathbb{R}^n \setminus \{0_n\}$
$\mathbb{S}^n$	set of $n \times n$ real symmetric matrices

Basic Functions and Operators

$\star$	lower triangular entries in symmetric matrices
$*$	generic element in a list
$\mathrm{Re}(\lambda)$	real part of $\lambda \in \mathbb{C}$
$\mathrm{Im}(\lambda)$	imaginary part of $\lambda \in \mathbb{C}$
X'	transpose of matrix X
$x > 0$	entry-wise positive vector x, i.e. $x_i > 0 \ \forall i$
$x \geq 0$	entry-wise nonnegative vector x, i.e. $x_i \geq 0 \ \forall i$
$X > 0$	symmetric positive definite matrix $X \in \mathbb{S}^n$
$X \geq 0$	symmetric positive semidefinite matrix $X \in \mathbb{S}^n$
$\lvert x \rvert$	1-norm of $x \in \mathbb{R}^n$, i.e. $\lvert x_1 \rvert + \ldots + \lvert x_n \rvert$
$\lVert x \rVert$	2-norm of $x \in \mathbb{R}^n$, i.e. $\lVert x \rVert = \sqrt{x'x}$
$\lVert x \rVert_\infty$	∞-norm of $x \in \mathbb{R}^n$, i.e. $\lVert x \rVert_\infty = \max\{\lvert x_1 \rvert, \ldots, \lvert x_n \rvert\}$
x^y	$x_1^{y_1} \cdots x_n^{y_n}$, with $x \in \mathbb{R}^n$, $y \in \mathbb{R}^n$
$\mathrm{sq}(x)$	$(x_1^2, \ldots, x_n^2)'$, with $x \in \mathbb{R}^n$
$\mathrm{sqr}(x)$	$(\sqrt{x_1}, \ldots, \sqrt{x_n})'$, with $x \in \mathbb{R}^n$

$\det(X)$	determinant of matrix X
$\text{rank}(X)$	rank of matrix X
$\text{diag}(X_1,\ldots,X_n)$	block diagonal matrix with square matrices $X_1,\ldots,X_n$ on the diagonal
$\text{img}(X)$	right image of matrix $X \in \mathbb{R}^{m\times n}$
$\ker(X)$	right null space of $X \in \mathbb{R}^{m\times n}$
$\text{spc}(X)$	set of eigenvalues of matrix X
$X \otimes Y$	Kronecker product of matrices X and Y, i.e.

$$X \otimes Y = \begin{pmatrix} x_{1,1}Y & x_{1,2}Y & \cdots \\ x_{2,1}Y & x_{2,2}Y & \cdots \\ \vdots & \vdots & \ddots \end{pmatrix}, \quad X = \begin{pmatrix} x_{1,1} & x_{1,2} & \cdots \\ x_{2,1} & x_{2,2} & \cdots \\ \vdots & \vdots & \ddots \end{pmatrix}$$

$\nabla f(x)$	vector of first derivatives of $f:\mathbb{R}^n \to \mathbb{R}$, i.e. $\nabla f(x) = (df(x)/dx_1, \ldots, df(x)/dx_n)'$
∂f	degree of polynomial $f(x)$, with $x \in \mathbb{R}^n$
$\partial^x f, \partial^y f$	degrees of polynomial $f(x,y)$ in x and y, respectively, with $x \in \mathbb{R}^n$ and $y \in \mathbb{R}^m$
$\partial\mathcal{V}$	boundary of set $\mathcal{V}$, with $\mathcal{V} \subset \mathbb{R}^n$
$\text{vol}(\mathcal{V})$	volume of set $\mathcal{V}$, with $\mathcal{V} \subset \mathbb{R}^n$

Abbreviations

BMI	Bilinear matrix inequality
DA	Domain of attraction
EVP	Eigenvalue problem
GEVP	Generalized eigenvalue problem
LEDA	Largest estimate of the domain of attraction
LERDA	Largest estimate of the robust domain of attraction
LF	Lyapunov function
LMI	Linear matrix inequality
PNS	Positive non sum of squares of polynomials
RDA	Robust domain of attraction
SDP	Semidefinite program, semidefinite programming
SMR	Square matricial representation
SOS	Sum of squares of polynomials

Part I: SOS Programming

Chapter 1
SOS Polynomials

This chapter introduces the heart of SOS programming, i.e. SOS polynomials, which are polynomials that can be expressed as sum of squares of polynomials. To this end, the chapter firstly recalls the standard representation of polynomials based on vector bases, which are denoted as power vectors. Then, it is shown how polynomials can be represented through a quadratic product in an extended space by using power vectors, which is known in the literature as Gram matrix method and SMR. SOS polynomials are hence introduced, showing in particular that a necessary and sufficient condition for a polynomial to be SOS is that the polynomial admits a positive semidefinite SMR matrix, which is an LMI feasibility test. This result is successively reformulated by introducing the SOS index, which measures how SOS is a polynomial and which can be found by solving an SDP. The chapter proceeds by introducing the representation of parameter-dependent polynomials and matrix polynomials via power vectors and the SMR, the concepts of SOS parameter-dependent polynomials and SOS matrix polynomials, and their characterization via the SMR and LMIs. Then, the problem of extracting power vectors from linear subspaces is addressed, which will play a key role in establishing optimality in SOS programming, and for which a simple solution based on computing the roots of a univariate polynomial is presented. Lastly, the gap between positive polynomials and SOS polynomials is briefly discussed, recalling classical and recent results. The studies above mentioned are presented for the case of general polynomials and, where appropriate, for the special cases of locally quadratic polynomials (i.e., polynomials without constant and linear monomials) and homogeneous polynomials (i.e., polynomials with all monomials of the same degree), which will be exploited throughout the book.

1.1 Polynomials and Power Vectors

This section introduces polynomials, locally quadratic polynomials and homogeneous polynomials, and their representation via power vectors.

G. Chesi: Domain of Attraction, LNCIS 415, pp. 3–44, 2011.
springerlink.com

1.1.1 General Case

Definition 1.1 (Polynomial). The function $f:\mathbb{R}^n \to \mathbb{R}$ is a *polynomial* of degree not greater than $d \in \mathbb{N}$ if

$$f(x) = \sum_{q\in\mathbb{N}^n,\ |q|\le d} a_q x^q \tag{1.1}$$

where $x \in \mathbb{R}^n$ and $a_q \in \mathbb{R}$. □

In (1.1), $a_q x^q$ is a *monomial* of degree $|q|$, and a_q is its coefficient. We denote the set of polynomials in an n-dimensional real variable as

$$\mathscr{P}_n = \{f:\ (1.1) \text{ holds for some } d \in \mathbb{N}\}. \tag{1.2}$$

Vector and matrices whose entries are polynomials are called *vector polynomials* and *matrix polynomials*, respectively. In particular, with the notation $\mathscr{P}_n^m$ and $\mathscr{P}_n^{m\times l}$ we denote the set of $m\times 1$ vectors and $m\times l$ matrices with entries in $\mathscr{P}_n$, respectively.

The degree of $f\in\mathscr{P}_n$, denoted by ∂f, is the maximum degree of the monomials of $f(x)$, i.e. by expressing $f(x)$ as in (1.1) one has that

$$\partial f = \max_{q\in\mathbb{N}^n,\ |q|\le d,\ a_q\neq 0} |q|. \tag{1.3}$$

The degrees of $g\in\mathscr{P}_n^m$ and $G\in\mathscr{P}_n^{m\times l}$, denoted by ∂g and ∂G, respectively, are the maximum of the degrees of their entries, i.e.

$$\begin{aligned} \partial g &= \max_{i=1,\dots,m} \partial g_i \\ \partial G &= \max_{i=1,\dots,m;j=1,\dots,l} \partial G_{i,j}. \end{aligned} \tag{1.4}$$

The maximum number of distinct monomials in a polynomial in $\mathscr{P}_n$ of degree d is given by

$$d_{pol}(n,d) = \frac{(n+d)!}{n!d!}. \tag{1.5}$$

Polynomials can be represented by vectors which contain their coefficients with respect to an appropriate base. The following definition introduces such bases.

Definition 1.2 (Power Vector). Let $b_{pol}(x,d)$ be a vector such that, for any $f\in\mathscr{P}_n$ with $\partial f\le d$, there exists $\mathbf{f}\in\mathbb{R}^{d_{pol}(n,d)}$ satisfying

$$f(x) = \mathbf{f}' b_{pol}(x,d). \tag{1.6}$$

Then, $b_{pol}(x,d)$ is called *power vector* for polynomials in $\mathscr{P}_n$ of degree not greater than d. □

Special choices for $b_{pol}(x,d)$ are those where each entry is a monomial. Among these, a typical choice is defined by the recursive rule

$$b_{pol}(x,d+1) = \begin{pmatrix} b_{pol}(x,d) \\ x_1 y_1(x,d) \\ x_2 y_2(x,d) \\ \vdots \\ x_n y_n(x,d) \end{pmatrix} \tag{1.7}$$

where $y_i(x,d)$, $i = 1,\dots,n$, is the lower subvector of $b_{pol}(x,d)$ starting with the monomial x_i^d, i.e.

$$b_{pol}(x,d) = \begin{pmatrix} 1 \\ x_1 \\ \vdots \\ x_n^{d-1} \\ y_1(x,d) \end{pmatrix} = \begin{pmatrix} 1 \\ x_1 \\ \vdots \\ x_1 x_n^{d-1} \\ y_2(x,d) \end{pmatrix} = \dots = \begin{pmatrix} 1 \\ x_1 \\ \vdots \\ x_{n-1} x_n^{d-1} \\ y_n(x,d) \end{pmatrix} \tag{1.8}$$

and the initialization is given by

$$\begin{cases} b_{pol}(x,0) = 1 \\ y_i(x,0) = 1 \;\; \forall i = 1,\dots,n. \end{cases} \tag{1.9}$$

For instance, with $n = 3$ and $d = 0,\dots,2$ one has

$$\begin{aligned} b_{pol}(x,0) &= 1 \\ b_{pol}(x,1) &= (1,x_1,x_2,x_3)' \\ b_{pol}(x,2) &= (1,x_1,x_2,x_3,x_1^2,x_1x_2,x_1x_3,x_2^2,x_2x_3,x_3^2)'. \end{aligned}$$

In the sequel we will assume that $b_{pol}(x,d)$ is defined according to (1.7) unless specified otherwise. Moreover, we will denote with the function $COE_{pol}(f)$ the vector $\mathbf{f}$ satisfying (1.6), i.e.

$$\begin{cases} \mathbf{f} = COE_{pol}(f) \\ \text{(1.6) holds.} \end{cases} \tag{1.10}$$

Example 1.1. Consider $f \in \mathcal{P}_1$ given by

$$f(x) = 4 - 3x + x^4. \tag{1.11}$$

Since $\partial f = 4$, $f(x)$ can be expressed according to (1.6) with $d = 4$ and

$$\begin{aligned} \mathbf{f} &= (4,-3,0,0,1)' \\ b_{pol}(x,d) &= \left(1,x,x^2,x^3,x^4\right)'. \end{aligned}$$

□

1.1.2 Special Cases

For several studies in control systems, one has often to consider polynomials that vanishes in the origin, i.e. such that $f(0_n) = 0$. In this case, a more appropriate representation consists of adopting a power vector that does not contain constant terms as explained in the following definition.

Definition 1.3 (Power Vector (continued)). Let $b_{lin}(x,d)$ be a vector such that, for any $f \in \mathscr{P}_n$ with $\partial f \leq d$ and $f(0_n) = 0$, there exists $\mathbf{f} \in \mathbb{R}^{d_{lin}(n,d)}$ satisfying

$$f(x) = \mathbf{f}' b_{lin}(x,d) \tag{1.12}$$

where

$$d_{lin}(n,d) = \frac{(n+d)!}{n!d!} - 1. \tag{1.13}$$

Then, $b_{lin}(x,d)$ is called *power vector* for polynomials $f \in \mathscr{P}_n$ with $\partial f \leq d$ and $f(0_n) = 0$. □

Similarly to $b_{pol}(x,d)$, $b_{lin}(x,d)$ admits different choices. In the sequel we will assume that $b_{lin}(x,d)$ is defined as $b_{pol}(x,d)$ in (1.7) by removing its first entry (i.e., the entry "1") unless specified otherwise. Moreover, we will denote with the function $COE_{lin}(f)$ the vector $\mathbf{f}$ satisfying (1.12), i.e.

$$\begin{cases} \mathbf{f} = COE_{lin}(f) \\ (1.12) \text{ holds.} \end{cases} \tag{1.14}$$

Polynomials that are locally definite, either positive or negative, represent another case of interest. These polynomials necessarily do not have constant and linear terms, and belong to the class of polynomials introduced in the following definition.

Definition 1.4 (Locally Quadratic Polynomial). Let $f \in \mathscr{P}_n$ satisfy

$$\begin{cases} f(0_n) = 0 \\ \nabla f(0_n) = 0_n. \end{cases} \tag{1.15}$$

Then, $f(x)$ is called *locally quadratic polynomial.* □

Locally quadratic polynomials can be represented by adopting a more compact power vector as explained hereafter.

Definition 1.5 (Power Vector (continued)). Let $b_{qua}(x,d)$ be a vector such that, for any locally quadratic polynomial $f \in \mathscr{P}_n$ with $\partial f \leq d$, there exists $\mathbf{f} \in \mathbb{R}^{d_{qua}(n,d)}$ satisfying

$$f(x) = \mathbf{f}' b_{qua}(x,d) \tag{1.16}$$

where

$$d_{qua}(n,d) = \frac{(n+d)!}{n!d!} - 1 - n. \tag{1.17}$$

Then, $b_{qua}(x,d)$ is called *power vector* for locally quadratic polynomials $f \in \mathscr{P}_n$ with $\partial f \leq d$. □

In the sequel we will assume that $b_{qua}(x,d)$ is defined as $b_{pol}(x,d)$ in (1.7) by removing its first $n+1$ entries (which are monomials of degree 0 and 1) unless specified otherwise. Moreover, we will denote with the function $COE_{qua}(f)$ the vector $\mathbf{f}$ satisfying (1.16), i.e.

$$\begin{cases} \mathbf{f} = COE_{qua}(f) \\ (1.16) \text{ holds.} \end{cases} \tag{1.18}$$

Another important class of polynomials is represented by the polynomials with monomials of the same degree, which are defined as follows.

Definition 1.6 (Homogeneous Polynomial (or Form)). Let $f \in \mathscr{P}_n$ with $\partial f = d$ satisfy

$$\left.\frac{d^{|q|} f(x)}{dx^q}\right|_{x=0_n} = 0 \ \forall q \in \mathbb{N}^n, \ |q| < d. \tag{1.19}$$

Then, $f(x)$ is called *homogeneous polynomial (or form)*. □

Definition 1.7 (Power Vector (continued)). Let $b_{hom}(x,d)$ be a vector such that, for any homogeneous polynomial $f \in \mathscr{P}_n$ with $\partial f = d$, there exists $\mathbf{f} \in \mathbb{R}^{d_{hom}(n,d)}$ satisfying

$$f(x) = \mathbf{f}' b_{hom}(x,d) \tag{1.20}$$

where

$$d_{hom}(n,d) = \frac{(n+d-1)!}{(n-1)!d!}. \tag{1.21}$$

Then, $b_{hom}(x,d)$ is called *power vector* for homogeneous polynomials $f \in \mathscr{P}_n$ with $\partial f = d$. □

In the sequel we will assume that $b_{hom}(x,d)$ is defined as $b_{pol}(x,d)$ in (1.7) by removing its first $d_{pol}(n,d-1)$ entries (which are monomials of degree smaller than d) unless specified otherwise. Moreover, we will denote with the function $COE_{hom}(f)$ the vector $\mathbf{f}$ satisfying (1.20), i.e.

$$\begin{cases} \mathbf{f} = COE_{hom}(f) \\ (1.20) \text{ holds.} \end{cases} \tag{1.22}$$

Example 1.2. Consider $f \in \mathscr{P}_2$ given by

$$f(x) = x_1^2 + 2x_2^2 - 4x_1^2 x_2. \tag{1.23}$$

Since $f(x)$ is a locally quadratic polynomial, we can write $f(x)$ according to (1.16) with

$$\begin{aligned} \mathbf{f} &= (1,0,2,0,-4,0,0)' \\ b_{qua}(x,d) &= \left(x_1^2, x_1x_2, x^2, x_1^3, x_1^2x_2, x_1x_2^2, x_2^3\right)'. \end{aligned}$$ □

Example 1.3. Consider $f \in \mathscr{P}_2$

$$f(x) = x_1^6 - 2x_1^5x_2 + 4x_1^3x_2^3 + x_2^6. \tag{1.24}$$

Since $f(x)$ is a homogeneous polynomial, we can write $f(x)$ according to (1.20) with

$$\begin{aligned} \mathbf{f} &= (1,-2,0,4,0,0,1)' \\ b_{hom}(x,d) &= \left(x_1^6, x_1^5x_2, x_1^4x_2^2, x_1^3x_2^3, x_1^2x_2^4, x_1x_2^5, x_2^6\right)'. \end{aligned}$$ □

1.2 SMR

This section introduces the representation of polynomials, locally quadratic polynomials and homogeneous polynomials via the SMR.

1.2.1 General Case

Definition 1.8 (SMR). Consider $f \in \mathscr{P}_n$ with $\partial f \leq 2m$, $m \in \mathbb{N}$, and let F be a symmetric matrix such that

$$f(x) = b_{pol}(x,m)'F\,b_{pol}(x,m). \tag{1.25}$$

Then, (1.25) is said an *SMR* of $f(x)$ with respect to $b_{pol}(x,m)$. □

In (1.25), the matrix F is said an *SMR matrix* of $f(x)$ with respect to $b_{pol}(x,m)$.

Theorem 1.1. *Consider $f \in \mathscr{P}_n$ with $\partial f \leq 2m$, $m \in \mathbb{N}$. Then, for any power vector $b_{pol}(x,m)$ there exists a symmetric matrix F such that* (1.25) *holds.*

Proof. Let us denote with $g_i(x)$, $i = 1,\ldots,d_{pol}(n,2m)$, the i-th monomial of $f(x)$. We have that $g_i(x)$ can be written as

$$g_i(x) = g_{i1}(x)g_{i2}(x)$$

where $g_{i1}(x)$ and $g_{i2}(x)$ are monomials with $\partial g_{ij} \leq m$ for $j = 1,2$. For definition of power vector, $g_{i1}(x)$ and $g_{i2}(x)$ can be written as

$$g_{ij}(x) = \mathbf{g}_{ij}'b_{pol}(x,m).$$

Thus, it follows that

$$\begin{aligned} g_i(x) &= g_{i1}(x)g_{i2}(x) \\ &= \mathbf{g}_{i1}'b_{pol}(x,m)\mathbf{g}_{i2}'b_{pol}(x,m) \\ &= b_{pol}(x,m)'\tilde{G}_i b_{pol}(x,m) \end{aligned}$$

where

$$\tilde{G}_i = \mathbf{g}_{i1}\mathbf{g}_{i2}'.$$

Let us observe that

$$g_i(x) = b_{pol}(x,m)'G_i b_{pol}(x,m)$$

where

$$G_i = \frac{\tilde{G}_i + \tilde{G}_i'}{2}.$$

Therefore, F in (1.25) can be chosen as

$$F = \sum_{i=1}^{d_{pol}(n,2m)} G_i,$$

i.e. the theorem holds. □

Let us consider the problem of characterizing the set of symmetric matrices F satisfying (1.25). We define such a set as

$$\mathscr{F} = \left\{F = F' :\ (1.25) \text{ holds}\right\}. \tag{1.26}$$

The following result states an important property of $\mathscr{F}$.

Theorem 1.2. *The set $\mathscr{F}$ in* (1.26) *is an affine subspace, i.e.*

$$y_1 F_1 + y_2 F_2 \in \mathscr{F}\ \ \forall F_1, F_2 \in \mathscr{F}\ \forall y_1, y_2 \in \mathbb{R},\ y_1 + y_2 = 1. \tag{1.27}$$

Proof. Let F_1 and F_2 be elements of $\mathscr{F}$. Then, for all $y_1, y_2 \in \mathbb{R}$ with $y_1 + y_2 = 1$ it follows that

$$\begin{aligned} & b_{pol}(x,m)'\left(y_1 F_1 + y_2 F_2\right) b_{pol}(x,m) \\ &= \textstyle\sum_{i=1}^{2} y_i b_{pol}(x,m)' F_i b_{pol}(x,m) \\ &= \textstyle\sum_{i=1}^{2} y_i f(x) \\ &= f(x) \end{aligned}$$

i.e. $y_1 F_1 + y_2 F_2 \in \mathscr{F}$. □

Since $\mathscr{F}$ is an affine space, its elements can be expressed as the sum between any element of $\mathscr{F}$ and an element in a linear subspace. Indeed, one has that

$$\mathscr{F} = \left\{F + L,\ L \in \mathscr{L}_{pol}(n,2m)\right\} \tag{1.28}$$

where F is any symmetric matrix such that (1.25) holds, and $\mathscr{L}_{pol}(n,2m)$ is the linear subspace

$$\mathscr{L}_{pol}(n,2m) = \left\{L = L' :\ b_{pol}(x,m)' L b_{pol}(x,m) = 0\ \ \forall x \in \mathbb{R}^n\right\}. \tag{1.29}$$

The following result characterizes the dimension of $\mathscr{L}_{pol}(n,2m)$.

Theorem 1.3. *The dimension of* $\mathscr{L}_{pol}(n,2m)$ *is given by*

$$l_{pol}(n,2m) = t_{pol}(n,2m) - d_{pol}(n,2m) \tag{1.30}$$

where

$$t_{pol}(n,2m) = \frac{1}{2} d_{pol}(n,m)\left(1 + d_{pol}(n,m)\right). \tag{1.31}$$

Proof. Let us define

$$a = \frac{1}{2} d_{pol}(n,m)\left(1 + d_{pol}(n,m)\right)$$

and let $l \in \mathbb{R}^a$ be a vector containing the free entries of a matrix $L \in \mathbb{S}^{d_{pol}(n,m)}$. Let $M : \mathbb{R}^a \to \mathbb{S}^{d_{pol}(n,m)}$ be the linear map from l to L, i.e.

$$M(l) = L.$$

We can write

$$b_{pol}(x,m)' M(l) b_{pol}(x,m) = (Hl)' b_{pol}(x,2m)$$

where $H \in \mathbb{R}^{d_{pol}(n,2m) \times a}$ is a suitable matrix. It follows that

$$\mathscr{L}_{pol}(n,2m) = \{M(l) : \ l \in \ker(H)\}.$$

This implies that

$$\begin{aligned} \dim\left(\mathscr{L}_{pol}(n,2m)\right) &= \dim\left(\ker(H)\right) \\ &= a - \operatorname{rank}(H). \end{aligned}$$

Now, let us prove that

$$\operatorname{rank}(H) = d_{pol}(n,2m).$$

In fact, let us suppose by contradiction that $\operatorname{rank}(H) \neq d_{pol}(n,2m)$. Since $d_{pol}(n,2m) \leq a$, this implies that $\operatorname{rank}(H) < d_{pol}(n,2m)$. Consequently, there exists $f \in \mathscr{P}_n$ with $\partial f \leq 2m$ such that

$$(Hl)' b_{pol}(x,2m) \neq f(x) \ \ \forall l \in \mathbb{R}^a$$

or, in other words

$$b_{pol}(x,m)' L b_{pol}(x,m) \neq f(x) \ \ \forall L \in \mathbb{S}^{d_{pol}(n,m)}$$

which contradicts Theorem 1.1. Therefore, $\operatorname{rank}(H) = d_{pol}(n,2m)$, and hence the theorem holds. □

Table 1.1 shows the dimension $l_{pol}(n,2m)$ of $\mathscr{L}_{pol}(n,2m)$ for some values of n and m. Let $L(\alpha)$ be a linear parametrization of $\mathscr{L}_{pol}(n,2m)$, i.e.

$$\begin{cases} \forall L \in \mathscr{L}_{pol}(n,2m) \ \exists \alpha \in \mathbb{R}^{l_{pol}(n,2m)} : \ L = L(\alpha) \\ L(\alpha) \in \mathscr{L}_{pol}(n,2m) \ \forall \alpha \in \mathbb{R}^{l_{pol}(n,2m)}. \end{cases} \tag{1.32}$$

The set $\mathscr{F}$ in (1.26) can be rewritten as

$$\mathscr{F} = \left\{F + L(\alpha),\ \alpha \in \mathbb{R}^{l_{pol}(n,2m)}\right\}. \tag{1.33}$$

This allows us to derive the following representation of polynomials.

Definition 1.9 (Complete SMR). Consider $f \in \mathscr{P}_n$ with $\partial f \leq 2m$, $m \in \mathbb{N}$. Let F be a symmetric matrix such that (1.25) holds, and $L(\alpha)$ be a linear parametrization of $\mathscr{L}_{pol}(n,2m)$. Then, for all $\alpha \in \mathbb{R}^{l_{pol}(n,2m)}$ one has that

$$f(x) = b_{pol}(x,m)' \left(F + L(\alpha)\right) b_{pol}(x,m). \tag{1.34}$$

The expression in (1.34) is called a *complete SMR* of $f(x)$ with respect to $b_{pol}(x,m)$. □

In (1.34), the matrix $F + L(\alpha)$ is said a *complete SMR matrix* of $f(x)$ with respect to $b_{pol}(x,m)$. Algorithms for the construction of the matrices F and $L(\alpha)$ in (1.34) are reported in Appendix C.

It is worthwhile to observe that the complete SMR matrix $F + L(\alpha)$ is not unique since the choices for F and $L(\alpha)$ are not unique. Nevertheless, for any of these choices, $F + L(\alpha)$ parametrizes the set of SMR matrices $\mathscr{F}$ of $f(x)$, i.e. (1.33) holds. Also, it is useful to observe that the SMR matrices of $f(x)$ depend on the chosen power vector $b_{pol}(x,m)$.

In the sequel we will denote with the functions $SMR_{pol}(f)$ and $CSMR_{pol}(f)$ an SMR matrix F and a complete SMR matrix $F + L(\alpha)$ of $f(x)$ with respect to $b_{pol}(x,m)$, i.e.

$$\begin{cases} F = SMR_{pol}(f) \\ \text{(1.25) holds} \end{cases} \tag{1.35}$$

and

$$\begin{cases} F + L(\alpha) = CSMR_{pol}(f) \\ \text{(1.34) holds} \end{cases} \tag{1.36}$$

where m is the smallest integer for which the SMR matrices can be built, i.e. $m = \lceil \partial f/2 \rceil$.

Table 1.1 Dimension $l_{pol}(n,2m)$ of $\mathscr{L}_{pol}(n,2m)$ for some values of n and m.

$n \setminus m$	1	2	3	4	5
1	0	1	3	6	10
2	0	6	27	75	165
3	0	20	126	465	1310
4	0	50	420	1990	7000
5	0	105	1134	6714	28875

Example 1.4. Consider $f(x) = 4 - 3x + x^4$ in (1.11). We can write $f(x)$ according to (1.34) with $m = 2$ and

$$b_{pol}(x,m) = \begin{pmatrix} 1 \\ x \\ x^2 \end{pmatrix}, \quad F = \begin{pmatrix} 4 & -1.5 & 0 \\ \star & 0 & 0 \\ \star & \star & 1 \end{pmatrix}, \quad L(\alpha) = \begin{pmatrix} 0 & 0 & -\alpha \\ \star & 2\alpha & 0 \\ \star & \star & 0 \end{pmatrix}.$$

□

1.2.2 *Special Cases*

Locally quadratic polynomials admit a more compact SMR, which consists of adopting the power vector $b_{lin}(x,m)$ instead of $b_{pol}(x,m)$. To this end, let us introduce the set

$$\mathscr{L}_{qua}(n,2m) = \left\{ L \in \mathbb{S}^{d_{lin}(n,m)} : \ b_{lin}(x,m)' L b_{lin}(x,m) = 0 \ \ \forall x \in \mathbb{R}^n \right\} \tag{1.37}$$

whose dimension (following the same reasoning of Theorem 1.3) is given by

$$l_{qua}(n,2m) = t_{qua}(n,2m) - d_{lin}(n,2m) + n \tag{1.38}$$

where

$$t_{qua}(n,2m) = \frac{1}{2} d_{lin}(n,m)\left(1 + d_{lin}(n,m)\right). \tag{1.39}$$

Definition 1.10 (Complete SMR (continued)). Let $f \in \mathscr{P}_n$ be a locally quadratic polynomial with $\partial f \leq 2m$, $m \in \mathbb{N}$. Let F be a symmetric matrix such that

$$f(x) = b_{lin}(x,m)' F b_{lin}(x,m) \tag{1.40}$$

and let $L(\alpha)$ be a linear parametrization of $\mathscr{L}_{qua}(n,2m)$. Then, for all $\alpha \in \mathbb{R}^{l_{quapol}(n,2m)}$ one has that

$$f(x) = b_{lin}(x,m)' \left(F + L(\alpha)\right) b_{lin}(x,m). \tag{1.41}$$

The expression in (1.41) is called a *complete SMR* of $f(x)$ with respect to $b_{lin}(x,m)$.

□

In the sequel we will denote with the functions $SMR_{qua}(f,m)$ and $CSMR_{qua}(f,m)$ an SMR matrix F and a complete SMR matrix $F + L(\alpha)$ of $f(x)$ with respect to $b_{lin}(x,m)$, i.e.

$$\begin{cases} F = SMR_{qua}(f) \\ \text{(1.15) and (1.40) hold} \end{cases} \tag{1.42}$$

and

$$\begin{cases} F + L(\alpha) = CSMR_{qua}(f) \\ \text{(1.15) and (1.41) hold} \end{cases} \tag{1.43}$$

where $m = \lceil \partial f / 2 \rceil$.

Lastly, a further more compact SMR can be introduced in order to represent homogeneous polynomials (of even degree) by adopting the power vector $b_{hom}(x,m)$ instead of $b_{pol}(x,m)$ or $b_{lin}(x,m)$. To this end, let us introduce the set

$$\mathscr{L}_{hom}(n,2m) = \left\{ L \in \mathbb{S}^{d_{hom}(n,m)} : \; b_{hom}(x,m)' L b_{hom}(x,m) = 0 \;\; \forall x \in \mathbb{R}^n \right\} \tag{1.44}$$

whose dimension (following the same reasoning of Theorem 1.3) is given by

$$l_{hom}(n,2m) = t_{hom}(n,2m) - d_{hom}(n,2m) \tag{1.45}$$

where

$$t_{hom}(n,2m) = \frac{1}{2} d_{hom}(n,m)\,(1 + d_{hom}(n,m)). \tag{1.46}$$

Definition 1.11 (Complete SMR (continued)). Let $f \in \mathscr{P}_n$ be a homogeneous polynomial with $\partial f = 2m$, $m \in \mathbb{N}$. Let F be a symmetric matrix such that

$$f(x) = b_{hom}(x,m)' F\, b_{hom}(x,m) \tag{1.47}$$

and let $L(\alpha)$ be a linear parametrization of $\mathscr{L}_{hom}(n,2m)$. Then, for all $\alpha \in \mathbb{R}^{l_{hom}(n,2m)}$ one has that

$$f(x) = b_{hom}(x,m)' \left(F + L(\alpha) \right) b_{hom}(x,m). \tag{1.48}$$

The expression in (1.48) is called a *complete SMR* of $f(x)$ with respect to $b_{hom}(x,m)$. □

In the sequel we will denote with the functions $SMR_{hom}(f)$ and $CSMR_{hom}(f)$ an SMR matrix F and a complete SMR matrix $F + L(\alpha)$ of $f(x)$ with respect to $b_{hom}(x,m)$, i.e.

$$\begin{cases} F = SMR_{hom}(f) \\ \text{(1.47) holds} \end{cases} \tag{1.49}$$

and

$$\begin{cases} F + L(\alpha) = CSMR_{hom}(f) \\ \text{(1.41) hold} \end{cases} \tag{1.50}$$

where $m = \partial f / 2$.

Example 1.5. Consider the locally quadratic polynomial $f(x) = x_1^2 + 2x_2^2 - 4x_1^2 x_2$ in (1.23). We can write $f(x)$ according to (1.41) with $m = 2$ and

$$b_{lin}(x,m) = \begin{pmatrix} x_1 \\ x_2 \\ x_1^2 \\ x_1 x_2 \\ x_2^2 \end{pmatrix}, \quad F = \begin{pmatrix} 1 & 0 & 0 & 0 & 0 \\ \star & 2 & -2 & 0 & 0 \\ \star & \star & 0 & 0 & 0 \\ \star & \star & \star & 0 & 0 \\ \star & \star & \star & \star & 0 \end{pmatrix}, \quad L(\alpha) = \begin{pmatrix} 0 & 0 & 0 & -\alpha_1 & -\alpha_2 \\ \star & 0 & \alpha_1 & \alpha_2 & 0 \\ \star & \star & 0 & 0 & -\alpha_3 \\ \star & \star & \star & 2\alpha_3 & 0 \\ \star & \star & \star & \star & 0 \end{pmatrix}.$$ □

Example 1.6. Consider the homogeneous polynomial $f(x) = x_1^6 - 2x_1^5 x_2 + 4x_1^3 x_2^3 + x_2^6$ in (1.24). We can write $f(x)$ according to (1.48) with $m = 3$ and

$$b_{hom}(x,m) = \begin{pmatrix} x_1^3 \\ x_1^2 x_2 \\ x_1 x_2^2 \\ x_2^3 \end{pmatrix}, \quad F = \begin{pmatrix} 1 & -1 & 0 & 2 \\ \star & 0 & 0 & 0 \\ \star & \star & 0 & 0 \\ \star & \star & \star & 1 \end{pmatrix}, \quad L(\alpha) = \begin{pmatrix} 0 & 0 & -\alpha_1 & -\alpha_2 \\ \star & 2\alpha_1 & \alpha_2 & -\alpha_3 \\ \star & \star & 2\alpha_3 & 0 \\ \star & \star & \star & 0 \end{pmatrix}.$$

□

1.3 SOS Polynomials

This section introduces the concepts of positivity and SOS for polynomials, locally quadratic polynomials and homogeneous polynomials, providing a characterization via the SMR and LMIs.

1.3.1 General Case

Definition 1.12 (Nonnegative/Positive Polynomial). Consider $f \in \mathscr{P}_n$. Then, $f(x)$ is *nonnegative* if $f(x) \geq 0$ for all $x \in \mathbb{R}^n$. Also, $f(x)$ is *positive* if $f(x) > 0$ for all $x \in \mathbb{R}^n$. □

As it will become clear in this section, nonnegative polynomials and positive polynomials can be studied with SOS polynomials, which are defined as follows.

Definition 1.13 (SOS Polynomial). Consider $f \in \mathscr{P}_n$. Then, $f(x)$ is called *SOS polynomial* if there exist $k \in \mathbb{N}$ and $g \in \mathscr{P}_n^k$ such that

$$f(x) = \sum_{i=1}^{k} g_i(x)^2. \tag{1.51}$$

□

We denote the set of SOS polynomials in $\mathscr{P}_n$ as

$$\mathscr{S}_n = \left\{ f \in \mathscr{P}_n : \text{ (1.51) holds for some } k \in \mathbb{N} \text{ and } g \in \mathscr{P}_n^k \right\}. \tag{1.52}$$

The SMR allows one to establish whether a polynomial is SOS through an LMI feasibility test. This is clarified in the following result.

Theorem 1.4. *Consider $f \in \mathscr{P}_n$, and define $F + L(\alpha) = CSMR_{pol}(f)$. Then, $f(x)$ is SOS if and only if there exists α satisfying the LMI*

$$F + L(\alpha) \geq 0. \tag{1.53}$$

Proof. "⇒" Suppose that $f(x)$ is SOS. Then, there exist $g \in \mathscr{P}_n^k$ such that (1.51) holds. Define $2m = \partial f$, and let us express the i-th entry of $g(x)$ according to (1.6) as

$$g_i(x) = \mathbf{g}_i' b_{pol}(x,m).$$

It follows that

$$\begin{aligned} f(x) &= \textstyle\sum_{i=1}^k g_i(x)^2 \\ &= \textstyle\sum_{i=1}^k \left(\mathbf{g}_i' b_{pol}(x,m)\right)^2 \\ &= \textstyle\sum_{i=1}^k b_{pol}(x,m)' \mathbf{g}_i \mathbf{g}_i' b_{pol}(x,m) \\ &= b_{pol}(x,m)' \bar{F} b_{pol}(x,m) \end{aligned}$$

where

$$\begin{aligned} \bar{F} &= \textstyle\sum_{i=1}^k F_i \\ F_i &= \mathbf{g}_i \mathbf{g}_i'. \end{aligned}$$

Since $F_i \geq 0$ for all $i = 1, \ldots, k$, it follows that $\bar{F} \geq 0$. Moreover, since $F + L(\alpha)$ parametrizes all the SMR matrices of $f(x)$, there exists α such that

$$F + L(\alpha) = \bar{F}.$$

"⇐" Suppose that there exists α such that (1.53) holds. By exploiting the Cholesky decomposition, one can write

$$F + L(\alpha) = H'H$$

for some $H \in \mathbb{R}^{d_{pol}(n,m) \times d_{pol}(n,m)}$. Hence, it follows that

$$\begin{aligned} f(x) &= b_{pol}(x,m)' \left(F + L(\alpha)\right) b_{pol}(x,m) \\ &= b_{pol}(x,m)' H' H b_{pol}(x,m) \\ &= \left(H b_{pol}(x,m)\right)^2 \\ &= \textstyle\sum_{i=1}^{d_{pol}(n,m)} g_i(x)^2 \end{aligned}$$

where

$$g_i(x) = H_i b_{pol}(x,m)$$

and H_i is the i-th row of H. □

Solving an LMI feasibility test amounts to solving a convex optimization problem whose solution can be systematically found, see Appendix A.

The following result characterizes the number of polynomials $g_i(x)$ that are needed to express $f(x)$ as in (1.51).

Theorem 1.5. *Consider $f \in \mathscr{P}_n$ with $\partial f \leq 2m$, $m \in \mathbb{N}$, and define $F + L(\alpha) = CSMR_{pol}(f)$. If $f(x)$ is SOS, then $f(x)$ can be written as in* (1.51) *with*

$$\begin{aligned} k &\leq d_{pol}(n,m) \\ g(x) &= G b_{pol}(x,m) \end{aligned} \tag{1.54}$$

where $G \in \mathbb{R}^{k \times d_{pol}(n,m)}$ *is any matrix such that*

$$G'G = F + L(\alpha) \tag{1.55}$$

and α *is any vector satisfying* (1.53).

Proof. Suppose that $f(x)$ is SOS. From Theorem 1.4 there exists α such that (1.53) holds. For such an α, one can write $F + L(\alpha) = H'H$ according to the Cholesky decomposition for some $H \in \mathbb{R}^{d_{pol}(n,m) \times d_{pol}(n,m)}$. Let $G \in \mathbb{R}^{k \times d_{pol}(n,m)}$ be the submatrix of H containing only the rows of H that are not null. It follows that $k \leq d_{pol}(n,m)$ and that $G'G = F + L(\alpha)$. Proceeding as in the second part of the proof of Theorem 1.4 with H replaced by G, we conclude that (1.51) holds with k and $g(x)$ as in (1.54). □

In order to further characterize SOS polynomials, we introduce the following index.

Definition 1.14 (SOS Index). Consider $f \in \mathscr{P}_n$, and define $\mathbf{f} = COE_{pol}(f)$ and $F + L(\alpha) = CSMR_{pol}(f)$. Let us introduce

$$\lambda_{pol}(f) = \begin{cases} \beta z^* & \text{if } \|\mathbf{f}\| > 0 \\ 0 & \text{otherwise} \end{cases} \tag{1.56}$$

where β is a normalizing factor chosen as

$$\beta = \frac{\sqrt{d_{pol}(n,m)}}{\|\mathbf{f}\|}, \quad m = \left\lceil \frac{\partial f}{2} \right\rceil \tag{1.57}$$

and z^* is the solution of the SDP

$$\begin{aligned} z^* = & \max_{z,\alpha} \ z \\ & \text{s.t.} \ \ F + L(\alpha) - zI \geq 0. \end{aligned} \tag{1.58}$$

Then, $\lambda_{pol}(f)$ is called *SOS index* of $f(x)$. □

As it can be observed from its definition and as it will become clearer in the sequel, the SOS index is a measure of how SOS is a polynomial. Computing the SOS index amounts to solving the SDP (1.58), which belongs to the class of convex optimization problems. See Appendix A for details about SDPs. Let us observe that β is introduced in (1.56) in order to normalize the SOS index. In particular:

1. $\lambda_{pol}(f) = \lambda_{pol}(\rho f)$ for all $\rho > 0$;
2. $\lambda_{pol}(f) = 1$ for $f(x) = \|b_{pol}(x,i)\|^2$ for all $i \in \mathbb{N}$.

The following result states that the SOS index always exists.

Theorem 1.6. *The SOS index* $\lambda_{pol}(f)$ *exists (and, hence, is bounded) for all* $f \in \mathscr{P}_n$.

Proof. First, let us observe that for any F there exists z and α such that

$$F + L(\alpha) - zI \geq 0.$$

This means that the feasible set of the SDP (1.58) is nonempty. Second, let us observe that for any F there exists z such that

$$F + L(\alpha) - zI \not\geq 0 \ \forall \alpha.$$

In fact, if one suppose for contradiction that this is not true, it follows that for all z there exists α such that

$$\begin{aligned} 0 &\leq b_{pol}(x,m)' \left(F + L(\alpha) - zI\right) b_{pol}(x,m) \\ &= f(x) - z\|b_{pol}(x,m)\| \end{aligned}$$

which is impossible since $\|b_{pol}(x,m)\|$ is positive definite and has the same degree of $f(x)$. This means that z^* cannot be unbounded. Lastly, let us observe that the feasible set of the SDP (1.58) is closed, i.e. z^* exists. □

The SOS index can be directly used to establish whether a polynomial is SOS as explained in the following corollary.

Corollary 1.1. *Consider $f \in \mathscr{P}_n$. Then, $f(x)$ is SOS if and only if*

$$\lambda_{pol}(f) \geq 0. \tag{1.59}$$

Proof. If $\|\mathbf{f}\| > 0$, then $\lambda_{pol}(f) \geq 0$ if and only if $z^* \geq 0$ where z^* is the solution of (1.58), and the result follows directly from Theorem 1.4 by observing that for $z = 0$ the LMI in (1.58) coincides with the LMI (1.53). Lastly, if $\|\mathbf{f}\| = 0$, one has that $\lambda_{pol}(f) = 0$ and $f(x)$ is SOS since $f(x) = 0$. □

Let us define the following positivity index for polynomials:

$$\mu_{pol}(f) = \begin{cases} \beta y \text{ if } \|\mathbf{f}\| > 0 \\ 0 \quad \text{otherwise} \end{cases} \tag{1.60}$$

where

$$y = \inf_x \frac{f(x)}{\|b_{pol}(x,m)\|^2} \tag{1.61}$$

and $\mathbf{f}$, β and m are as in Definition 1.14. Clearly:

1. $f(x)$ is nonnegative if and only if $\mu_{pol}(f) \geq 0$.
2. $f(x)$ is positive if $\mu_{pol}(f) > 0$.

The following result provides a relationship between the SOS index and the positivity index.

Theorem 1.7. *Consider $f \in \mathscr{P}_n$. Then,*

$$\lambda_{pol}(f) \leq \mu_{pol}(f). \tag{1.62}$$

Proof. If $||\mathbf{f}|| > 0$, let us pre- and post-multiply the LMI in (1.58) by $b_{pol}(x,m)'$ and $b_{pol}(x,m)$, respectively. It follows that

$$\begin{aligned} 0 &\leq b_{pol}(x,m)' (F + L(\alpha) - zI) b_{pol}(x,m) \\ &= f(x) - z||b_{pol}(x,m)||^2 \end{aligned}$$

and hence

$$z \leq \frac{f(x)}{||b_{pol}(x,m)||^2}$$

whenever the LMI in (1.58) holds. Lastly, if $||\mathbf{f}|| = 0$, one simply has that $\lambda_{pol}(f) = \mu_{pol}(f) = 0$. □

Consequently, the SOS index can be used to investigate positivity of a polynomial as explained hereafter.

Corollary 1.2. *Consider $f \in \mathscr{P}_n$. The following statements hold.*

1. *$f(x)$ is nonnegative if $\lambda_{pol}(f) \geq 0$.*
2. *$f(x)$ is positive if $\lambda_{pol}(f) > 0$.*

Proof. Direct consequence of Theorem 1.7 and Corollary 1.1. □

Example 1.7. Consider $f(x) = 4 - 3x + x^4$ in (1.11). By using the complete SMR matrix $F + L(\alpha)$ of $f(x)$ shown in Example 1.4, it turns out that $\lambda_{pol}(f) = 0.181$. This implies from Corollary 1.1 that $f(x)$ is SOS. The optimal values of z and α in (1.58) are

$$z = 0.532, \quad \alpha = 0.829.$$

Moreover, from Theorem 1.5 we find that $f(x)$ can be decomposed as $f(x) = \sum_{i=1}^{k} g_i(x)^2$ with $k = 3$ and

$$g(x) = \begin{pmatrix} 2 - 0.75x - 0.415x^2 \\ 1.047x - 0.297x^2 \\ 0.86x^2 \end{pmatrix}.$$

□

1.3.2 Special Cases

For locally quadratic polynomials, the results just presented for characterizing SOS properties can be simplified. In the following we present some of these results that will be exploited throughout the book.

Definition 1.15 (Positive semidefinite/definite polynomial). Let $f \in \mathscr{P}_n$ be either a locally quadratic polynomial or a homogeneous polynomial. Then, $f(x)$ is *positive semidefinite* if $f(x) \geq 0$ for all $x \in \mathbb{R}^n$. Also, $f(x)$ is *positive definite* if $f(x) > 0$ for all $x \in \mathbb{R}_0^n$. □

Positive semidefiniteness and definiteness can be investigated through the SOS index, which in the case of locally quadratic polynomials is defined as follows.

Definition 1.16 (SOS Index (continued)). Let $f \in \mathscr{P}_n$ be a locally quadratic polynomial, and define $\mathbf{f} = COE_{qua}(f)$ and $F + L(\alpha) = CSMR_{qua}(f)$. Let us introduce

$$\lambda_{qua}(f) = \begin{cases} \beta z^* & \text{if } \|\mathbf{f}\| > 0 \\ 0 & \text{otherwise} \end{cases} \tag{1.63}$$

where β is a normalizing factor chosen as

$$\beta = \frac{\sqrt{d_{lin}(n,m)}}{\|\mathbf{f}\|}, \quad m = \left\lceil \frac{\partial f}{2} \right\rceil \tag{1.64}$$

and z^* is the solution of the SDP

$$\begin{aligned} z^* &= \max_{z,\alpha} \ z \\ &\quad \text{s.t.} \ \ F + L(\alpha) - zI \geq 0. \end{aligned} \tag{1.65}$$

Then, $\lambda_{qua}(f)$ is called *SOS index* of $f(x)$ (for locally quadratic polynomials). □

Let us define the following positivity index for locally quadratic polynomials:

$$\mu_{qua}(f) = \begin{cases} \beta y & \text{if } \|\mathbf{f}\| > 0 \\ 0 & \text{otherwise} \end{cases} \tag{1.66}$$

where

$$y = \inf_x \frac{f(x)}{\|b_{lin}(x,m)\|^2} \tag{1.67}$$

and $\mathbf{f}$, β and m are as in Definition 1.16. Clearly:

1. a locally quadratic polynomial $f(x)$ is positive semidefinite if and only if $\mu_{qua}(f) \geq 0$.
2. a locally quadratic polynomial $f(x)$ is positive definite if $\mu_{qua}(f) > 0$.

Corollary 1.3. *Let $f \in \mathscr{P}_n$ be a locally quadratic polynomial. The following statements hold.*

1. *$f(x)$ is SOS if and only if there exists α satisfying the LMI $F+L(\alpha)\geq 0$ where $F+L(\alpha)=CSMR_{qua}(f)$.*
2. *The SOS index $\lambda_{qua}(f)$ exists (and, hence, is bounded).*
3. *$f(x)$ is SOS if and only if $\lambda_{qua}(f)\geq 0$.*
4. *$f(x)$ is positive semidefinite if $\lambda_{qua}(f)\geq 0$.*
5. *$f(x)$ is positive definite if $\lambda_{qua}(f)>0$.*
6. *$\lambda_{qua}(f)\leq \mu_{qua}(f)$.*

Proof. Analogous to the proofs of Theorems 1.6 and 1.7, and to the proofs of Corollaries 1.1 and 1.2. □

The same results can be obtained for the class of homogeneous polynomials as explained hereafter. Observe that positive semidefiniteness and definiteness for homogeneous polynomials are defined as in Definition 1.15.

Definition 1.17 (SOS Index (continued)). Let $f\in\mathscr{P}_n$ be a homogeneous polynomial of even degree, and define $\mathbf{f}=COE_{hom}(f)$ and $F+L(\alpha)=CSMR_{hom}(f)$. Let us introduce

$$\lambda_{hom}(f)=\begin{cases}\beta z^* & \text{if } \|\mathbf{f}\|>0\\ 0 & \text{otherwise}\end{cases} \tag{1.68}$$

where β is a normalizing factor chosen as

$$\beta=\frac{\sqrt{d_{hom}(n,m)}}{\|\mathbf{f}\|},\quad m=\frac{\partial f}{2} \tag{1.69}$$

and z^* is the solution of the SDP

$$\begin{aligned} z^* &= \max_{z,\alpha}\ z\\ &\quad \text{s.t. } F+L(\alpha)-zI\geq 0. \end{aligned} \tag{1.70}$$

Then, $\lambda_{hom}(f)$ is called *SOS index* of $f(x)$ (for homogeneous polynomials). □

Let us define the following positivity index for homogeneous polynomials:

$$\mu_{hom}(f)=\begin{cases}\beta y & \text{if } \|\mathbf{f}\|>0\\ 0 & \text{otherwise}\end{cases} \tag{1.71}$$

where

$$y=\inf_x \frac{f(x)}{\|b_{hom}(x,m)\|^2} \tag{1.72}$$

and $\mathbf{f}$, β and m are as in Definition 1.17. Clearly:

1. a homogeneous polynomial $f(x)$ is positive semidefinite if and only if $\mu_{hom}(f)\geq 0$.
2. a homogeneous polynomial $f(x)$ is positive definite if and only if $\mu_{hom}(f)>0$.

Corollary 1.4. *Let $f \in \mathscr{P}_n$ be a homogeneous polynomial of even degree. The following statements hold.*

1. *$f(x)$ is SOS if and only if there exists α satisfying the LMI $F + L(\alpha) \geq 0$ where $F + L(\alpha) = CSMR_{hom}(f)$.*
2. *The SOS index $\lambda_{hom}(f)$ exists (and, hence, is bounded).*
3. *$f(x)$ is SOS if and only if $\lambda_{hom}(f) \geq 0$.*
4. *$f(x)$ is positive semidefinite if $\lambda_{hom}(f) \geq 0$.*
5. *$f(x)$ is positive definite if $\lambda_{hom}(f) > 0$.*
6. *$\lambda_{hom}(f) \leq \mu_{hom}(f)$.*

Proof. Analogous to the proofs of Theorems 1.6 and 1.7, and to the proofs of Corollaries 1.1 and 1.2. □

Example 1.8. Consider the locally quadratic polynomial $f(x) = x_1^2 + 2x_2^2 - 4x_1^2x_2$ in (1.23). By using the complete SMR matrix of $f(x)$ shown in Example 1.5, it turns out that $\lambda_{qua}(f) = -0.211$. This means that $f(x)$ is not SOS, which is consistent with the fact that the degree of $f(x)$ is odd. The optimal values of z and α in (1.65) are

$$z = -0.432, \quad \alpha = (1.101, 0.000, 0.208)'.$$

Also, we consider another locally quadratic polynomial obtained by adding $x_1^4 + x_2^4$ to $f(x)$, hence obtaining

$$g(x) = x_1^2 + 2x_2^2 - 4x_1^2x_2 + x_1^4 + x_2^4. \tag{1.73}$$

It turns out that $\lambda_{qua}(g) = 0.044$, and hence $g(x)$ is SOS from Corollary 1.3. The optimal values of z and α in (1.65) are

$$z = 0.095, \quad \alpha = (0.915, 0.000, 0.510)'.$$ □

Example 1.9. Consider the homogeneous polynomial $f(x) = x_1^6 - 2x_1^5x_2 + 4x_1^3x_2^3 + x_2^6$ in (1.24). By using the complete SMR matrix of $f(x)$ shown in Example 1.6, it turns out that $\lambda_{hom}(f) = -0.032$, and hence $f(x)$ is not SOS from Corollary 1.4. The optimal values of z and α in (1.70) are

$$z = -0.074, \quad \alpha = (0.832, 1.285, 0.827)'.$$

Also, we consider another homogeneous polynomial obtained by replacing $4x_1^3x_2^3$ in $f(x)$ with $3x_1^3x_2^3$, hence obtaining

$$g(x) = x_1^6 - 2x_1^5x_2 + 3x_1^3x_2^3 + x_2^6. \tag{1.74}$$

It turns out that $\lambda_{hom}(g) = 0.083$, and hence $g(x)$ is SOS from Corollary 1.4. The optimal values of z and α in (1.70) are

$$z = 0.160, \quad \alpha = (0.759, 0.946, 0.740)'. \qquad \square$$

1.4 SOS Parameter-Dependent Polynomials and SOS Matrix Polynomials

This section introduces the representation of parameter-dependent polynomials and matrix polynomials via power vectors and via the SMR, the concepts of SOS parameter-dependent polynomials and SOS matrix polynomials, and their characterization via the SMR and LMIs.

1.4.1 SOS Parameter-Dependent Polynomials: General Case

Polynomials can be parameter-dependent, i.e. with coefficients affected by some parameters. In this book we consider the case where the parameters affect polynomially the coefficients, i.e.

$$f(x,\theta) = \sum_{q \in \mathbb{N}^n,\ |q| \le d} a_q(\theta) x^q \tag{1.75}$$

where $x \in \mathbb{R}^n$, $\theta \in \mathbb{R}^{n_\theta}$ and $a_q \in \mathscr{P}_{n_\theta}$ with $\partial a_q \le d_\theta$. It follows that $f(x,\theta)$ can be expressed as

$$f(x,\theta) = \sum_{q \in \mathbb{N}^n,\ |q| \le d,\ r \in \mathbb{N}^{n_\theta},\ |r| \le d_\theta} a_{q,r} x^q \theta^r \tag{1.76}$$

where $a_{q,r} \in \mathbb{R}$. In (1.76), $a_{q,r} x^q \theta^r$ is a monomial of degree $|q|$ in x and of degree $|r|$ in θ, and $a_{q,r}$ is its coefficient. The degree of $f(x,\theta)$ in x, denoted by $\partial^x f$, is given by

$$\partial^x f = \max_{q \in \mathbb{N}^n,\ |q| \le d,\ r \in \mathbb{N}^{n_\theta},\ |r| \le d_\theta,\ a_{q,r} \ne 0} |q| \tag{1.77}$$

while the degree of in θ, denoted by $\partial^\theta f$, is

$$\partial^\theta f = \max_{q \in \mathbb{N}^n,\ |q| \le d,\ r \in \mathbb{N}^{n_\theta},\ |r| \le d_\theta,\ a_{q,r} \ne 0} |r|. \tag{1.78}$$

We denote the set of parameter-dependent polynomials in a n-dimensional real variable with a n_θ-dimensional real parameter as

$$\mathscr{P}_{n,n_\theta} = \{f : \text{ (1.76) holds for some } d, d_\theta \in \mathbb{N}\}. \tag{1.79}$$

Hence, with the notation $\mathscr{P}^m_{n,n_\theta}$ and $\mathscr{P}^{m\times l}_{n,n_\theta}$ we denote the set of $m \times 1$ vectors and $m \times l$ matrices with entries in $\mathscr{P}_{n,n_\theta}$, respectively.

By using power vectors, $f(x,\theta)$ can be written as

$$f(x,\theta) = \mathbf{f}' \left(b_{pol}(\theta,d_\theta) \otimes b_{pol}(x,d)\right) \tag{1.80}$$

for a unique $\mathbf{f} \in \mathbb{R}^{d_{polpol}(n,n_\theta,d,d_\theta)}$ where

$$d_{polpol}(n,n_\theta,d,d_\theta) = d_{pol}(n,d)d_{pol}(n_\theta,d_\theta). \tag{1.81}$$

In the sequel we will denote with the function $COE_{polpol}(f)$ the vector $\mathbf{f}$ satisfying (1.80), i.e.

$$\begin{cases} \mathbf{f} = COE_{polpol}(f) \\ \text{(1.80) holds.} \end{cases} \tag{1.82}$$

Parameter-dependent polynomials can be written also via the SMR. Specifically, let us introduce the notation

$$\Delta\left(A,b,c\right) = (b \otimes c)' A \left(b \otimes c\right). \tag{1.83}$$

Then, a parameter-dependent polynomial $f(x,\theta)$ with $\partial^x f \le 2m$ and $\partial^\theta f \le 2m_\theta$, $m, m_\theta \in \mathbb{N}$, can be written as

$$f(x,\theta) = \Delta\left(F, b_{pol}(\theta,m_\theta), b_{pol}(x,m)\right) \tag{1.84}$$

for some symmetric matrix F. The matrix F is not unique, and all such matrices can be parametrized via the complete SMR. Indeed, let $L(\alpha)$ be a linear parametrization of the linear subspace

$$\begin{aligned}\mathscr{L}_{polpol}(n,n_\theta,2m,2m_\theta) = \{L = L' :\ & \Delta\left(F, b_{pol}(\theta,m_\theta), b_{pol}(x,m)\right) = 0 \\ & \forall x \in \mathbb{R}^n\ \forall \theta \in \mathbb{R}^{n_\theta}\}.\end{aligned} \tag{1.85}$$

Then, the complete SMR of $f(x,\theta)$ is given by

$$f(x,\theta) = \Delta\left(F + L(\alpha), b_{pol}(\theta,m_\theta), b_{pol}(x,m)\right). \tag{1.86}$$

It turns out that the dimension of $\mathscr{L}_{polpol}(n,n_\theta,2m,2m_\theta)$ is given by

$$l_{polpol}(n,n_\theta,2m,2m_\theta) = t_{polpol}(n,n_\theta,2m,2m_\theta) - d_{polpol}(n,n_\theta,2m,2m_\theta) \tag{1.87}$$

where

$$t_{polpol}(n,n_\theta,2m,2m_\theta) = \frac{1}{2} d_{polpol}(n,n_\theta,m,m_\theta)\left(d_{polpol}(n,n_\theta,m,m_\theta)+1\right). \tag{1.88}$$

In order to simplify the presentation, in the sequel we will denote with the functions $SMR_{polpol}(f)$ and $CSMR_{polpol}(f)$ an SMR matrix F and a complete SMR matrix $F + L(\alpha)$ of $f(x,\theta)$ with respect to $b_{pol}(\theta,m_\theta) \otimes b_{pol}(x,m)$, i.e.

$$\begin{cases} F = SMR_{polpol}(f) \\ \text{(1.84) holds} \end{cases} \tag{1.89}$$

and

$$\begin{cases} F+L(\alpha)=CSMR_{polpol}(f) \\ (1.86) \text{ holds} \end{cases} \quad (1.90)$$

where m and m_θ are the smallest integers for which the SMR matrices can be built, i.e. $m=\lceil \partial^x f/2 \rceil$ and $m_\theta=\lceil \partial^\theta f/2 \rceil$.

A parameter-dependent polynomial $f(x,\theta)$ is said SOS if there exist $k\in\mathbb{N}$ and $g\in\mathscr{P}^k_{n,n_\theta}$ such that

$$f(x,\theta)=\sum_{i=1}^{k} g_i(x,\theta)^2. \quad (1.91)$$

We denote the set of SOS parameter-dependent polynomials in $\mathscr{P}_{n,n_\theta}$ as

$$\mathscr{S}_{n,n_\theta}=\left\{f\in\mathscr{P}_{n,n_\theta}:\ (1.91) \text{ holds for some } k\in\mathbb{N} \text{ and } g\in\mathscr{P}^k_{n,n_\theta}\right\}. \quad (1.92)$$

The SMR allows one to establish whether a parameter-dependent polynomial is SOS. Indeed, similarly to Theorem 1.4 for the case of polynomials, one has that $f(x,\theta)$ is SOS if and only if there exists α satisfying the LMI

$$F+L(\alpha)\geq 0. \quad (1.93)$$

SOS parameter-dependent polynomials can be decomposed similarly to what done in Theorem 1.5 for the case of SOS polynomials. The SOS index for parameter-dependent polynomials is defined as follows.

Definition 1.18 (SOS Index (continued)). Let $f\in\mathscr{P}_{n,n_\theta}$, and define $\mathbf{f}=COE_{polpol}(f)$ and $F+L(\alpha)=CSMR_{polpol}(f)$. Let us introduce

$$\lambda_{polpol}(f)=\begin{cases} \beta z^* & \text{if } \|\mathbf{f}\|>0 \\ 0 & \text{otherwise} \end{cases} \quad (1.94)$$

where β is a normalizing factor chosen as

$$\beta=\frac{\sqrt{d_{polpol}(n,n_\theta,m,m_\theta)}}{\|\mathbf{f}\|},\quad m=\left\lceil\frac{\partial^x f}{2}\right\rceil,\quad m_\theta=\left\lceil\frac{\partial^\theta f}{2}\right\rceil \quad (1.95)$$

and z^* is the solution of the SDP

$$\begin{aligned} z^* &= \max_{z,\alpha}\ z \\ &\text{s.t. } F+L(\alpha)-zI\geq 0. \end{aligned} \quad (1.96)$$

Then, $\lambda_{polpol}(f)$ is called *SOS index* of $f(x,\theta)$ (for parameter-dependent polynomials). □

Corollary 1.5. *Consider $f\in\mathscr{P}_{n,n_\theta}$. The following statements hold.*

1. *The SOS index $\lambda_{polpol}(f)$ exists (and, hence, is bounded).*
2. *$f(x,\theta)$ is SOS if and only if $\lambda_{polpol}(f)\geq 0$.*

Proof. Analogous to the proofs of Theorem 1.6 and Corollary 1.1. □

1.4.2 SOS Parameter-Dependent Polynomials: Special Cases

If $f \in \mathscr{P}_{n,n_\theta}$ is such that $f(0_n,\theta) = 0$, a more appropriate representation can be introduced by using the power vector $b_{lin}(x,d)$ instead of $b_{pol}(x,d)$, indeed

$$f(x,\theta) = \mathbf{f}' \left(b_{pol}(\theta,m_\theta) \otimes b_{lin}(x,m)\right) \tag{1.97}$$

for a unique $\mathbf{f} \in \mathbb{R}^{d_{linpol}(n,n_\theta,d,d_\theta)}$ where

$$d_{linpol}(n,n_\theta,d,d_\theta) = d_{lin}(n,d)d_{pol}(n_\theta,d_\theta). \tag{1.98}$$

In the sequel we will denote with the function $COE_{linpol}(f)$ the vector $\mathbf{f}$ satisfying (1.97), i.e.

$$\begin{cases} \mathbf{f} = COE_{linpol}(f) \\ \text{(1.97) holds.} \end{cases} \tag{1.99}$$

Moreover, if $f(x,\theta)$ is a locally quadratic polynomial in x for all fixed θ, i.e.

$$\forall \theta \in \mathbb{R}^{n_\theta} \quad \begin{cases} f(0_n,\theta) = 0 \\ \nabla f(0_n,\theta) = 0_n \end{cases} \tag{1.100}$$

then $f(x,\theta)$ can be written as

$$f(x,\theta) = \mathbf{f}' \left(b_{pol}(\theta,m_\theta) \otimes b_{qua}(x,m)\right) \tag{1.101}$$

for a unique $\mathbf{f} \in \mathbb{R}^{d_{quapol}(n,n_\theta,d,d_\theta)}$ where

$$d_{quapol}(n,n_\theta,d,d_\theta) = d_{qua}(n,d)d_{pol}(n_\theta,d_\theta). \tag{1.102}$$

In the sequel we will denote with the function $COE_{quapol}(f)$ the vector $\mathbf{f}$ satisfying (1.102), i.e.

$$\begin{cases} \mathbf{f} = COE_{quapol}(f) \\ \text{(1.102) holds.} \end{cases} \tag{1.103}$$

For $f(x,\theta)$ satisfying (1.100) with $\partial^x f \le 2m$ and $\partial^\theta f \le 2m_\theta$, $m, m_\theta \in \mathbb{N}$, a more appropriate SMR can be obtained as

$$f(x,\theta) = \Delta\left(F, b_{pol}(\theta,m_\theta), b_{lin}(x,m)\right) \tag{1.104}$$

where F is a symmetric matrix. Hence, the complete SMR is defined as

$$f(x,\theta) = \Delta\left(F + L(\alpha), b_{pol}(\theta,m_\theta), b_{lin}(x,m)\right) \tag{1.105}$$

where $L(\alpha)$ is a linear parametrization of the linear subspace

$$\mathscr{L}_{quapol}(n,n_\theta,2m,2m_\theta) = \{L = L' : \Delta\left(F, b_{pol}(\theta,m_\theta), b_{lin}(x,m)\right) = 0 \\ \forall x \in \mathbb{R}^n \ \forall \theta \in \mathbb{R}^{n_\theta}\} \tag{1.106}$$

whose dimension is given by

$$l_{quapol}(n,n_\theta,2m,2m_\theta) = t_{quapol}(n,n_\theta,2m,2m_\theta) - d_{qua}(n,2m)d_{pol}(n_\theta,2m_\theta) \tag{1.107}$$

where

$$t_{quapol}(n,n_\theta,2m,2m_\theta) = \frac{1}{2}d_{linpol}(n,n_\theta,m,m_\theta)\left(d_{linpol}(n,n_\theta,m,m_\theta)+1\right). \tag{1.108}$$

In the sequel we will denote with the functions $SMR_{quapol}(f)$ and $CSMR_{quapol}(f)$ an SMR matrix F and a complete SMR matrix $F+L(\alpha)$ of $f(x,\theta)$ with respect to $b_{pol}(\theta,m_\theta)\otimes b_{lin}(x,m)$, i.e.

$$\begin{cases} F = SMR_{quapol}(f) \\ (1.104) \text{ holds} \end{cases} \tag{1.109}$$

and

$$\begin{cases} F+L(\alpha) = CSMR_{quapol}(f) \\ (1.105) \text{ holds} \end{cases} \tag{1.110}$$

where m and m_θ are the smallest integers for which the SMR matrices can be built, i.e. $m = \lceil \partial^x f/2 \rceil$ and $m_\theta = \lceil \partial^\theta f/2 \rceil$

Similarly to Theorem 1.4 for the case of polynomials, one has that $f \in \mathscr{P}_{n,n_\theta}$ satisfying (1.100) is SOS if and only if there exists α satisfying the LMI

$$F + L(\alpha) \geq 0. \tag{1.111}$$

Definition 1.19 (SOS Index (continued)). Let $f \in \mathscr{P}_{n,n_\theta}$ satisfy (1.100), and define $\mathbf{f} = COE_{quapol}(f)$ and $F+L(\alpha) = CSMR_{quapol}(f)$. Let us introduce

$$\lambda_{quapol}(f) = \begin{cases} \beta z^* & \text{if } \|\mathbf{f}\| > 0 \\ 0 & \text{otherwise} \end{cases} \tag{1.112}$$

where β is a normalizing factor chosen as

$$\beta = \frac{\sqrt{d_{linpol}(n,n_\theta,m,m_\theta)}}{\|\mathbf{f}\|}, \quad m = \left\lceil \frac{\partial^x f}{2} \right\rceil, \quad m_\theta = \left\lceil \frac{\partial^\theta f}{2} \right\rceil \tag{1.113}$$

and z^* is the solution of the SDP

$$\begin{aligned} z^* &= \max_{z,\alpha} \ z \\ &\text{s.t. } F + L(\alpha) - zI \geq 0. \end{aligned} \tag{1.114}$$

Then, $\lambda_{quapol}(f)$ is called *SOS index* of $f(x,\theta)$ (for parameter-dependent locally quadratic polynomials). □

Corollary 1.6. *Let $f \in \mathscr{P}_{n,n_\theta}$ satisfy* (1.100). *The following statements hold.*

1. *The SOS index $\lambda_{quapol}(f)$ exists (and, hence, is bounded).*
2. *$f(x,\theta)$ is SOS if and only if $\lambda_{quapol}(f) \geq 0$.*

Proof. Analogous to the proofs of Theorem 1.6 and Corollary 1.1. □

Similarly, one can consider the case where $f(x,\theta)$ is a homogeneous polynomial in x for all fixed $\theta \in \mathbb{R}^{n_\theta}$. The details of this extension are omitted for conciseness.

1.4.3 SOS Matrix Polynomials

As explained in Section 1.1.1, a matrix polynomial is a matrix whose entries are polynomials. Matrix polynomials can be represented by using power vectors analogously to polynomials. Specifically, for $F \in \mathscr{P}_{n_\theta}^{n\times p}$ with $\partial F = d$ one can write

$$F(\theta) = \mathbf{F}\left(b_{pol}(\theta,d)\otimes I_p\right) \tag{1.115}$$

for a unique $\mathbf{F} \in \mathbb{R}^{n\times p d_{pol}(n_\theta,d)}$. In the sequel we will denote with the function $COE_{mat}(F)$ the matrix $\mathbf{F}$ satisfying (1.115), i.e.

$$\begin{cases} \mathbf{F} = COE_{mat}(F) \\ (1.115) \text{ holds.} \end{cases} \tag{1.116}$$

Also, symmetric matrix polynomials can be represented through the SMR. Specifically, $F = F' \in \mathscr{P}_{n_\theta}^{n\times n}$ with $\partial F \leq 2m$, $m \in \mathbb{N}$, can be written as

$$F(\theta) = \Delta\left(G, b_{pol}(\theta,m), I_n\right) \tag{1.117}$$

where G is a symmetric matrix. The matrix G is not unique, and all such matrices can be parametrized via the complete SMR. Indeed, let $L(\alpha)$ be a linear parametrization of the linear subspace

$$\mathscr{L}_{mat}(n_\theta,2m,n) = \left\{L = L' : \Delta\left(G, b_{pol}(\theta,m), I_n\right) = 0 \;\; \forall\theta \in \mathbb{R}^{n_\theta}\right\}. \tag{1.118}$$

Then, the complete SMR of $F(\theta)$ is given by

$$F(\theta) = \Delta\left(G + L(\alpha), b_{pol}(\theta,m), I_n\right). \tag{1.119}$$

It turns out that the dimension of $\mathscr{L}_{mat}(n_\theta,2m,n)$ is given by

$$\mathscr{L}_{mat}(n_\theta,2m,n) = t_{mat}(n_\theta,2m,n) - d_{qua}(n,2)d_{pol}(n_\theta,2m) \tag{1.120}$$

where

$$t_{mat}(n_\theta, 2m, n) = \frac{1}{2} d_{linpol}(n, n_\theta, 1, m)\left(d_{linpol}(n, n_\theta, 1, m) + 1\right). \quad (1.121)$$

In order to simplify the presentation, in the sequel we will denote with the functions $SMR_{mat}(F)$ and $CSMR_{mat}(F)$ an SMR matrix G and a complete SMR matrix $G + L(\alpha)$ of $F(\theta)$, i.e.

$$\begin{cases} G = SMR_{mat}(F) \\ (1.117) \text{ holds} \end{cases} \quad (1.122)$$

and

$$\begin{cases} G + L(\alpha) = CSMR_{mat}(F) \\ (1.119) \text{ holds} \end{cases} \quad (1.123)$$

where m is the smallest integer for which the SMR matrices can be built, i.e. $m = \lceil \partial F/2 \rceil$.

A matrix polynomial $F = F' \in \mathscr{P}_{n_\theta}^{n\times n}$ is said SOS if there exist $k \in \mathbb{N}$ and $H_1, \ldots, H_k \in \mathscr{P}_{n_\theta}^{n\times n}$ such that

$$F(\theta) = \sum_{i=1}^{k} H_i(\theta)' H_i(\theta). \quad (1.124)$$

The SMR allows one to establish whether a matrix polynomial is SOS. Indeed, $F(\theta)$ is SOS if and only if there exists α satisfying the LMI

$$G + L(\alpha) \geq 0. \quad (1.125)$$

SOS matrix polynomials can be decomposed according to (1.124) similarly to what done in Theorem 1.5 for the case of SOS polynomials. Specifically, consider $F = F' \in \mathscr{P}_{n_\theta}^{n\times n}$ with $\partial F \leq 2m$, $m \in \mathbb{N}$, and define $G + L(\alpha) = CSMR_{mat}(F)$. If $F(\theta)$ is SOS, then $F(\theta)$ can be written as in (1.124) with

$$\begin{array}{c} k \leq d_{pol}(n_\theta, m) \\ H_i(\theta) = Q_i\left(b_{pol}(\theta, m) \otimes I_n\right) \end{array} \quad (1.126)$$

where $Q_1, \ldots, Q_k \in \mathbb{R}^{n\times n d_{pol}(n_\theta, m)}$ are submatrices according to $Q = (Q_1', \ldots, Q_k')'$, and Q is any matrix such that

$$Q'Q = G + L(\alpha) \quad (1.127)$$

where α is any vector satisfying (1.125).

Definition 1.20 (SOS Index (continued)). Let $F = F' \in \mathscr{P}_{n_\theta}^{n\times n}$, and define $\mathbf{F} = COE_{mat}(F)$ and $G + L(\alpha) = CSMR_{mat}(F)$. Let us introduce

$$\lambda_{mat}(F) = \begin{cases} \beta z^* & \text{if } \|\mathbf{F}\| > 0 \\ 0 & \text{otherwise} \end{cases} \quad (1.128)$$

where β is a normalizing factor chosen as

$$\beta = \frac{\sqrt{d_{pol}(n_\theta, m)}}{\|\mathbf{F}\|}, \quad m = \left\lceil \frac{\partial F}{2} \right\rceil \tag{1.129}$$

and z^* is the solution of the SDP

$$\begin{aligned} z^* &= \max_{z,\alpha} \; z \\ &\quad \text{s.t. } \; G + L(\alpha) - zI \geq 0. \end{aligned} \tag{1.130}$$

Then, $\lambda_{mat}(f)$ is called *SOS index* of $F(\theta)$ (for matrix polynomials). □

Corollary 1.7. *Consider $F = F' \in \mathscr{P}_{n_\theta}^{n\times n}$. The following statements hold.*

1. *The SOS index $\lambda_{mat}(F)$ exists (and, hence, is bounded).*
2. *$F(\theta)$ is SOS if and only if $\lambda_{mat}(F) \geq 0$.*

Proof. Analogous to the proofs of Theorem 1.6 and Corollary 1.1. □

It is useful to observe that the SMR of matrix polynomials is related to the SMR of parameter-dependent polynomials. Specifically, consider $F = F' \in \mathscr{P}_{n_\theta}^{n\times n}$ and define

$$h(x,\theta) = x'F(\theta)x \tag{1.131}$$

where $x \in \mathbb{R}^n$. Then,

$$CSMR_{mat}(F) = CSMR_{quapol}(h). \tag{1.132}$$

Consequently, one can establish whether a matrix polynomial is SOS also by establishing whether a parameter-dependent polynomial is SOS, i.e.

$$F(\theta) \text{ is SOS} \iff h(x,\theta) \text{ is SOS}. \tag{1.133}$$

Example 1.10. Consider $F = F' \in \mathscr{P}_2^{2\times 2}$ given by

$$F(\theta) = \begin{pmatrix} 2 - \theta_1 + \theta_1^2 + 3\theta_2^2 & -1 + \theta_2 + 2\theta_1\theta_2 \\ \star & 1 + 3\theta_1^2 + \theta_1\theta_2 + \theta_2^2 \end{pmatrix}.$$

Let us rewrite $F(\theta)$ as

$$F(\theta) = F_{00} + \theta_1 F_{10} + \theta_2 F_{01} + \theta_1^2 F_{20} + \theta_1\theta_2 F_{11} + \theta_2^2 F_{02}$$

where

$$F_{00} = \begin{pmatrix} 2 & -1 \\ \star & 1 \end{pmatrix}, \quad F_{10} = \begin{pmatrix} -1 & 0 \\ \star & 0 \end{pmatrix}, \quad F_{01} = \begin{pmatrix} 0 & 1 \\ \star & 0 \end{pmatrix}$$
$$F_{20} = \begin{pmatrix} 1 & 0 \\ \star & 3 \end{pmatrix}, \quad F_{11} = \begin{pmatrix} 0 & 2 \\ \star & 1 \end{pmatrix}, \quad F_{02} = \begin{pmatrix} 3 & 0 \\ \star & 1 \end{pmatrix}.$$

We can express $F(\theta)$ according to (1.119) with $m=1$, $n=2$ and

$$b_{pol}(\theta,m)=\begin{pmatrix}1\\ \theta_1\\ \theta_2\end{pmatrix},\quad G=\begin{pmatrix}F_{00} & F_{10}/2 & F_{01}/2\\ \star & F_{20} & F_{11}/2\\ \star & \star & F_{02}\end{pmatrix}$$

$$L(\alpha)=\begin{pmatrix}0 & 0 & 0 & \alpha_1 & 0 & \alpha_2\\ \star & 0 & -\alpha_1 & 0 & -\alpha_2 & 0\\ \star & \star & 0 & 0 & 0 & \alpha_3\\ \star & \star & \star & 0 & -\alpha_3 & 0\\ \star & \star & \star & \star & 0 & 0\\ \star & \star & \star & \star & \star & 0\end{pmatrix}.$$

By solving the SDP (1.130) we find that $\lambda_{mat}(F)=0.012$, hence implying that $F(\theta)$ is SOS. The optimal values of z and α in (1.130) are

$$z=0.032,\quad \alpha=(0.212,-0.390,1.758)'.$$

Moreover, according to (1.126)–(1.127), we find that $F(\theta)$ can be decomposed as $F(\theta)=\sum_{i=1}^{k}H_i(\theta)'H_i(\theta)$ with $k=3$ and

$$H_1(\theta)=\begin{pmatrix}1.414-0.354\theta_1 & -0.707-0.162\theta_1+0.267\theta_2\\ -0.030\theta_1+0.880\theta_2 & 0.707-0.162\theta_1+0.267\theta_2\end{pmatrix}$$
$$H_2(\theta)=\begin{pmatrix}0.935\theta_1+0.028\theta_2 & -0.066\theta_1+0.399\theta_2\\ 1.092\theta_2 & 1.716\theta_1+0.357\theta_2\end{pmatrix}$$
$$H_3(\theta)=\begin{pmatrix}1.016\theta_2 & -0.627\theta_2\\ 0 & 0.422\theta_2\end{pmatrix}.$$

□

1.5 Extracting Power Vectors from Linear Subspaces

This section addresses the problem of finding power vectors in a linear subspace, which will play a key role in establishing tightness of bounds provided by SOS programming. In particular, Section 1.5.1 considers the power vector $b_{pol}(x,m)$ (used in the SMR of polynomials), while Section 1.5.1 considers the power vectors $b_{lin}(x,m)$ (used in the SMR of locally quadratic polynomials), $b_{hom}(x,m)$ (used in the SMR of homogeneous polynomials), and $b_{pol}(\theta,m_\theta)\otimes b_{lin}(x,m)$ (used in the SMR of locally parameter-dependent polynomials and matrix polynomials).

1.5.1 General Case

First, let us consider the power vector $b_{pol}(x,m)$, for which the problem can be formulated as

$$\mathscr{X}_{pol}=\left\{x\in\mathbb{R}^n:\ b_{pol}(x,m)\in\mathscr{V}\right\}\tag{1.134}$$

where $\mathscr{V} \subseteq \mathbb{R}^{d_{pol}(n,m)}$ is defined by

$$\mathscr{V} = \operatorname{img}(V) \tag{1.135}$$

where $V \in \mathbb{R}^{d_{pol}(n,m)\times u}$ is a matrix with full column rank, i.e.

$$\operatorname{rank}(V) = u. \tag{1.136}$$

This problem amounts to solving a system of polynomial equations. Indeed, let us observe that the condition $b_{pol}(x,m) \in \mathscr{V}$ can be written as

$$b_{pol}(x,m) = Vs \tag{1.137}$$

for some $s \in \mathbb{R}^u$, and hence $\mathscr{X}_{pol}$ can be found by solving the system of polynomial equations (1.137) in the variables x and s. Although various methods have been proposed in the literature for solving system of polynomial equations (such as resultants, homotopy methods, etc), such a task is indeed nontrivial.

In the sequel we present a strategy that, whenever possible, exploits the structure of (1.137) in order to find $\mathscr{X}_{pol}$ by computing the roots of a polynomial in one scalar variable only. We consider the following four cases depending on n, m and u.

Case I. Let us consider the case where u satisfies

$$u \leq m. \tag{1.138}$$

For an integer i satisfying $1 \leq i \leq n$, let $U \in \mathbb{R}^{u\times u}$ be the submatrix of V such that

$$y(x_i) = Us \tag{1.139}$$

where

$$y(x_i) = (1, x_i, \ldots, x_i^{u-1})' \tag{1.140}$$

and s is as in (1.137). Let us suppose that we have chosen an integer i such that U is invertible (we will consider that such an i does not exist in Case III). We can rewrite (1.137) as

$$b_{pol}(x,m) = Wy(x_i) \tag{1.141}$$

where

$$W = VU^{-1}, \tag{1.142}$$

i.e. expressing $b_{pol}(x,m)$ as a linear combination of the columns of W with weights depending on powers of x_i only. It follows that x_i satisfies

$$x_i^u = ay(x_i) \tag{1.143}$$

where a is the row of W corresponding to the monomial x_i^u, or in other words x_i is a root of the polynomial

$$q(x_i) = x_i^u - \sum_{k=0}^{u-1} a_{k+1} x_i^k. \tag{1.144}$$

Let us define the set of real roots of $q(x_i)$ as

$$\hat{\mathscr{X}_i} = \{x_i \in \mathbb{R} : \; q(x_i) = 0\} \tag{1.145}$$

and the set of corresponding x as

$$\hat{\mathscr{X}} = \left\{CWy(x_i), \; x_i \in \hat{\mathscr{X}_i}\right\} \tag{1.146}$$

where $C \in \mathbb{R}^{n \times d_{pol}(n,m)}$ is the matrix that extracts x from $b_{pol}(x,m)$, i.e.

$$x = Cb_{pol}(x,m). \tag{1.147}$$

Then, the sought set $\mathscr{X}_{pol}$ is a subset of $\hat{\mathscr{X}}$ according to

$$\mathscr{X}_{pol} = \left\{x \in \hat{X} : \; b_{pol}(x,m) \in \mathscr{V}\right\}. \tag{1.148}$$

Case II. Consider now the case where u satisfies

$$u = m + 1. \tag{1.149}$$

Observe that, if $n = 1$, then it directly follows that $\mathscr{X}_{pol} = \mathbb{R}$. Moreover, if $m = 1$, either $\mathscr{X}_{pol} = \emptyset$ (if the row of V corresponding to the monomial 1 is null) or $\mathscr{X}_{pol}$ is infinite (if the row of V corresponding to the monomial 1 is not null). Hence, we consider that $n > 1$ and $m > 1$.

Let us define the matrix U according to (1.139). Let us suppose that we have chosen an integer i such that U is invertible (we will consider that such an i does not exist in Case III). Let us define the matrix W as in (1.142). It follows that x_i satisfies

$$ay(x_i)x_i = by(x_i) \tag{1.150}$$

where a and b are the rows of W corresponding to the monomials x_j and $x_i x_j$ for some integer j satisfying $1 \leq j \leq n$ and $j \neq i$, or in other words x_i is a root of the polynomial

$$q(x_i) = a_u x_i^u + \sum_{k=1}^{u-1} (a_k - b_{k+1}) x_i^k - b_1. \tag{1.151}$$

Hence, the sought set $\mathscr{X}_{pol}$ is found according to (1.145)–(1.148).

Case III. Here we complete Cases I and II by considering that the matrix U defined by (1.139) is singular for all i, i.e.

$$\begin{cases} u \leq m+1 \\ \nexists i : \; \det(U) \neq 0. \end{cases} \tag{1.152}$$

For a given i denote with U_j the submatrix of V such that

$$y_j(x_i) = U_j s \tag{1.153}$$

where

$$y_j(x_i) = (1, x_i, \ldots, x_i^{j-1})' \tag{1.154}$$

and let k be the largest j such that U_j has full rank. Since U is not invertible, it follows that $k < u$, i.e. $k \leq m$. Let a be the row of U corresponding to x_i^k, i.e.

$$x_i^k = as. \tag{1.155}$$

Define

$$b = aU_k' \left(U_k U_k'\right)^{-1}. \tag{1.156}$$

Since U_{k+1} has not full rank, it follows that $a' \in \mathrm{img}(U_k')$, which leads us to

$$a = bU_k. \tag{1.157}$$

This implies that $as = by_k(x_i)$, and hence

$$x_i^k = by_k(x_i). \tag{1.158}$$

The solutions for x_i are the roots of the polynomial

$$q_i(x_i) = x_i^k - \sum_{j=0}^{k-1} b_{j+1} x_i^j \tag{1.159}$$

that are gathered into

$$\hat{\mathscr{X}_i} = \{x_i \in \mathbb{R} : \; q_i(x_i) = 0\}. \tag{1.160}$$

By repeating the procedure (1.153)–(1.160) for all $i = 1, \ldots, n$, one obtain the set of candidate solutions

$$\hat{\mathscr{X}} = \left\{x \in \mathbb{R}^n : \; x_i \in \hat{\mathscr{X}_i}\right\} \tag{1.161}$$

and the sought set $\mathscr{X}_{pol}$ is found as in (1.148).

Case IV. Next, let us consider the case where u satisfies

$$m + 2 \leq u \leq n(m-1) + 2. \tag{1.162}$$

Clearly, this implies that $n \geq 2$. Let us write the condition $b_{pol}(x,m) \in \mathscr{V}$ as in (1.137) for some $s \in \mathbb{R}^u$, and for an integer i satisfying $1 \leq i \leq n$, let $X_1 \in \mathbb{R}^{u \times u}$ be the submatrix of V such that

$$y(x) = X_1 s \tag{1.163}$$

where $y(x) \in \mathbb{R}^u$ is a vector whose entries are the first u monomials in the list

$$\begin{array}{c} 1, x_i, \ldots, x_i^m \\ x_{j_1}, x_{j_1}x_i, \ldots, x_{j_1}x_i^{m-2} \\ \vdots \\ x_{j_k}, x_{j_k}x_i, \ldots, x_{j_k}x_i^{m-2} \end{array} \tag{1.164}$$

where $j_1, \ldots, j_k$ are distinct integers in $[1,n]$ different from i, and k is the smallest integer for which the above list contains at least u entries. Let us define

$$z(x) = (x_{j_1}, \ldots, x_{j_k})'. \tag{1.165}$$

If X_1 is invertible, one can rewrite (1.137) as

$$b_{pol}(x,m) = X_2 y(x) \tag{1.166}$$

where

$$X_2 = V X_1^{-1}. \tag{1.167}$$

Let $X_3 \in \mathbb{R}^{k \times u}$ be the submatrix of X_2 such that

$$x_i^{m-1} z(x) = X_3 y(x). \tag{1.168}$$

We can write

$$X_3 y(x) = X_4(x_i) z(x) + X_5(x_i) \tag{1.169}$$

where $X_4(x_i)$ and $X_5(x_i)$ are polynomial functions of x_i only. Hence, one gets

$$z(x) = X_6(x_i)^{-1} X_5(x_i) \tag{1.170}$$

where

$$X_6(x_i) = x_i^{m-1} I - X_4(x_i) \tag{1.171}$$

which means that $x_{j_1}, \ldots, x_{j_k}$ are rational functions of x_1. This implies that the entries of $y(x)$ are rational functions of x_1 as well, and hence we can write

$$y(x) = w(x_i) \tag{1.172}$$

where the function $w(x_i)$ is rational. It follows that x_i must satisfy

$$bw(x_i) - (aw(x_i))^2 = 0 \tag{1.173}$$

where a and b be the rows of X_2 corresponding to the monomials x_l and x_l^2 for an integer $l \in \{j_1, \ldots, j_k\}$, i.e.

$$\begin{aligned} x_l &= aw(x_i) \\ x_l^2 &= bw(x_i), \end{aligned} \tag{1.174}$$

or in other words x_i must be a zero of the rational function

$$q(x_i) = bw(x_i) - (aw(x_i))^2. \tag{1.175}$$

Let us define the set of these zeros as

$$\hat{\mathscr{X}_i} = \{x_i \in \mathbb{R} : \; q(x_i) = 0\} \tag{1.176}$$

and the set of corresponding x as

$$\hat{\mathscr{X}} = \left\{ CX_2 w(x_i), \; x_i \in \hat{\mathscr{X}_i} \right\}. \tag{1.177}$$

The sought set $\mathscr{X}_{pol}$ is a subset of $\hat{\mathscr{X}}$ according to (1.148).

Example 1.11. Let us consider the computation of $\mathscr{X}_{pol}$ in (1.134) with

$$b_{pol}(x,m) = \begin{pmatrix} 1 \\ x_1 \\ x_2 \\ x_1^2 \\ x_1 x_2 \\ x_2^2 \end{pmatrix}, \; V = \begin{pmatrix} 2 & -1 \\ 0 & 3 \\ 2 & -4 \\ 2 & -1 \\ -2 & 4 \\ 4 & -8 \end{pmatrix}.$$

We have that $n = 2$, $m = 2$ and $u = 2$. Hence, (1.138) is satisfied, and let us proceed as in Case I. The matrix U is invertible with $i = 1$, and we obtain

$$W = \begin{pmatrix} 1 & 0 \\ 0 & 1 \\ 1 & -1 \\ 1 & 0 \\ -1 & 1 \\ 2 & -2 \end{pmatrix}, \; y(x_1) = \begin{pmatrix} 1 \\ x_1 \end{pmatrix}.$$

It follows that

$$q(x_1) = x_1^2 - 1$$

which provides according to (1.145)

$$\hat{\mathscr{X}_1} = \{-1, 1\}$$

and, from (1.146), the set of candidates

$$\hat{\mathscr{X}} = \left\{ \begin{pmatrix} -1 \\ 2 \end{pmatrix}, \begin{pmatrix} 1 \\ 0 \end{pmatrix} \right\}.$$

From $\hat{\mathscr{X}}$ we directly obtain the sought set $\mathscr{X}_{pol}$ according to (1.148) by simply establishing whether $b_{pol}(x,m) \in \mathscr{V}$ for each x in $\hat{\mathscr{X}}$. We find that all the candidates in $\hat{\mathscr{X}}$ belong to $\mathscr{X}_{pol}$, hence concluding that $\mathscr{X}_{pol} = \hat{\mathscr{X}}$. □

Example 1.12. Let us consider the computation of $\mathscr{X}_{pol}$ in (1.134) with

$$b_{pol}(x,m) = \begin{pmatrix} 1 \\ x_1 \\ x_2 \\ x_1^2 \\ x_1x_2 \\ x_2^2 \end{pmatrix}, \quad V = \begin{pmatrix} 2 & 0 & 2 & 2 \\ -2 & -1 & 2 & -1 \\ 2 & 3 & 2 & 1 \\ 10 & -1 & 4 & 13 \\ -6 & 0 & -2 & -8 \\ 4 & 9 & 10 & 5 \end{pmatrix}.$$

We have that $n = 2$, $m = 2$ and $u = 4$. Hence, (1.162) is satisfied, and let us proceed as in Case IV. The matrix U is invertible with $i = 1$ and $j_1 = 2$, and we obtain

$$X_2 = \begin{pmatrix} 1.000 & 0.000 & 0.000 & 0.000 \\ -0.000 & 1.000 & 0.000 & 0.000 \\ 0.000 & 0.000 & 0.000 & 1.000 \\ 0.000 & 0.000 & 1.000 & 0.000 \\ 0.818 & -0.091 & -0.727 & -0.273 \\ -3.545 & 2.727 & 0.818 & 4.182 \end{pmatrix}, \quad y(x) = \begin{pmatrix} 1 \\ x_1 \\ x_1^2 \\ x_2 \end{pmatrix}.$$

From this we obtain that

$$w(x_1) = \begin{pmatrix} 1 \\ x_1 \\ x_1^2 \\ \dfrac{-8x_1^2 - x_1 + 9}{11x_1 + 3} \end{pmatrix}.$$

It follows that

$$q(x_1) = 35x_1^4 - 245x_1^2 + 210x_1$$

which provides according to (1.145)

$$\hat{\mathscr{X}_1} = \{-3, 0, 1, 2\}$$

and, from (1.146), the set of candidates

$$\hat{\mathscr{X}} = \left\{ \begin{pmatrix} -3 \\ 2 \end{pmatrix}, \begin{pmatrix} 0 \\ -3 \end{pmatrix}, \begin{pmatrix} 1 \\ 0 \end{pmatrix}, \begin{pmatrix} 2 \\ -1 \end{pmatrix} \right\}.$$

According to (1.148) we simply find that $\mathscr{X}_{pol} = \hat{\mathscr{X}}$. □

1.5.2 Special Cases

Next, we consider the power vector $b_{lin}(x,m)$. Let us observe that, since $b_{lin}(x,m)$ vanishes for $x = 0_n$ and since the null vector belongs to $\mathscr{V}$, we can focus our attention on the case $x \neq 0_n$. This means that the problem can be formulated as

$$\mathscr{X}_{lin} = \{x \in \mathbb{R}_0^n : \ b_{lin}(x,m) \in \mathscr{V}\}. \tag{1.178}$$

As in Section 1.5.1, this problem amounts to solving a system of polynomial equations. Hereafter we explain how Cases I–IV derived in Section 1.5.1 should be modified in order to compute $\mathscr{X}_{lin}$.

Case V. Let us consider (1.178) when u satisfies

$$u \leq m-1. \tag{1.179}$$

Let us proceed as in (1.137)–(1.142) by replacing the definition of $y(x_i)$ in (1.140) with

$$y(x_i) = (x_i, \ldots, x_i^u)'. \tag{1.180}$$

Let us suppose that we have chosen an integer i such that U is invertible (we will consider that such an i does not exist in Case VII). By defining W as in (1.142), it follows that x_i satisfies

$$x_i^{u+1} = ay(x_i) \tag{1.181}$$

where a is the row of W corresponding to the monomial x_i^{u+1}, or in other words x_i is a root of the polynomial $q(x_i)$ in (1.144). We define $\hat{\mathscr{X}_i}$ as in (1.145), and

$$\hat{\mathscr{X}} = \left\{CWy(x_i), \ x_i \in \hat{\mathscr{X}_i}\right\} \tag{1.182}$$

where $C \in \mathbb{R}^{n \times d_{lin}(n,m)}$ is the matrix that extracts x from $b_{lin}(x,m)$, i.e.

$$x = Cb_{lin}(x,m). \tag{1.183}$$

Then, the sought set $\mathscr{X}_{lin}$ is a subset of $\hat{\mathscr{X}}$ according to

$$\mathscr{X}_{lin} = \left\{x \in \hat{X} : \ b_{lin}(x,m) \in \mathscr{V}\right\}. \tag{1.184}$$

Case VI. Consider (1.178) when u satisfies

$$u = m. \tag{1.185}$$

If $n = 1$, then it directly follows that $\mathscr{X}_{lin} = \mathbb{R}$. Moreover, if $m = 1$, it follows that $\mathscr{X}_{lin}$ is infinite. Hence, we consider that $n > 1$ and $m > 1$.

Let us define the matrix U as in Case V. Let us suppose that we have chosen an integer i such that U is invertible (we will consider that such an i does not exist in Case VII). Let us define the matrix W as in (1.142). It follows that x_i satisfies (1.150) where a and b are the rows of W corresponding to the monomials x_j and $x_i x_j$ for some integer j satisfying $1 \leq j \leq n$ and $j \neq i$, or in other words x_i is a root of the polynomial $q(x_i)$ in (1.151). Hence, the sought set $\mathscr{X}_{lin}$ is found according to (1.145) and (1.182)–(1.184).

Case VII. Here we complete Cases V and VI by considering that the matrix U defined in Case V is singular for all i, i.e.

$$\begin{cases} u \leq m \\ \nexists i: \ \det(U) \neq 0. \end{cases} \tag{1.186}$$

For a given i let us define the matrix U_k and the vector b according to (1.153)–(1.156) by replacing the definition of $y_j(x_i)$ in (1.154) with

$$y_j(x_i) = (x_i, \ldots, x_i^j)'. \tag{1.187}$$

It follows that $k < m$ and

$$x_i^{k+1} = by_k(x_i). \tag{1.188}$$

The solutions for x_i are the roots of the polynomial $q(x_i)$ in (1.159) that are gathered into $\hat{\mathscr{X}}_i$ in (1.160). By repeating this procedure for all $i = 1, \ldots, n$, one obtain the set of candidate solutions $\hat{\mathscr{X}}$ in (1.161), and the sought set $\mathscr{X}_{lin}$ is found as in (1.184).

Case VIII. Lastly, consider (1.178) when u satisfies

$$m+1 \leq u \leq n(m-1)+1. \tag{1.189}$$

Clearly, this implies that $n \geq 2$. We proceed as in Case IV till the definition of $\hat{\mathscr{X}}$ in (1.177) by replacing the list of monomials in (1.164) with

$$\begin{array}{c} x_i, \ldots, x_i^m \\ x_{j_1}, x_{j_1}x_i, \ldots, x_{j_1}x_i^{m-2} \\ \vdots \\ x_{j_k}, x_{j_k}x_i, \ldots, x_{j_k}x_i^{m-2} \end{array} \tag{1.190}$$

and the matrix C used in (1.177) with the one defined in (1.183). Then, the sought set $\mathscr{X}_{lin}$ is found according to (1.184).

Example 1.13. Let us consider the computation of $\mathscr{X}_{lin}$ in (1.178) with

$$b_{lin}(x,m) = \begin{pmatrix} x_1 \\ x_2 \\ x_1^2 \\ x_1x_2 \\ x_2^2 \end{pmatrix}, \ V = \begin{pmatrix} 1 & 5 \\ 3 & -3 \\ 5 & 7 \\ -3 & 3 \\ 9 & -9 \end{pmatrix}.$$

We have that $n = 2$, $m = 2$ and $u = 2$. Hence, (1.179) is satisfied, and let us proceed as in Case V. The matrix U is invertible with $i = 1$, and we obtain

$$q(x_1) = x_1^2 - x_1 - 2$$

which provides according to (1.145)

$$\hat{\mathscr{X}}_1 = \{-1, 2\}$$

and, from (1.182), the set of candidates

$$\hat{\mathscr{X}} = \left\{ \begin{pmatrix} -1 \\ 3 \end{pmatrix}, \begin{pmatrix} 2 \\ 0 \end{pmatrix} \right\}.$$

From $\hat{\mathscr{X}}$ we directly obtain the sought set $\mathscr{X}_{lin}$ according to (1.184) by simply establishing whether $b_{lin}(x,m) \in \mathscr{V}$ for each x in $\hat{\mathscr{X}}$. We find that all the candidates in $\hat{\mathscr{X}}$ belong to $\mathscr{X}_{lin}$, hence concluding that $\mathscr{X}_{lin} = \hat{\mathscr{X}}$. □

Example 1.14. Let us consider the computation of $\mathscr{X}_{lin}$ in (1.178) with

$$b_{lin}(x,m) = \begin{pmatrix} x_1 \\ x_2 \\ x_3 \\ x_1^2 \\ x_1x_2 \\ x_1x_3 \\ x_2^2 \\ x_2x_3 \\ x_3^2 \end{pmatrix}, \; V = \begin{pmatrix} 4 & 1 \\ -5 & 4 \\ 4 & -2 \\ 6 & 3 \\ -4 & 5 \\ 4 & -2 \\ 19 & -8 \\ -12 & 6 \\ 8 & -4 \end{pmatrix}.$$

We have that $n = 3$, $m = 2$ and $u = 2$. Hence, we proceed as in Case V. The matrix U is invertible with $i = 2$, and we obtain

$$q(x_2) = 0.583x_2^2 + 1.167x_2 - 1.750$$

which provides according to (1.145)

$$\hat{\mathscr{X}}_2 = \{-3, 1\}$$

and, from (1.182), the set of candidates

$$\hat{\mathscr{X}} = \left\{ \begin{pmatrix} 1 \\ -3 \\ 2 \end{pmatrix}, \begin{pmatrix} 2 \\ 1 \\ 0 \end{pmatrix} \right\}.$$

We find that all the candidates in $\hat{\mathscr{X}}$ belong to $\mathscr{X}_{lin}$, hence concluding that $\mathscr{X}_{lin} = \hat{\mathscr{X}}$. □

Next, we consider the power vector $b_{hom}(x,m)$. Let us observe that, since $b_{hom}(x,m)$ contains monomials with the same degree, namely m, it follows that

$$b_{hom}(x,m) \in \mathscr{V} \iff b_{hom}(\rho x,m) \in \mathscr{V} \;\; \forall \rho \in \mathbb{R}_0. \tag{1.191}$$

This means that the search can be reduced by considering vectors x with fixed norm and fixed direction, e.g. according to

$$\begin{cases} \|x\|_\infty = 1 \\ x_j > 0, \ j = \min\{i : \ |x_i| = 1\}. \end{cases} \tag{1.192}$$

Hence, the problem can be formulated as

$$\mathscr{X}_{hom} = \{x \in \mathbb{R}^n : \ b_{hom}(x,m) \in \mathscr{V} \ \text{ and (1.192) holds}\}. \tag{1.193}$$

One way to determine $\mathscr{X}_{hom}$ in (1.193) consists of exploiting the procedure previously described for the case of power vector $b_{pol}(x,m)$. Specifically, consider $x \in \mathscr{X}_{hom}$ and let i be such that $x_i \neq 0$. We can define

$$z(y) = \frac{x}{x_i} \tag{1.194}$$

where

$$y = (x_1, \dots, x_{i-1}, x_{i+1}, \dots, x_n)'. \tag{1.195}$$

Indeed, the i-th entry of $z(y)$ is 1, and $z(y)$ is a function of $n-1$ scalar variables, namely $x_1, \dots, x_{i-1}, x_{i+1}, \dots, x_n$ which are gathered into y. It follows that

$$b_{hom}(x,m) \in \mathscr{V} \ \text{ and } \ x_i \neq 0 \iff b_{pol}(z(y),m) \in \mathscr{V}. \tag{1.196}$$

Hence, let us define

$$\mathscr{Y}_i = \left\{y \in \mathbb{R}^{n-1} : \ b_{pol}(z(y),m) \in \mathscr{V}\right\} \tag{1.197}$$

and

$$\mathscr{X}_i = \left\{x = \gamma \frac{w(y)}{\|w(y)\|_\infty}, \ y \in \mathscr{Y}_i, \ \gamma = \operatorname{sgn}(w_j(y)), \ j = \min\{i : \ |w_i(y)_i| = \|w(y)\|_\infty\}\right\} \tag{1.198}$$

where $w(y)$ is the function that from y recovers the corresponding x, i.e.

$$w(y) = (y_1, \dots, y_{i-1}, 1, y_i, \dots, y_{n-1})'. \tag{1.199}$$

It follows that

$$x \in \mathscr{X}_{hom} \ \text{ and } \ x_i \neq 0 \iff x \in \mathscr{X}_i. \tag{1.200}$$

Therefore, the sought set $\mathscr{X}_{hom}$ is finally found by repeating this procedure for all $i = 1, \dots, n$, hence obtaining

$$\mathscr{X}_{hom} = \bigcup_{i=1}^{n} \mathscr{X}_i. \tag{1.201}$$

Example 1.15. Let us consider the computation of $\mathscr{X}_{hom}$ in (1.193) with

$$b_{hom}(x,m) = \begin{pmatrix} x_1^3 \\ x_1^2 x_2 \\ x_1 x_2^2 \\ x_2^3 \end{pmatrix}, \ V = \begin{pmatrix} 0 & 2 \\ -1 & 3 \\ -3 & 5 \\ -7 & 9 \end{pmatrix}.$$

We have that $n = 2$, $m = 3$ and $u = 2$. We compute the sets $\mathscr{X}_1$ and $\mathscr{X}_2$ defined in (1.198) by proceeding as in Case I. We find that

$$\mathscr{X}_1 = \mathscr{X}_2 = \mathscr{X}_{hom} = \left\{ \begin{pmatrix} 0.500 \\ 1.000 \end{pmatrix}, \begin{pmatrix} 1.000 \\ 1.000 \end{pmatrix} \right\}. \qquad \square$$

Lastly, we consider the problem of finding the Kronecker product of two power vectors in a linear subspace, which arises when dealing with parameter-dependent polynomials and matrix polynomials. In particular, we consider the problem

$$\mathscr{W} = \left\{ (x, \theta) \in \mathbb{R}_0^n \times \mathbb{R}^{n_\theta} : \; b_{pol}(\theta, m_\theta) \otimes b_{lin}(x, m) \in \mathscr{V} \right\} \tag{1.202}$$

where $\mathscr{V} \subseteq \mathbb{R}^{d_{linpol}(n, n_\theta, m, m_\theta)}$ is a linear subspace.

One way to address this problem consists of observing that

$$b_{pol}(\theta, m_\theta) \otimes b_{lin}(x, m) = (b_{lin}(x, m)', \ldots,)' \tag{1.203}$$

i.e. the first $d_{lin}(n, m)$ entries of $b_{pol}(\theta, m_\theta) \otimes b_{lin}(x, m)$ correspond to $b_{lin}(x, m)$. This means that one can address the determination of the x-part of $\mathscr{W}$ with the method in Section 1.5.2 for computing (1.178). Once that the x-part is determined, the θ-part of $\mathscr{W}$ can be found with the method in Section 1.5.1 for computing (1.134) by substituting in $b_{pol}(\theta, m_\theta) \otimes b_{lin}(x, m)$ the found values of x. See also [19] for other ways of exploiting the structure of $b_{pol}(\theta, m_\theta) \otimes b_{lin}(x, m)$.

1.6 Gap between Positive Polynomials and SOS Polynomials

While SOS polynomials are nonnegative, nonnegative polynomials may be not SOS. This fact was discovered by Hilbert in 1888 via a non-constructive proof [92]. The nonnegative polynomials that are not SOS are called PNS [21]. An example of PNS polynomial was provided by Motzkin in 1967. This polynomial has degree 6 in 2 scalar variables, and is given by

$$f_{Mot}(x) = x_1^4 x_2^2 + x_1^2 x_2^4 - 3x_1^2 x_2^2 + 1. \tag{1.204}$$

Indeed, it can be verified that $f_{Mot}(x)$ is nonnegative, in particular

$$\min_x f_{Mot}(x) = 0. \tag{1.205}$$

However, $f_{Mot}(x)$ is not SOS, which can be proved either analytically as shown in [94], or numerically by calculating its SOS index, which turns out to be $\lambda_{pol}(f_{Mot}) = -0.006$.

Another example was found by Choi and Lam, which is a polynomial of degree 4 in 4 scalar variables given by

$$f_{ChoLam}(x) = x_1^2 x_2^2 + x_1^2 x_3^2 + x_2^2 x_3^2 - 4x_1 x_2 x_3 + 1. \tag{1.206}$$

Also this polynomial is nonnegative and not SOS, in particular $\lambda_{pol}(f_{ChoLam}) = -0.030$.

Another example was found with the parametrization of PNS polynomials proposed in [21],

$$\begin{aligned} f_{Che}(x) = {} & 4(x_1^6 + x_2^6 + 1) - 19(x_1^4 x_2^2 + x_2^4 + x_1^2) - 30x_1^2 x_2^2 \\ & + 29(x_1^4 + x_2^4 x_1^2 + x_2^2) - \delta(x_1^2 + x_2^2 + 1)^3 \end{aligned} \quad (1.207)$$

which is a PNS polynomial for all $\delta \in (0, 0.436)$. Indeed, for such values of δ, $f_{Che}(x)$ is positive and not SOS. For instance, with $\delta = 0.4$ we obtain $\lambda_{pol}(f_{Che}) = -0.037$.

Although nonnegative polynomials may not be SOS, it is true that nonnegative polynomials are the ratio of two SOS polynomials, or equivalently sum of squares of rational functions. This was proved by Artin in 1927, providing a positive answer to Hilbert 17th problem as explained hereafter.

Theorem 1.8 ([51]). *Consider $f \in \mathscr{P}_n$. Then, $f(x)$ is nonnegative if and only if there exist $g_1, g_2 \in \mathscr{S}_n$ with $g_2(x)$ not identically zero such that*

$$f(x) = \frac{g_1(x)}{g_2(x)}. \quad (1.208)$$

Depending on the number of variables and degree of a polynomial, nonnegativity coincides with the property to be SOS, or equivalently a polynomial cannot be SOS, as explained in the following result.

Theorem 1.9 ([51]). *Consider $f \in \mathscr{P}_n$ and suppose that $(n, \partial f) \in \mathscr{E}$ where*

$$\mathscr{E} = \{(n,2), n \in \mathbb{N}\} \cup \{(1,2k), k \in \mathbb{N}\} \cup \{(2,4)\}. \quad (1.209)$$

Then, $f(x)$ is nonnegative if and only if $f(x)$ is SOS.

This result can be extended to the case of positive polynomials as explained hereafter.

Theorem 1.10. *Consider $f \in \mathscr{P}_n$ and suppose that $(n, \partial f) \in \mathscr{E}$. Suppose also that $f(x)$ is positive and that its highest degree form is positive definite. Then, $f(x)$ admits a positive definite SMR matrix, i.e.*

$$\exists F > 0 : \ f(x) = b_{pol}(x,m)' F b_{pol}(x,m). \quad (1.210)$$

Proof. Suppose that $f(x)$ is positive and that its highest degree form is positive definite. Define

$$g(x) = f(x) - a \|b_{pol}(x,m)\|^2$$

where

$$a = \min_x \frac{f(x)}{\|b_{pol}(x,m)\|^2}.$$

We have that $a > 0$ and that $g(x)$ is nonnegative. Moreover, since $(n,2m) \in \mathscr{E}$, it follows from Theorem 1.9 that $g(x)$ is SOS. Hence, there exists $G \geq 0$ such that

$$g(x) = b_{pol}(x,m)'Gb_{pol}(x,m).$$

Now, let us express $f(x)$ as $f(x) = b_{pol}(x,m)'Fb_{pol}(x,m)$. It follows that

$$F = G + aI$$

which implies that $F > 0$. □

The following result states that any PNS polynomial is the vertex of a cone of PNS polynomials.

Theorem 1.11 ([21]). *Consider $f \in \mathscr{P}_n$ with $\partial f = 2m$, $m \in \mathbb{N}$, and suppose that $f(x)$ is PNS. Define*

$$M = F + L(\alpha^*) \tag{1.211}$$

where $F + L(\alpha^*)$ is the found optimal value of $F + L(\alpha)$ in (1.58). *Let V be an orthonormal matrix whose columns are the eigenvectors of the eigenvalues of M strictly greater than the minimum eigenvalue of M. Define the cone of polynomials*

$$\mathscr{C}(f) = \{g \in \mathscr{P}_n : \ g(x) = f(x) + s(x,u,V), \ \ u > 0\} \tag{1.212}$$

where

$$s(x,u,V) = b_{pol}(x,m)'V\mathrm{diag}(u)V'b_{pol}(x,m). \tag{1.213}$$

Then, $\mathscr{C}(f)$ contains only PNS polynomials.

1.7 Notes and References

The representation of polynomials via the SMR described in Section 1.2 was introduced in [36]. In the literature, this representation is also known as Gram matrix, see e.g. [40] and references therein. The idea of representing polynomials through a quadratic form in a vector of monomials was proposed several times and in various contexts, see e.g. [9, 12, 4]. The SMR for parameter-dependent polynomials (or, equivalently, matrix polynomials) described in Section 1.4 was introduced in [30]. Algorithms for the computation of the SMR were provided in [15, 31, 33].

The LMI test for recognizing SOS polynomials described in Section 1.3 was introduced in [36, 83, 112]. For SOS parameter-dependent polynomials and SOS matrix polynomials, the LMI tests in Section 1.4 were introduced in [30, 66, 17, 91, 57]. The SOS indexes described in these sections extend the definition proposed in [33] for the case of homogeneous polynomials.

Alternative techniques for investigating positivity of polynomials via LMIs were proposed for example via slack variables [84], also in the context of robust analysis [79, 99, 80].

The extraction of power vectors from linear spaces described in Section 1.5 was proposed for the case of homogeneous polynomials in [28, 29]. An alternative technique for extracting power vectors from linear spaces was presented in [44, 55].

Hilbert 17th problem and the gap between positive polynomials and SOS polynomials discussed in Section 1.6 have attracted much interest, see e.g. the classical book [51] and recent contributions such as [94, 7, 69, 95, 21, 100].

Lastly, there exist various results in the literature that further characterize SOS polynomials, such as [93] which explains that the monomials required to write a polynomial as SOS can be chosen without loss of generality in the Newton polytope. This may allow to reduce the complexity of the LMI feasibility tests built in Section 1.3 for establishing if a polynomial is SOS.

Chapter 2
Optimization with SOS Polynomials

This chapter describes the use of SOS polynomials in typical optimization problems over polynomials. First, the case of unconstrained optimization is considered. It is shown how one can establish whether a polynomial is either positive or non-positive by introducing a generalized SOS index which measures how a polynomial is the ratio of two SOS polynomials and which can be found by solving an SDP. Then, the problem of determining the minimum of a rational function is considered, showing how one can obtain a lower bound through the generalized SOS index. This lower bound converges to the sought minimum as the order of the generalized SOS index increases, moreover a necessary and sufficient condition is provided for establishing its tightness. Second, the case of constrained optimization is considered. Stengle Positivstellensatz and some of its re-elaborations are reviewed, which provide strategies for establishing positivity of a polynomial over semialgebraic sets based on SOS polynomials. Then, the problem of determining the minimum of a rational function over semialgebraic sets is considered. For this problem it is shown that a direct use of the Positivstellensatz allows one to obtain a lower bound through SDPs, moreover a necessary and sufficient condition is provided for establishing its tightness. Moreover, the problem of solving systems of polynomial equations and inequalities is addressed, showing how a combination of the generalized SOS index and the Positivstellensatz can be used to compute the sought solutions through an SDP. Also, the problem of establishing whether a matrix polynomial is positive definite over semialgebraic sets is addressed by introducing a matrix version of the Positivstellensatz. Third, the case of optimization over special sets is considered. In particular, the chapter addresses the problems of establishing positivity of a polynomial over an ellipsoid and over the simplex. It is shown that these problems are equivalent to establishing positivity of a homogeneous polynomial constructed without introducing unknown multipliers. Lastly, the problem of searching for a positive semidefinite SMR matrix of a SOS polynomial with the largest rank is considered, showing that it can be exactly solved with a finite sequence of LMI feasibility tests.

G. Chesi: Domain of Attraction, LNCIS 415, pp. 45–85, 2011.
springerlink.com

2.1 Unconstrained Optimization

This section addresses the problems of establishing positivity or non-positivity of a polynomial, determining the minimum of a rational function, and solving systems of polynomial equations.

2.1.1 *Positivity and Non-positivity: General Case*

As seen in Chapter 1, one can investigate positivity of a polynomial by using the SOS index. Exploiting Theorem 1.8, one can introduce a generalized SOS index that can provide less conservative results in this investigation. Such a new index is defined as follows.

Definition 2.1 (Generalized SOS Index). Consider $f \in \mathscr{P}_n$ and $k \in \mathbb{N}$, and define $\mathbf{f} = COE_{pol}(f)$. Let us introduce

$$\lambda_{pol}(f,k) = \begin{cases} \beta z^* & \text{if } \|\mathbf{f}\| > 0 \\ 0 & \text{otherwise} \end{cases} \tag{2.1}$$

where β is a normalizing factor chosen as

$$\beta = \frac{\sqrt{d_{pol}(n,m_0)}}{\|\mathbf{f}\|}, \quad m_0 = \left\lceil \frac{\partial f}{2} \right\rceil \tag{2.2}$$

and z^* is the solution of the SDP

$$\begin{aligned} z^* &= \max_{Q,z,\alpha} z \\ &\text{s.t.} \begin{cases} P(Q)+L(\alpha)-zI \geq 0 \\ Q \geq 0 \\ \text{trace}(Q) = 1 \end{cases} \end{aligned} \tag{2.3}$$

and

$$\begin{aligned} P(Q)+L(\alpha) &= CSMR_{pol}(p) \\ p(x) &= q(x)f(x) \\ q(x) &= b_{pol}(x,k)'Qb_{pol}(x,k). \end{aligned} \tag{2.4}$$

Then, $\lambda_{pol}(f,k)$ is called *generalized SOS index of* $f(x)$. □

As it can be observed from its definition and as it will become clearer in the sequel, the generalized SOS index is a measure of how a polynomial is the ratio of two SOS polynomials. Indeed, the solution z^* of the SDP (2.3) corresponds to the maximum achievable z under the condition that $q(x)f(x) - z\|b_{pol}(x,m)\|^2$ and $q(x)$ are SOS, with $q(x)$ not identically zero and $m = m_0 + k = \lceil \partial qf/2 \rceil$. The integer k bounds the degree of $q(x)$ and $p(x)$, in particular from (2.4) it follows that $\partial q \leq 2k$

and $\partial p \leq 2m$. The equation $\text{trace}(Q) = 1$ normalizes the variables involved in (2.3): in fact,

$$\begin{array}{c} P(Q) + L(\alpha) - zI \geq 0 \text{ and } Q \geq 0 \text{ hold} \\ \Updownarrow \\ P(\rho Q) + L(\rho\alpha) - \rho zI \geq 0 \text{ and } \rho Q \geq 0 \text{ hold for all } \rho > 0. \end{array}$$

Let us also observe that the generalized SOS index coincides with the SOS index for $k = 0$, i.e.

$$\lambda_{pol}(f, 0) = \lambda_{pol}(f). \tag{2.5}$$

The generalized SOS index owns a monotonicity property as explained in the following result.

Theorem 2.1. *Consider $f \in \mathscr{P}_n$ and $k \in \mathbb{N}$. The following statements hold.*

1. *$\lambda_{pol}(f, k) \geq 0$ implies that $\lambda_{pol}(f, \tilde{k}) \geq 0$ for all $\tilde{k} \geq k$.*
2. *$\lambda_{pol}(f, k) > 0$ implies that $\lambda_{pol}(f, \tilde{k}) > 0$ for all $\tilde{k} \geq k$.*

Proof. Obviously we can just prove the theorem for $\tilde{k} = k + 1$. Let us select

$$\tilde{q}(x) = aq(x)(1 + \|x\|^2)$$

for some $a \in \mathbb{R}$. It follows that $\tilde{q}(x)$ can be written as

$$\tilde{q}(x) = b_{pol}(x, k+1)' \tilde{Q} b_{pol}(x, k+1)$$

with

$$\tilde{Q} = C'_{k+1}(Q \otimes aI_{n+1})C_{k+1}$$

where C_{k+1} is the matrix satisfying

$$b_{pol}(x, k) \otimes (1, x')' = C_{k+1} b_{pol}(x, k+1).$$

Then, it follows that $\tilde{p}(x) = \tilde{q}(x) f(x)$ can be written as

$$\tilde{p}(x) = b_{pol}(x, b+k+1)' \tilde{P}(\tilde{Q}) b_{pol}(x, b+k+1)$$

with

$$\tilde{P}(\tilde{Q}) = C'_{b+k+1}((P(Q) + L(\alpha)) \otimes aI_{n+1})C_{b+k+1}$$

where b is the smallest integer such that $2b \geq \partial f$.

Hence, let us suppose that $\lambda_{pol}(f, k) \geq 0$. This means that $P(Q) + L(\alpha) \geq 0$, $Q \geq 0$ and $\text{trace}(Q) = 1$. Since C_{k+1} and C_{b+k+1} are full column rank matrices, it follows that

$$\begin{cases} 0 \leq \tilde{P}(\tilde{Q}) \\ 0 \leq \tilde{Q} \\ 1 = \text{trace}(Q) \end{cases}$$

by simply selecting

$$a = \left(\text{trace}\left(C'_{k+1}(Q \otimes I_{n+1})C_{k+1}\right)\right)^{-1}$$

which implies that $\lambda_{pol}(f,k+1) \geq 0$. Then, let us suppose that $\lambda_{pol}(f,k) > 0$. This means that $P(Q)+L(\alpha) > 0$, $Q \geq 0$ and $\mathrm{trace}(Q) = 1$. Similarly it follows that

$$\begin{cases} 0 < \tilde{P}(\tilde{Q}) \\ 0 \leq \tilde{Q} \\ 1 = \mathrm{trace}(Q) \end{cases}$$

for the same value of a, hence implying that $\lambda_{pol}(f,k+1) > 0$. □

The following result explains how the generalized SOS index can be used to investigate positivity of a polynomial.

Theorem 2.2. *Consider $f \in \mathscr{P}_n$. The following statements hold.*

1. *$f(x)$ is nonnegative if and only if there exists k such that $\lambda_{pol}(f,k) \geq 0$.*
2. *$f(x)$ is positive if there exists k such that $\lambda_{pol}(f,k) > 0$.*

Proof. Let us consider the first statement. Suppose that there exists k such that $\lambda_{pol}(f,k) \geq 0$. This means that there exists a not identically zero SOS polynomial $q(x)$ such that $q(x)f(x)$ is SOS, and hence $f(x)$ is nonnegative from Theorem 1.8. Then, suppose that $f(x)$ is nonnegative. From Theorem 1.8 it follows that there exist two SOS polynomials $g_1(x)$ and $g_2(x)$ such that $g_2(x)f(x) = g_1(x)$ with $g_2(x)$ not identically zero. Let us define $q(x) = ag_2(x)$ where $a > 0$. Since $g_2(x)$ is SOS and $a > 0$ it follows that $q(x)$ admits a positive semidefinite SMR matrix Q, moreover a can be chosen so that $\mathrm{trace}(Q) = 1$. Then, one has that $p(x) = q(x)f(x) = ag_2(x)f(x) = ag_1(x)$. Since $g_1(x)$ is SOS and $a > 0$ it follows that $p(x)$ admits a positive semidefinite matrix, i.e. there exists α such that the LMIs in (2.3) hold with $z = 0$. This means that $\lambda_{pol}(f,k) \geq 0$.

Next, let us consider the second statement. Suppose that there exists k such that $\lambda_{pol}(f,k) > 0$. This means that there exists a SOS polynomial $q(x)$ such that $q(x)f(x)$ is positive, and hence $f(x)$ is positive because

$$q(x) \geq 0 \text{ and } q(x)f(x) > 0 \;\Rightarrow\; q(x) > 0 \text{ and } f(x) > 0.$$

□

Depending on the number of variables and degree of $f(x)$, a stronger version of the conditions in Theorem 2.2 can be derived as explained hereafter.

Theorem 2.3. *Consider $f \in \mathscr{P}_n$ and suppose that $(n, \partial f) \in \mathscr{E}$ with $\mathscr{E}$ as in* (1.209). *Then, the following statements hold.*

1. *$f(x)$ is nonnegative if and only if $\lambda_{pol}(f,k) \geq 0$ for an arbitrarily chosen $k \in \mathbb{N}$.*
2. *$f(x)$ is positive and its highest degree form is positive definite if and only if $\lambda_{pol}(f,k) > 0$ for an arbitrarily chosen $k \in \mathbb{N}$.*

Proof. We prove the theorem for $k = 0$, which implies that the theorem holds for any $k \in \mathbb{N}$ from Theorems 2.1 and 2.2. Suppose that $(n, \partial f) \in \mathscr{E}$. Then, it follows from Theorem 1.9 that $f(x)$ is nonnegative if and only if $f(x)$ is SOS, i.e. $\lambda_{pol}(f) \geq 0$. Lastly, it follows from Theorem 1.10 that $f(x)$ is positive and its largest degree form is positive definite if and only if $f(x)$ admits a positive definite SMR matrix, i.e. $\lambda_{pol}(f) > 0$. □

At this point, the question is what can be said about nonnegativity of a polynomial if its generalized SOS index is negative. In fact, this can be the case of a nonnegative polynomial, hence a PNS polynomial, or of a polynomial that is negative for some values of the variable. The following result provides an answer to this question.

Theorem 2.4. *Consider $f \in \mathscr{P}_n$ and suppose that $\lambda_{pol}(f,k) < 0$ for some $k \in \mathbb{N}$. Define*

$$\mathscr{M} = \left\{x \in \mathbb{R}^n : \; b_{pol}(x,m) \in \ker(M)\right\} \tag{2.6}$$

where $m = \lceil \partial f/2 \rceil + k$ and

$$M = P(Q^*) + L(\alpha^*) - z^* I \tag{2.7}$$

where $P(Q^*) + L(\alpha^*) - z^* I$ is the found optimal value of $P(Q) + L(\alpha) - zI$ in (2.3). *Then, there exists $x \in \mathbb{R}^n$ such that $f(x) < 0$ if $\mathscr{M}$ is nonempty. In particular,*

$$f(x) < 0 \;\; \forall x \in \mathscr{M}. \tag{2.8}$$

Proof. Let $x \in \mathscr{M}$. It follows that

$$\begin{aligned} 0 &= b_{pol}(x,m)' M b_{pol}(x,m) \\ &= q(x) f(x) - \lambda_{pol}(f,k) \|b_{pol}(x,m)\|^2 \end{aligned}$$

where $q(x) \geq 0$ due to the second LMI in (2.3). Since $\lambda_{pol}(f,k) < 0$ and $\|b_{pol}(x,m)\|$ > 0, it follows that $q(x)f(x) \neq 0$. This means that $q(x) > 0$ and $f(x) < 0$. □

Theorem 2.4 provides a condition for establishing whether a polynomial $f(x)$ is negative for some x. In particular, this happens if the set $\mathscr{M}$ in (2.6) is nonempty. This set can be computed as explained in Section 1.5.1.

We conclude this section observing that, while the condition for nonnegativity provided by Theorem 2.2 is sufficient and necessary, the one provided for positivity is only sufficient. Indeed, a polynomial needs to have its highest degree form positive definite in order to have a positive generalized SOS index, while such a requirement is not necessary for the polynomial in order to be positive. For instance,

$$f(x) = 1 + x_1^2 + x_2^4 \text{ is positive, but } \lambda_{pol}(f,k) = 0 \;\; \forall k \in \mathbb{N}.$$

One way to cope with this problem was proposed in [81] in the context of systems analysis and consists of establishing whether the polynomial under investigation is not smaller than a known positive polynomial. In our case, this idea can be exploited as follows. For $k \in \mathbb{N}$ and $\varepsilon \in \mathbb{R}$, $\varepsilon > 0$, define the SDP

$$z_{pol} = \sup_{Q,z,\Psi,\alpha} z$$
$$\text{s.t.} \begin{cases} \tilde{P}(Q,\Psi) + L(\alpha) - zI \geq 0 \\ Q \geq 0 \\ \text{trace}(Q) = 1 \\ \sum_{j=1}^{m} \Psi_{i,j} \geq \varepsilon \ \forall i = 1,\dots,n \\ \Psi_{i,j} \geq 0 \ \forall i = 1,\dots,n \ \forall j = 1,\dots,m \end{cases} \tag{2.9}$$

where $m = \lceil \partial f/2 \rceil + k$ and

$$\begin{aligned} \tilde{P}(Q,\Psi) + L(\alpha) &= CSMR_{pol}(\tilde{p}) \\ \tilde{p}(x) &= q(x)f(x) - \sum_{(i,j)=(1,1)}^{(n,m)} \Psi_{i,j} x_i^{2j} \\ q(x) &= b_{pol}(x,k)' Q b_{pol}(x,k). \end{aligned} \tag{2.10}$$

Then,

$$z_{pol} \geq 0 \ \Rightarrow \ f(x) \text{ is positive} \tag{2.11}$$

because the LMIs in (2.9) impose that $q(x)f(x) - z\|b_{pol}(x,m)\|^2 - \sum_{(i,j)=(1,1)}^{(n,m)} \Psi_{i,j} x_i^{2j}$ and $q(x)$ are SOS with $\Psi_{i,j} \geq 0$ such that $\sum_{j=1}^{m} \Psi_{i,j} \geq \varepsilon$. In other words, z_{pol} is an alternative generalized SOS index defined for investigating positivity (rather than nonnegativity).

Example 2.1. Consider the PNS polynomial in (1.204) (Motzkin polynomial). While $\lambda_{pol}(f_{Mot}) = -0.006$, we find $\lambda_{pol}(f_{Mot}, 1) = 0.000$. In particular, the optimal value of Q in (2.3) provides that $q(x) f_{Mot}(x)$ is SOS with

$$q(x) = 0.451 + 0.278x_1^2 + 0.272x_2^2.$$

Similarly, for the PNS polynomial in (1.206) (Choi and Lam polynomial), we have $\lambda_{pol}(f_{ChoLam}) = -0.030$ and $\lambda_{pol}(f_{ChoLam}, 1) = 0.000$. In particular, the optimal value of Q in (2.3) provides that $q(x) f_{ChoLam}(x)$ is SOS with

$$q(x) = 0.226 + 0.255x_1^2 + 0.250x_2^2 + 0.268x_3^2.$$

Lastly, for the PNS polynomial in (1.207) with $\delta = 0.4$, we have $\lambda_{pol}(f_{Che}) = -0.037$ and $\lambda_{pol}(f_{Che}, 1) = 0.001$, hence proving that $f_{Che}(x)$ is positive. In particular, the optimal value of Q in (2.3) provides that $q(x) f_{Che}(x)$ is SOS with

$$q(x) = 0.333 + 0.333x_1^2 + 0.333x_2^2.$$

□

Example 2.2. Consider the polynomial

$$f(x) = 3.3 - 2x_1 + 3x_2 - x_1x_2 + x_1^2x_2 - x_1x_2^3 + x_1^3x_2^2 + x_1^6 + x_2^6.$$

It turns out that $\lambda_{pol}(f) = -5.787 \cdot 10^{-5}$. Let us use Theorem 2.4. We find that $\mathscr{M}$ in (2.6) for $k = 0$ is given by

$$\mathscr{M} = \{(-1.039, -1.144)'\}$$

and hence $f(x)$ is non-positive since $\mathscr{M} \neq \emptyset$. Indeed, $f(x) = -0.002$ for $x \in \mathscr{M}$ (in this case, $\dim(\ker(M)) = 1$). It is worth noticing that the region where $f(x)$ is negative is rather small as shown by Figure 2.1.

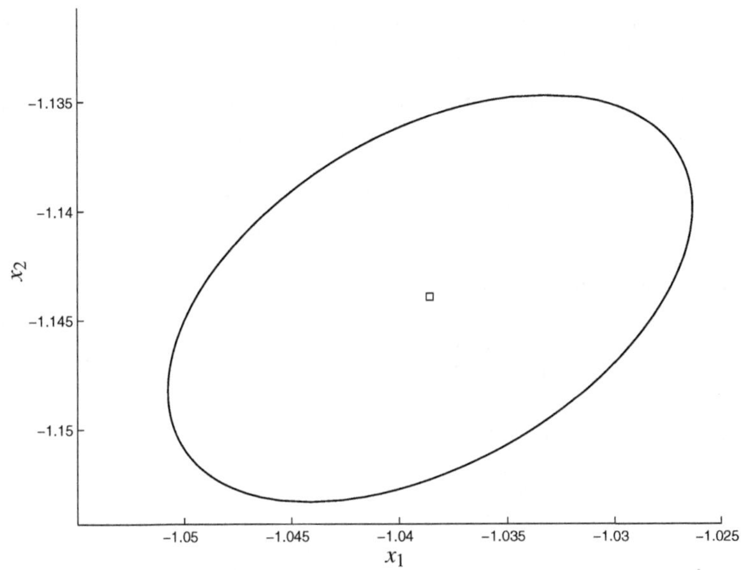

Fig. 2.1 Example 2.2. Curve $f(x) = 0$ (solid) and point in $\mathscr{M}$ ("□" marker). The polynomial is negative inside the region delimited by the curve. □

Example 2.3. Consider the polynomial

$$f(x) = 1 + 1.4x_1x_2x_3 + x_1^6 + 2.8x_1^3x_2^2x_3 + x_2^6 + x_3^6.$$

It turns out that $\lambda_{pol}(f) = -0.039$. Let us use Theorem 2.4. We find that $\mathscr{M}$ in (2.16) for $k = 0$ is given by

$$\mathscr{M} = \{(-1.323, 1.240, 1.114)', (1.323, 1.240, -1.114)'\}$$

and hence $f(x)$ is non-positive since $\mathscr{M} \neq \emptyset$. Indeed, $f(x) = -1.755$ for all $x \in \mathscr{M}$ (in this case, $\dim(\ker(M)) = 2$). It is worth noticing that $f(x)$ is unbounded from below. □

Example 2.4. Consider the polynomial

$$f(x) = 1 + x_1x_2 + x_2^2 + x_1^4.$$

With $k = 0$ and $\varepsilon = 0.001$, (2.9) provides $z_{pol} = 0.000$. Therefore, we conclude that $f(x)$ is positive from (2.11). □

2.1.2 Positivity and Non-positivity: Special Cases

If $f(x)$ is a locally quadratic polynomial, then the results just presented can be simplified. In the following we present some of them.

Definition 2.2 (Generalized SOS Index (continued)). Let $f \in \mathscr{P}_n$ be a locally quadratic polynomial, and let $k \in \mathbb{N}$. Let us define $\mathbf{f} = COE_{lin}(f)$ and

$$\lambda_{qua}(f,k) = \begin{cases} \beta z^* & \text{if } \|\mathbf{f}\| > 0 \\ 0 & \text{otherwise} \end{cases} \tag{2.12}$$

where β is a normalizing factor chosen as

$$\beta = \frac{\sqrt{d_{lin}(n,m_0)}}{\|\mathbf{f}\|}, \quad m_0 = \left\lceil \frac{\partial f}{2} \right\rceil \tag{2.13}$$

and z^* is the solution of the SDP

$$\begin{aligned} z^* &= \max_{Q,z,\alpha} z \\ &\text{s.t.} \begin{cases} P(Q) + L(\alpha) - zI \geq 0 \\ Q \geq 0 \\ \text{trace}(Q) = 1 \end{cases} \end{aligned} \tag{2.14}$$

and

$$\begin{aligned} P(Q) + L(\alpha) &= CSMR_{qua}(p) \\ p(x) &= q(x)f(x) \\ q(x) &= b_{pol}(x,k)'Qb_{pol}(x,k). \end{aligned} \tag{2.15}$$

Then, $\lambda_{qua}(f,k)$ is called *generalized SOS index of* $f(x)$ *(for locally quadratic polynomials).* □

The generalized SOS index for locally quadratic polynomials coincides with the SOS index for such polynomials for $k = 0$, i.e. $\lambda_{qua}(f) = \lambda_{qua}(f,0)$. Moreover, $\lambda_{qua}(f,k)$ owns a monotonicity property as explained in the following result.

Corollary 2.1. *Let* $f \in \mathscr{P}_n$ *be a locally quadratic polynomial, and let* $k \in \mathbb{N}$. *The following statements hold.*

1. $\lambda_{qua}(f,k) \geq 0$ *implies that* $\lambda_{qua}(f,\tilde{k}) \geq 0$ *for all* $\tilde{k} \geq k$.
2. $\lambda_{qua}(f,k) > 0$ *implies that* $\lambda_{qua}(f,\tilde{k}) > 0$ *for all* $\tilde{k} \geq k$.

Proof. Analogous to the proof of Theorem 2.1. □

The generalized SOS index $\lambda_{qua}(f,k)$ allows one to investigate positive semidefiniteness and positive definiteness as explained hereafter.

Corollary 2.2. *Let $f \in \mathscr{P}_n$ be a locally quadratic polynomial. The following statements hold.*

1. *$f(x)$ is positive semidefinite if and only if there exists k such that $\lambda_{qua}(f,k) \geq 0$.*
2. *$f(x)$ is positive definite if there exists k such that $\lambda_{qua}(f,k) > 0$.*

Proof. Analogous to the proof of Theorem 2.2. □

A stronger version of the previous result can be obtained depending on the number of variables and degree of $f(x)$ as reported in the following corollary.

Corollary 2.3. *Let $f \in \mathscr{P}_n$ be a locally quadratic polynomial, and suppose that $(n, \partial f) \in \mathscr{E}$ with $\mathscr{E}$ as in* (1.209). *Then, the following statements hold.*

1. *$f(x)$ is positive semidefinite if and only if $\lambda_{qua}(f,k) \geq 0$ for an arbitrarily chosen $k \in \mathbb{N}$.*
2. *$f(x)$ is positive definite and its largest degree form is positive definite if and only if $\lambda_{qua}(f,k) > 0$ for an arbitrarily chosen $k \in \mathbb{N}$.*

Proof. Analogous to the proof of Theorem 2.3. □

Also, one can establish whether a locally quadratic polynomial is negative for some value of the variable as explained hereafter.

Corollary 2.4. *Let $f \in \mathscr{P}_n$ be a locally quadratic polynomial, and suppose that $\lambda_{qua}(f,k) < 0$ for some $k \in \mathbb{N}$. Define*

$$\mathscr{M} = \{x \in \mathbb{R}_0^n : \ b_{lin}(x,m) \in \ker(M)\} \tag{2.16}$$

where $m = \lceil \partial f/2 \rceil + k$ and

$$M = P(Q^*) + L(\alpha^*) - \lambda_{qua}(f,k) I \tag{2.17}$$

where $P(Q^*) + L(\alpha^*)$ is the found optimal value of $P(Q) + L(\alpha)$ in (2.14). *Then, there exists $x \in \mathbb{R}_0^n$ such that $f(x) < 0$ if $\mathscr{M}$ is nonempty. In particular,*

$$f(x) < 0 \ \ \forall x \in \mathscr{M}. \tag{2.18}$$

Proof. Analogous to the proof of Theorem 2.4 by observing that $0_n \notin \mathscr{M}$ and hence $\|b_{lin}(x,m)\| > 0$ for all $x \in \mathscr{M}$. □

Corollary 2.4 provides a condition for establishing whether a locally quadratic polynomial $f(x)$ is negative for some x. In particular, this happens if the set $\mathscr{M}$ in (2.16) is nonempty. This set can be computed as explained in Section 1.5.2. Observe that the origin is excluded from the set $\mathscr{M}$ in (2.16). This is in accordance with the fact that $f(x)$ is a locally quadratic polynomial, and hence cannot be negative at the origin.

Let us observe that, while the condition for positive semidefiniteness provided by Corollary 2.2 is sufficient and necessary, the one provided for positive definiteness is only sufficient. Indeed, a locally quadratic polynomial needs to have its second degree and highest degree forms positive definite in order to have a positive generalized SOS index, while such a requirement is not necessary for the polynomial in order to be positive definite. For instance,

$$\left.\begin{array}{l} f(x) = x_2^2 + x_1^4 + x_2^4 \\ \text{and} \\ f(x) = x_1^2 + x_2^2 + x_1^4 \end{array}\right\} \text{ are positive definite, but } \lambda_{qua}(f,k) = 0 \ \forall k \in \mathbb{N}.$$

One way to cope with this problem is to follow the technique exploited in (2.9). Specifically, for $k \in \mathbb{N}$ and $\varepsilon \in \mathbb{R}$, $\varepsilon > 0$, define the SDP

$$z_{qua} = \sup_{Q,z,\Psi,\alpha} \ z$$
$$\text{s.t.} \left\{\begin{array}{r} \tilde{P}(Q,\Psi) + L(\alpha) - zI \geq 0 \\ Q \geq 0 \\ \text{trace}(Q) = 1 \\ \sum_{j=1}^{m} \Psi_{i,j} \geq \varepsilon \ \forall i = 1,\ldots,n \\ \Psi_{i,j} \geq 0 \ \forall i = 1,\ldots,n \ \forall j = 1,\ldots,m \end{array}\right. \tag{2.19}$$

where $m = \lceil \partial f/2 \rceil + k$ and

$$\begin{aligned} \tilde{P}(Q,\Psi) + L(\alpha) &= CSMR_{qua}(\tilde{p}) \\ \tilde{p}(x) &= q(x)f(x) - \textstyle\sum_{(i,j)=(1,1)}^{(n,m)} \Psi_{i,j} x_i^{2j} \\ q(x) &= b_{pol}(x,k)' Q b_{pol}(x,k). \end{aligned} \tag{2.20}$$

Then, it directly follows that

$$z_{qua} \geq 0 \ \Rightarrow \ f(x) \text{ is positive definite.} \tag{2.21}$$

Lastly, the previous results can be further simplified for the case of homogeneous polynomials.

Definition 2.3 (Generalized SOS Index (continued)). Let $f \in \mathscr{P}_n$ be a homogeneous polynomial of even degree, and let $k \in \mathbb{N}$. Let us define $\mathbf{f} = COE_{hom}(f)$ and

$$\lambda_{hom}(f,k) = \begin{cases} \beta z^* & \text{if } \|\mathbf{f}\| > 0 \\ 0 & \text{otherwise} \end{cases} \tag{2.22}$$

where β is a normalizing factor chosen as

$$\beta = \frac{\sqrt{d_{hom}(n,m_0)}}{\|\mathbf{f}\|}, \quad m_0 = \frac{\partial f}{2} \tag{2.23}$$

and z^* is the solution of the SDP

$$z^* = \max_{Q,z,\alpha} z \quad \text{s.t.} \begin{cases} P(Q)+L(\alpha)-zI \geq 0 \\ Q \geq 0 \\ \text{trace}(Q) = 1 \end{cases} \tag{2.24}$$

and

$$\begin{aligned} P(Q)+L(\alpha) &= CSMR_{hom}(p) \\ p(x) &= q(x)f(x) \\ q(x) &= b_{hom}(x,k)'Qb_{hom}(x,k). \end{aligned} \tag{2.25}$$

Then, $\lambda_{hom}(f,k)$ is called *generalized SOS index of $f(x)$ (for homogeneous polynomials).* □

The generalized SOS index for homogeneous polynomials coincides with the SOS index for such polynomials for $k=0$, i.e. $\lambda_{hom}(f) = \lambda_{hom}(f,0)$. Moreover, $\lambda_{hom}(f,k)$ owns the following monotonicity property.

Corollary 2.5. *Let $f \in \mathscr{P}_n$ be a homogeneous polynomial, and let $k \in \mathbb{N}$. The following statements hold.*

1. *$\lambda_{hom}(f,k) \geq 0$ implies that $\lambda_{hom}(f,\tilde{k}) \geq 0$ for all $\tilde{k} \geq k$.*
2. *$\lambda_{hom}(f,k) > 0$ implies that $\lambda_{hom}(f,\tilde{k}) > 0$ for all $\tilde{k} \geq k$.*

Proof. Analogous to the proof of Theorem 2.1. □

The generalized SOS index $\lambda_{hom}(f,k)$ allows one to investigate positive semidefiniteness and positive definiteness as explained hereafter. Interesting, the condition is sufficient and necessary not only for positive semidefiniteness but also for positive definiteness contrary to the case of polynomials and locally quadratic polynomials.

Theorem 2.5. *Consider $f \in \mathscr{P}_n$ be a homogeneous polynomial. The following statements hold.*

1. *$f(x)$ is positive semidefinite if and only if there exists k such that $\lambda_{hom}(f,k) \geq 0$.*
2. *$f(x)$ is positive definite if and only if there exists k such that $\lambda_{hom}(f,k) > 0$.*

Proof. Analogous to the proof of Theorem 2.2 for the first statement and for the sufficiency of the second statement. The necessity of the second statement follows from a result in [100] which states that a homogeneous polynomial $f(x)$ is positive definite if and only if for all positive definite homogeneous polynomials $g(x)$ such that ∂g divides ∂f there exists $i \in \mathbb{N}$ such that $g(x)^i f(x)$ is SOS. □

Corollary 2.6. *Let $f \in \mathscr{P}_n$ be a homogeneous polynomial, and suppose that $(n-1, \partial f) \in \mathscr{E}$ with $\mathscr{E}$ as in* (1.209). *Then, the following statements hold.*

1. *$f(x)$ is positive semidefinite if and only if $\lambda_{hom}(f,k) \geq 0$ for an arbitrarily chosen $k \in \mathbb{N}$.*
2. *$f(x)$ is positive definite if and only if $\lambda_{hom}(f,k) > 0$ for an arbitrarily chosen $k \in \mathbb{N}$.*

Proof. Analogous to the proof of Theorem 2.3 by observing that a homogeneous polynomial in n scalar variables is equivalent to a polynomial in $n-1$ scalar variables. □

Corollary 2.7. *Consider $f \in \mathscr{P}_n$ homogeneous, and suppose that $\lambda_{hom}(f,k) < 0$ for some $k \in \mathbb{N}$. Define*

$$\mathscr{M} = \{x \in \mathbb{R}_0^n : \ b_{hom}(x,m) \in \ker(M) \ \text{ and } (1.192) \text{ holds}\} \tag{2.26}$$

where $m = \partial f/2 + k$ and

$$M = P(Q^*) + L(\alpha^*) - \lambda_{hom}(f,k)I \tag{2.27}$$

where $P(Q^*) + L(\alpha^*)$ is the found optimal value of $P(Q) + L(\alpha)$ in (2.24). *Then, there exists $x \in \mathbb{R}_0^n$ such that $f(x) < 0$ if $\mathscr{M}$ is nonempty. In particular,*

$$f(x) < 0 \ \ \forall x \in \mathscr{M}. \tag{2.28}$$

Proof. Analogous to the proof of Theorem 2.4 by observing that $0_n \notin \mathscr{M}$ and hence $\|b_{hom}(x,m)\| > 0$ for all $x \in \mathscr{M}$. □

Corollary 2.7 provides a condition for establishing whether a homogeneous polynomial $f(x)$ is negative for some x. In particular, this happens if the set $\mathscr{M}$ in (2.26) is nonempty. This set can be computed as explained in Section 1.5.2 after (1.193).

Example 2.5. Consider the locally quadratic polynomial $f(x) = x_1^2 + 2x_2^2 - 4x_1^2 x_2$ in (1.23). As seen in Example 1.8, it turns out that $\lambda_{qua}(f) = -0.211$. Clearly, $f(x)$ is non-positive since its degree is odd. Another way of proving this fact consists of using Corollary 2.4: indeed, the set $\mathscr{M}$ in (2.16) for $k = 0$ is given by

$$\mathscr{M} = \{(-1.876, 1.300)', (1.876, 1.300)'\}$$

and it turns out that $f(x) = -11.399$ for all $x \in \mathscr{M}$ (in this case, $\dim(\ker(M)) = 2$). □

Example 2.6. Consider the locally quadratic polynomial $g(x) = x_1^2 + 2x_2^2 - 4x_1^2 x_2 + x_1^4 + x_2^4$ in (1.73). As seen in Example 1.8, it turns out that $\lambda_{qua}(g) = 0.044$, and hence $g(x)$ is positive definite from Corollary 2.2.

It is worth observing that, by using the SOS index $\lambda_{pol}(g)$ (which considers the case of generic polynomials) instead of $\lambda_{qua}(g)$, one gets $\lambda_{pol}(g) = 0.000$, which proves that $g(x)$ is positive semidefinite but does not prove that $g(x)$ is positive definite. □

Example 2.7. Consider the homogeneous polynomial $f(x) = x_1^6 - 2x_1^5 x_2 + 4x_1^3 x_2^3 + x_2^6$ in (1.24). As seen in Example 1.9, it turns out that $\lambda_{hom}(f) = -0.032$. From Corollary 2.6 it follows that $f(x)$ can be negative for some values of x. This can be proved also by using Corollary 2.7: indeed, the set $\mathscr{M}$ in (2.26) for $k = 0$ is given by

$$\mathscr{M} = \left\{(-0.863, 1.000)'\right\}.$$

and it turns out that $f(x) = -0.200$ for $x \in \mathscr{M}$ (in this case, $\dim(\ker(M)) = 1$). □

Example 2.8. Consider the homogeneous polynomial $g(x) = x_1^6 - 2x_1^5 x_2 + 3x_1^3 x_2^3 + x_2^6$ in (1.74). As seen in Example 1.9, it turns out that $\lambda_{hom}(g) = 0.083$, and hence $g(x)$ is positive definite from Theorem 2.5.

It is worth observing that, by using the SOS indexes $\lambda_{pol}(g)$ and $\lambda_{qua}(g)$ (which consider the cases of generic polynomials and locally quadratic polynomials) instead of $\lambda_{hom}(f)$, one gets $\lambda_{pol}(g) = 0.000$, which proves that $g(x)$ is positive semidefinite but does not prove that $g(x)$ is positive definite. □

2.1.3 *Minimum of Rational Functions*

One of the most direct applications of SOS polynomials is in unconstrained optimization of polynomials and rational functions, which can be formulated as

$$\gamma = \inf_x \frac{f(x)}{g(x)} \tag{2.29}$$

for $f, g \in \mathscr{P}_n$. We consider that $g(x)$ satisfies

$$g(x) > 0 \;\; \forall x \in \mathbb{R}^n. \tag{2.30}$$

Condition (2.30) is a mild assumption because, if not satisfied, one has that:

1. either $-g(x)$ is positive, and hence one can just redefine the problem changing the sign of $f(x)$ and $g(x)$;
2. or $g(x)$ can change sign, which means that γ is in general $-\infty$ (unless $f(x)$ vanishes whenever $g(x)$ does).

In order to solve (2.29), let us exploit the generalized SOS index. Specifically, for $k \in \mathbb{N}$ let us define

$$\hat{\gamma} = \sup_{z} z \quad \text{s.t.} \quad \begin{cases} \lambda_{pol}(h,k) \geq 0 \\ h(x) = f(x) - zg(x). \end{cases} \tag{2.31}$$

Since $\lambda_{pol}(h,k) \geq 0$ implies that $f(x) - zg(x)$ is nonnegative, it directly follows that $\hat{\gamma}$ is a lower bound of the sought γ, i.e.

$$\hat{\gamma} \leq \gamma. \tag{2.32}$$

From the definition of generalized SOS index, (2.31) can be rewritten as

$$\hat{\gamma} = \sup_{q,z} z \quad \text{s.t.} \quad \begin{cases} p \in \mathscr{S}_n, \ \partial p \leq 2m \\ q \in \mathscr{S}_n, \ q(x) \text{ not identically zero} \end{cases} \tag{2.33}$$

where

$$p(x) = q(x)(f(x) - zg(x)) \tag{2.34}$$

and

$$m = \left\lceil \frac{\max\{\partial f, \partial g\}}{2} \right\rceil + k. \tag{2.35}$$

By exploiting the SMR, (2.33) can be solved as explained in the following result.

Corollary 2.8. *Let $f, g \in \mathscr{P}_n$ and suppose that* (2.30) *holds. Then, the lower bound $\hat{\gamma}$ in* (2.31) *is given by*

$$\hat{\gamma} = \sup_{Q,z,\alpha} z \quad s.t. \quad \begin{cases} P(Q,z) + L(\alpha) \geq 0 \\ Q \geq 0 \\ \operatorname{trace}(Q) = 1 \end{cases} \tag{2.36}$$

where

$$\begin{cases} P(Q,z) + L(\alpha) = CSMR_{pol}(p) \\ p(x) = q(x)(f(x) - zg(x)) \\ q(x) = b_{pol}(x,k)' Q b_{pol}(x,k). \end{cases} \tag{2.37}$$

Proof. Direct consequence of the definition of SOS polynomials. □

Let us observe that the optimization problem (2.36) might be nonconvex due to the presence of a BMI, specifically the inequality $P(Q,z) + L(\alpha) \geq 0$ where $P(Q,z)$ is a bilinear matrix function of Q and z.

Nevertheless, the lower bound $\hat{\gamma}$ can be found via a one-parameter sequence of convex optimization problems, namely SDPs, by solving (2.36) for fixed values of z. Moreover, this search can be performed via a bisection algorithm, hence speeding up the convergence.

Let us observe that, for $k = 0$, $\hat{\gamma}$ can be found via an SDP only: in fact, in such a case, $Q = 1$, i.e. Q is constant.

Clearly, $\hat{\gamma}$ can be used as an alternative generalized SOS index of a polynomial (in the case $g(x) = constant$) being a lower bound of its minimum. However, contrary to the generalized SOS index, $\hat{\gamma}$ might not exist. For instance, this is the case when $f(x)$ is unbounded from below.

The lower bound $\hat{\gamma}$ owns the following monotonicity property.

Theorem 2.6. *Let $\hat{\gamma}$ be as in* (2.36). *For all $\tilde{k} \in \mathbb{N}$ one has*

$$\hat{\gamma}|_{k=\tilde{k}} \leq \hat{\gamma}|_{k=\tilde{k}+1} \,. \tag{2.38}$$

Proof. Suppose that $\lambda_{pol}(p,k) \geq 0$. This means that there exists a not identically zero SOS polynomial $q(x)$ with $\partial q \leq 2k$ such that $q(x)p(x)$ is a SOS polynomial. Define

$$\tilde{q}(x) = a(1 + \|x\|^2)q(x)$$

where $a > 0$. It follows that $\tilde{q}(x)$ is a not identically zero SOS polynomial with $\partial\tilde{q} = 2(k+1)$. Moreover, $\tilde{q}(x)p(x)$ is a SOS polynomial since $1 + \|x\|^2$ is. Then, a can be chosen so that $\tilde{q}(x)$ has a positive semidefinite SMR matrix $\tilde{Q}$ satisfying $\mathrm{trace}(\tilde{Q}) = 1$. This implies that $\lambda_{pol}(p,k+1) \geq 0$, and hence (2.38) holds. □

The following result provides a necessary and sufficient condition for establishing tightness of the found $\hat{\gamma}$.

Theorem 2.7. *Let γ and $\hat{\gamma}$ be as in* (2.29) *and* (2.36), *respectively. Suppose that the minimizers of* (2.36) *exist. Define*

$$\mathscr{M} = \left\{x \in \mathbb{R}^n : \, b_{pol}(x,m) \in \ker(M)\right\} \tag{2.39}$$

where m is as in (2.35) *and*

$$M = P(Q^*,\hat{\gamma}) + L(\alpha^*) \tag{2.40}$$

where $P(Q^*,\hat{\gamma}) + L(\alpha^*)$ is the found optimal value of $P(Q,z) + L(\alpha)$ in (2.36). *Define also*

$$\mathscr{M}_1 = \left\{x \in \mathscr{M} : \, \frac{f(x)}{g(x)} = \hat{\gamma}\right\}. \tag{2.41}$$

Then,

$$\mathscr{M}_1 \neq \emptyset \iff \begin{cases} \hat{\gamma} = \gamma \\ \textit{the minimizer of } (2.29) \textit{ exists.} \end{cases} \tag{2.42}$$

Moreover, for $k = 0$, one has that

$$\mathscr{M} = \mathscr{M}_1. \tag{2.43}$$

Proof. "$\Leftarrow$" Suppose that $\hat{\gamma} = \gamma$, and let x^* be the minimizer of (2.29). Pre- and post-multiplying M by $b_{pol}(x^*,m)'$ and $b_{pol}(x^*,m)$, respectively, we get

$$\begin{aligned} & b_{pol}(x^*,m)' M b_{pol}(x^*,m) \\ & = q(x^*)(f(x^*) - \gamma g(x^*)) \\ & = 0. \end{aligned}$$

Moreover, M is positive semidefinite for definition of $\hat{\gamma}$. This implies that

$$b_{pol}(x^*,m) \in \ker(M)$$

i.e. $x^* \in \mathscr{M}$. Consequently, $x^* \in \mathscr{M}_1$.

"$\Rightarrow$" Suppose that $\mathscr{M}_1 \neq \emptyset$ and let $x \in \mathscr{M}_1$. It follows that there exists x such that $f(x)/g(x) = \hat{\gamma}$, and hence $\hat{\gamma}$ is an upper bound of γ. But $\hat{\gamma}$ is also a lower bound of γ. Consequently, $\hat{\gamma} = \gamma$ and $f(x)/g(x) = \gamma$.

Lastly, let us consider the case $k = 0$. Suppose that $\mathscr{M} \neq \emptyset$ and let $x \in \mathscr{M}$. It follows that

$$\begin{aligned} 0 & = b_{pol}(x,m)' M b_{pol}(x,m) \\ & = f(x) - \hat{\gamma}_0 g(x) \end{aligned}$$

i.e. there exists x such that $f(x)/g(x) = \hat{\gamma}_0$, and hence $\hat{\gamma}_0$ is an upper bound of γ. But $\hat{\gamma}_0$ is also a lower bound of γ. Consequently, $\hat{\gamma}_0 = \gamma$ and $f(x)/g(x) = \gamma$. This means that $x \in \mathscr{M}_1$ and, hence, that $\mathscr{M}_1 \subseteq \mathscr{M}$. But for definition of N_1 one has that $\mathscr{M} \subseteq \mathscr{M}_1$. Therefore, $\mathscr{M} = \mathscr{M}_1$. □

Theorem 2.7 provides a condition for establishing whether the lower bound provided by (2.36) is tight. In particular, this happens if the set $\mathscr{M}_1$ in (2.41) is nonempty. This set can be found via trivial substitution from the set $\mathscr{M}$ in (2.39) (at least whenever $\mathscr{M}$ is finite), which can be computed as explained in Section 1.5.1.

As explained in the following result, the lower bound $\hat{\gamma}$ is nonconservative for sufficiently large k.

Theorem 2.8. *Let γ and $\hat{\gamma}$ be as in* (2.29) *and* (2.36), *respectively. Then, there exists $\tilde{k}$ such that $\hat{\gamma} = \gamma$ for all $k \geq \tilde{k}$.*

Proof. Consider $p(x) = f(x) - zg(x)$ for $z = \gamma$. We have that $p(x)$ is nonnegative. Moreover, from Theorem 2.2, there exists $\tilde{k}$ such that $\lambda_{pol}(p,\tilde{k}) \geq 0$. Consequently, from (2.31) it follows that $\hat{\gamma}$ is an upper bound of γ for $k = \tilde{k}$. But from (2.32) it follows that $\hat{\gamma}$ is a lower bound of γ for all $k \in \mathbb{N}$. Hence, $\hat{\gamma} = \gamma$ for $k = \tilde{k}$. Therefore, from Theorem 2.6 we conclude that $\hat{\gamma} = \gamma$ for all $k \geq \tilde{k}$. □

Depending on the number of variables and degree of $f(x)$, the tightness of the lower bound $\hat{\gamma}$ can be ensured a priori as explained hereafter.

Theorem 2.9. *Let γ and $\hat{\gamma}$ be as in* (2.29) *and* (2.36), *respectively. Suppose that $(n, 2m_0) \in \mathscr{E}$ with $\mathscr{E}$ as in* (1.209) *and*

$$m_0 = \left\lceil \frac{\max\{\partial f, \partial g\}}{2} \right\rceil. \tag{2.44}$$

Then, $\hat{\gamma} = \gamma$ for an arbitrarily chosen $k \in \mathbb{N}$.

Proof. Suppose that $(n, 2m_0) \in \mathscr{E}$. Then, it follows from Theorem 1.9 that $p(x) = f(x) - zg(x)$ is nonnegative if and only if $p(x)$ is SOS, i.e. $\lambda_{pol}(p) \geq 0$. Therefore, $\hat{\gamma} = \gamma$ for $k = 0$. From Theorem 2.6 we conclude that $\hat{\gamma} = \gamma$ for all $k \in \mathbb{N}$. □

Example 2.9. Consider (2.29) with

$$\begin{aligned} f(x) &= -2x_2x_3 - 3x_1^2x_2 + x_1^4 + x_2^4 + x_3^4 \\ g(x) &= 1 + x_1 - x_3 + 2x_1^2 + 2x_2^2 + 2x_3^2. \end{aligned}$$

We have that $g(x)$ satisfies (2.30) (for instance, this can be verified with the SOS index, indeed $\lambda_{pol}(g) = 0.327$ and hence $g(x)$ is positive). With $k = 0$ Corollary 2.8 provides the lower bound $\hat{\gamma} = -0.549$ of γ. In order to establish whether this lower bound is tight, let us use Theorem 2.7. We have that the set $\mathscr{M}$ in (2.39) is

$$\mathscr{M} = \left\{(-0.864, 0.757, 0.582)'\right\}$$

(in this case, $\dim(\ker(M)) = 1$) and hence $\hat{\gamma}$ is tight since $\mathscr{M}$ is nonempty. Moreover, according to Theorem 2.7, the vector in $\mathscr{M}$ is a minimizer of $f(x)/g(x)$, indeed $f(x)/g(x) = -0.549$ for $x \in \mathscr{M}$. □

Example 2.10. Consider (2.29) with

$$\begin{aligned} f(x) &= f_{Che}(x) \\ g(x) &= 1 \end{aligned}$$

where $f_{Che}(x)$ is the polynomial in (1.207) for $\delta = 0.4$. With $k = 0$ Corollary 2.8 provides the lower bound $\hat{\gamma} = -11.710$ of γ. This lower bound can be improved by increasing k, indeed with $k = 1$ we find $\hat{\gamma} = 0.087$. In order to establish whether this lower bound is tight, let us use Theorem 2.7. We have that the set $\mathscr{M}$ in (2.39) is

$$\mathscr{M} = \left\{\pm(0.584, 0.000)'\right\}$$

(in this case, $\dim(\ker(M)) = 2$) and hence $\hat{\gamma}$ is tight since $\mathscr{M}$ is nonempty. Moreover, according to Theorem 2.7, the vectors in $\mathscr{M}$ are minimizers of $f_{Che}(x)$, indeed $f_{Che}(x) = 0.087$ for all $x \in \mathscr{M}$. □

2.2 Constrained Optimization

This section addresses the Positivstellensatz and the problems of determining the minimum of a rational function and establishing positivity of a matrix polynomial over semialgebraic sets.

2.2.1 Positivstellensatz

Some fundamental techniques for investigating constrained positivity of polynomials are known as *Positivstellensatz*. In order to describe these techniques, let us introduce the following basic definitions, see e.g. [8, 83, 70] for details.

The *multiplicative monoid* generated by $p \in \mathscr{P}_n^k$ is the set of finite products of $p_1(x), \ldots, p_k(x)$, i.e.

$$mulmon(p) = \left\{ \bar{p}(x) = p(x)^d : d \in \{0,1\}^k \right\}. \tag{2.45}$$

The *cone* generated by $p \in \mathscr{P}_n^k$ is the set

$$cone(p) = \left\{ \bar{p}(x) = s(x)'q(x) : s \in \mathscr{S}_n^l,\ q_1, \ldots, q_l \in mulmon(p) \right\}. \tag{2.46}$$

Lastly, the *ideal* generated by $p \in \mathscr{P}_n^k$ is the set

$$ideal(p) = \left\{ \bar{p}(x) = r(x)'p(x) : r \in \mathscr{P}_n^k \right\}. \tag{2.47}$$

The following Positivstellensatz was proposed by Stengle.

Theorem 2.10 ([108]). *Let $f \in \mathscr{P}_n^k$, $g \in \mathscr{P}_n^l$ and $h \in \mathscr{P}_n^m$. Define the set*

$$\mathscr{K} = \{x \in \mathbb{R}^n : f(x) \geq 0 \text{ and } g(x) = 0 \text{ and } h_i(x) \neq 0 \ \forall i = 1, \ldots, m\}. \tag{2.48}$$

Then,

$$\begin{array}{lll} \mathscr{K} = \emptyset & \iff & \exists \bar{f} \in cone(f),\ \bar{g} \in ideal(g),\ \bar{h} \in mulmon(h) : \\ & & \bar{f}(x) + \bar{g}(x) + \bar{h}(x)^2 = 0. \end{array} \tag{2.49}$$

A stronger version of the Positivstellensatz was proposed by Schmuedgen as follows.

Theorem 2.11 ([103]). *Let $f \in \mathscr{P}_n^k$ be such that the set*

$$\mathscr{X} = \{x \in \mathbb{R}^n : f(x) \geq 0\} \tag{2.50}$$

is compact. Let $p \in \mathscr{P}_n$. Then,

$$p(x) > 0 \ \forall x \in \mathscr{X} \iff p \in cone(f). \tag{2.51}$$

A further stronger version was proposed by Putinar as stated in the following result.

Theorem 2.12 ([92]). *Consider $p \in \mathscr{P}_n$ and $f \in \mathscr{P}_n^k$. Suppose that $f_1(x),\ldots,f_k(x)$ have even degree, the highest degree forms of $f_1(x),\ldots,f_k(x)$ do not have common zeros in $\mathbb{R}_0^n$, and $\mathscr{X}$ in (2.50) is compact. Then,*

$$p(x) > 0 \ \forall x \in \mathscr{X} \quad \Longleftrightarrow \quad \exists q \in \mathscr{S}_n, s \in \mathscr{S}_n^k : \ p(x) = q(x) + s(x)'f(x). \quad (2.52)$$

It is worth mentioning that the Positivstellensatz can be regarded as a generalization of the *S-procedure*, where the polynomials defining the problem are quadratic and the multipliers are constants [10].

2.2.2 Minimum of Rational Functions

The problem we consider consists of determining the minimum of a rational function in the presence of constraints expressed as polynomial inequalities and polynomial equalities. Specifically, we consider

$$\begin{aligned} \gamma = \inf_x \ & \frac{f(x)}{g(x)} \\ \text{s.t.} \ & x \in \Theta \end{aligned} \quad (2.53)$$

where $f,g \in \mathscr{P}_n$, and $\Theta \subset \mathbb{R}^n$ is a semialgebraic set defined as

$$\Theta = \{x \in \mathbb{R}^n : \ a(x) \geq 0, \ b(x) = 0\} \quad (2.54)$$

for $a \in \mathscr{P}_n^{n_a}$ and $b \in \mathscr{P}_n^{n_b}$. Similarly to the case of unconstrained optimization, we assume that $g(x)$ satisfies (2.30).

A lower bound of γ can be found by exploiting the Positivstellensatz techniques presented in Section 2.2.1. Specifically, extending the approach introduced for the case of unconstrained minimization addressed in Section 2.1.3, for $k \in \mathbb{N}$ let us define

$$\begin{aligned} \hat{\gamma} = \ & \sup_{q,r,s,z} \ z \\ & \text{s.t.} \ \begin{cases} p \in \mathscr{S}_n, \ \partial p \leq 2m \\ q \in \mathscr{S}_n, \ q(x) \neq 0 \ \forall x \in \mathbb{R}^n \\ r \in \mathscr{S}_n^{n_a} \\ s \in \mathscr{P}_n^{n_b} \end{cases} \end{aligned} \quad (2.55)$$

where

$$p(x) = q(x)(f(x) - zg(x)) - r(x)'a(x) - s(x)'b(x) \quad (2.56)$$

and

$$m = \left\lceil \frac{\max\{\partial f, \partial g, \partial a, \partial b\}}{2} \right\rceil + k. \quad (2.57)$$

The following theorem explains that (2.55) provides a lower bound of γ.

Theorem 2.13. *Let $f,g \in \mathscr{P}_n$ and suppose that* (2.30) *holds. Let γ and $\hat{\gamma}$ be as in* (2.53) *and* (2.55)*, respectively. Then,*

$$\hat{\gamma} \leq \gamma. \tag{2.58}$$

Proof. Suppose that the constraints in (2.55) hold. Consider $x \in \Theta$. It follows that

$$\begin{aligned} 0 &\leq p(x) \\ &= q(x)(f(x) - zg(x)) - r(x)'a(x) - s(x)'b(x) \\ &\leq q(x)(f(x) - zg(x)) \end{aligned}$$

since $p(x) \geq 0$, $r(x) \geq 0$, $a(x) \geq 0$ and $b(x) = 0$. Then, since $q(x) > 0$, it follows that

$$0 \leq f(x) - zg(x)$$

i.e. $z \leq \gamma$. □

In (2.55), the integer k bounds the degrees of the polynomials $q(x), r_1(x), \ldots, r_{n_a}(x)$ and $s_1(x), \ldots, s_{n_b}(x)$. Indeed, since $\partial p \leq 2m$, it follows that

$$\begin{aligned} \partial q &= 2k \\ \partial r_i &= 2m - 2\lceil \partial a_i / 2 \rceil \\ \partial s_i &= 2m - \partial b_i. \end{aligned} \tag{2.59}$$

By exploiting the SMR, $\hat{\gamma}$ in (2.55) can be computed as explained in the following result.

Corollary 2.9. *Let $f,g \in \mathscr{P}_n$ and suppose that* (2.30) *holds. Then, the lower bound $\hat{\gamma}$ in* (2.55) *is given by*

$$\begin{aligned} \hat{\gamma} = &\sup_{Q,R,\mathbf{s},z,\alpha} \; z \\ &\text{s.t.} \begin{cases} P(Q,R,\mathbf{s},z) + L(\alpha) \geq 0 \\ Q > 0 \\ R \geq 0 \\ \operatorname{trace}(Q) = 1 \end{cases} \end{aligned} \tag{2.60}$$

where

$$\begin{cases} P(Q,R,\mathbf{s},z) + L(\alpha) = CSMR_{pol}(p) \\ q(x) = b_{pol}(x,k)' Q b_{pol}(x,k) \\ R = \operatorname{diag}(R_1, \ldots, R_{n_a}) \\ R_i = SMR_{pol}(r_i) \;\; \forall i = 1, \ldots, n_a \\ \mathbf{s} = (\mathbf{s}_1', \ldots, \mathbf{s}_{n_b}')' \\ \mathbf{s}_i = COE_{pol}(s_i) \;\; \forall i = 1, \ldots, n_b \\ \partial p \leq 2m \end{cases} \tag{2.61}$$

and $p(x)$ and m are as in (2.56)–(2.57).

Proof. Direct consequence of the definition of SOS polynomials. □

For $k = 0$, (2.60) amounts to solving an SDP. Otherwise, if $k > 0$, (2.60) can be solved through a one-parameter sequence of SDPs.

Let us observe that Q has to be positive definite in (2.60). This because $q(x)$ has to be positive in the proof of Theorem 2.13.

Let us also observe that Corollary 2.9 can be used also in the case of unconstrained x, simply by choosing $n_a = n_b = 0$ in (2.54) (which yields $\Theta = \mathbb{R}^n$). In this case, $q(x)$ can be allowed to be nonnegative rather than positive, i.e. one can allow Q to be positive semidefinite in (2.60) analogously to Corollary 2.8.

The following result provides a necessary and sufficient condition for establishing tightness of the found $\hat{\gamma}$.

Theorem 2.14. *Let γ and $\hat{\gamma}$ be as in* (2.53) *and* (2.60), *respectively. Suppose that the minimizers of* (2.60) *exist. Define*

$$\mathscr{M} = \left\{x \in \mathbb{R}^n : \; b_{pol}(x,m) \in \ker(M)\right\} \tag{2.62}$$

where m is as in (2.57) *and*

$$M = P(Q^*, R^*, \mathbf{s}^*, \hat{\gamma}) + L(\alpha^*) \tag{2.63}$$

where $P(Q^*, R^*, \mathbf{s}^*, \hat{\gamma}) + L(\alpha^*)$ is the found optimal value of $P(Q, R, \mathbf{s}, z) + L(\alpha)$ in (2.60). *Define also*

$$\mathscr{M}_1 = \left\{x \in \mathscr{M} : \; x \in \Theta \;\; and \;\; \frac{f(x)}{g(x)} = \hat{\gamma}\right\}. \tag{2.64}$$

Then,

$$\mathscr{M}_1 \neq \emptyset \quad \Longleftrightarrow \quad \begin{cases} \hat{\gamma} = \gamma \\ \textit{the minimizer of (2.53) exists.} \end{cases} \tag{2.65}$$

Proof. "$\Leftarrow$" Suppose that $\hat{\gamma} = \gamma$, and let x^* be the minimizer of (2.53). Pre- and post-multiplying M by $b_{pol}(x^*, m)'$ and $b_{pol}(x^*, m)$, respectively, we get

$$\begin{aligned} & b_{pol}(x^*, m)' M b_{pol}(x^*, m) \\ & = q(x^*)(f(x^*) - \gamma g(x^*)) - r(x^*)' a(x^*) - s(x^*)' b(x^*) \\ & \leq q(x^*)(f(x^*) - \gamma g(x^*)) \\ & = 0 \end{aligned}$$

since $r(x^*) \geq 0$, $a(x^*) \geq 0$, $b(x^*) = 0$ and $\gamma = f(x^*)/g(x^*)$. Moreover, M is positive semidefinite for definition of $\hat{\gamma}$. This implies that

$$\begin{aligned} 0 & \leq b_{pol}(x^*, m)' M b_{pol}(x^*, m) \\ & \leq 0 \end{aligned}$$

and hence that

$$b_{pol}(x^*, m) \in \ker(M)$$

i.e. $x^* \in \mathscr{M}$. Consequently, $x^* \in \mathscr{M}_1$.

"⇒" Suppose that $\mathscr{M}_1 \neq \emptyset$ and let $x \in \mathscr{M}_1$. It follows that there exists $x \in \Theta$ such that $f(x)/g(x) = \hat{\gamma}$, and hence $\hat{\gamma}$ is an upper bound of γ. But $\hat{\gamma}$ is also a lower bound of γ. Consequently, $\hat{\gamma} = \gamma$ and $f(x)/g(x) = \gamma$. □

Theorem 2.14 provides a condition for establishing whether the lower bound provided by (2.60) is tight. In particular, this happens if the set $\mathscr{M}_1$ in (2.64) is nonempty. This set can be found via trivial substitution from the set $\mathscr{M}$ in (2.62) (at least whenever $\mathscr{M}$ is finite), which can be computed as explained in Section 1.5.1.

Example 2.11. Consider (2.53) with

$$\begin{aligned} f(x) &= 7 + x_1x_2^2 - x_2^3 + x_1^4 \\ g(x) &= 1 + x_2 + x_2^2 \\ a(x) &= 4 - x_1^2 - 2x_2^2 \\ b(x) &= -1 + x_1 + x_2 + x_1^2 + x_1x_2 + x_2^3. \end{aligned}$$

With $k = 0$ Corollary 2.9 provides the lower bound $\hat{\gamma} = 1.913$ of γ (observe that, according to (2.59), this lower bound has been obtained with $\partial q = 0$, $\partial r = 2$ and $\partial s = 1$). In order to establish whether this lower bound is tight, let us use Theorem 2.14. We find that

$$\mathscr{M} = \{(-0.814, 0.989)'\}$$

(in this case, $\dim(\ker(M)) = 1$). Moreover, $a(x) = 1.380$ and $b(x) = 0.000$ for $x \in \mathscr{M}$, and hence $\mathscr{M}_1 = \mathscr{M}$. Therefore, $\hat{\gamma}$ is tight since $\mathscr{M}_1$ is nonempty. Furthermore, the points in $\mathscr{M}_1$ are minimizers of (2.53), indeed $f(x)/g(x) = 1.913$ and $x \in \Theta$ for $x \in \mathscr{M}_1$. □

2.2.3 Solving Systems of Polynomial Equations and Inequalities

Another application of SOS polynomials is in solving systems of polynomial equations. Specifically, let us consider the problem of computing the set

$$\mathscr{X} = \{x \in \Theta : \; f(x) = 0\} \tag{2.66}$$

where $f \in \mathscr{P}_n^{n_f}$ and Θ is a semialgebraic set described as

$$\Theta = \{x \in \mathbb{R}^n : \; a(x) \geq 0\} \tag{2.67}$$

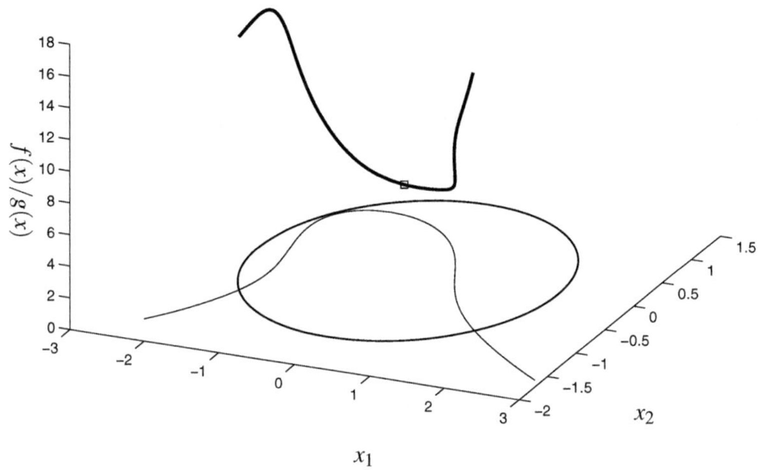

Fig. 2.2 Example 2.11. Curves $a(x) = 0$ and $b(x) = 0$ (solid) on the plane $x_3 = 0$, function $f(x)/g(x)$ for $x \in \Theta$ (thick solid), and $f(x)/g(x)$ for $x \in \mathscr{M}_1$ ("□" marker).

for $a \in \mathscr{P}_n^{n_a}$. Whenever nonempty, $\mathscr{X}$ corresponds to the set of minimizers of the optimization problem

$$\begin{aligned} \gamma = \inf_x \; & f(x)'f(x) \\ \text{s.t. } & x \in \Theta. \end{aligned} \tag{2.68}$$

This suggests that $\mathscr{X}$ can be found through Theorem 2.14 when $f(x)$ and $g(x)$ in (2.53) are replaced by $f(x)'f(x)$ and 1, respectively. Indeed, for $k \in \mathbb{N}$ let us introduce the following SDP:

$$\begin{aligned} \hat{\gamma} = \sup_{Q,R,z,\alpha} \; & z \\ \text{s.t. } & \begin{cases} P(Q,R,z) + L(\alpha) \geq 0 \\ Q > 0 \\ R \geq 0 \\ \operatorname{trace}(Q) = 1 \end{cases} \end{aligned} \tag{2.69}$$

where

$$\begin{cases} P(Q,R,z) + L(\alpha) = CSMR_{pol}(p) \\ q(x) = b_{pol}(x,k)'Qb_{pol}(x,k) \\ R = \operatorname{diag}(R_1,\ldots,R_{n_a}) \\ R_i = SMR_{pol}(r_i) \;\; \forall i = 1,\ldots,n_a \\ \partial p \leq 2m \end{cases} \tag{2.70}$$

with

$$p(x) = q(x)f(x)'f(x) - z - r(x)'a(x) \tag{2.71}$$

and

$$m = \left\lceil \frac{\max\{2\partial f, \partial a\}}{2} \right\rceil + k. \quad (2.72)$$

The following result explains the derivation of $\mathscr{X}$ from the SDP (2.69).

Theorem 2.15. *Let $\hat{\gamma}$ be as in* (2.69). *Then,*

$$\hat{\gamma} \geq 0. \quad (2.73)$$

Moreover,

$$\hat{\gamma} > 0 \;\Rightarrow\; \mathscr{X} = \emptyset. \quad (2.74)$$

Lastly,

$$\hat{\gamma} = 0 \;\Rightarrow\; \mathscr{X} = \mathscr{M}_1 \quad (2.75)$$

where

$$\mathscr{M}_1 = \{x \in \mathscr{M} : \; x \in \Theta \;\; and \;\; r_i(x)a_i(x) = 0 \;\; \forall i = 1, \ldots, n_a\} \quad (2.76)$$

and

$$\mathscr{M} = \left\{x \in \mathbb{R}^n : \; b_{pol}(x,m) \in \ker(M)\right\} \quad (2.77)$$

where m is as in (2.72) *and*

$$M = P(Q^*, R^*, 0) + L(\alpha^*) \quad (2.78)$$

where $P(Q^*,R^*,0)+L(\alpha^*)$ is the found optimal value of $P(Q,R,z)+L(\alpha)$ in (2.69).

Proof. First, let us observe that $\hat{\gamma} \geq 0$ since, for $z = 0$, there exists $q(x) = 1$ and $r(x) = 0$ such that $q(x)$ and $p(x) = f(x)'f(x)$ are SOS polynomials, and hence there exist Q, R and α such that the LMIs in (2.69) hold for $z = 0$.

Second, let us observe that $\hat{\gamma} > 0$ implies that $p(x) = q(x)f(x)'f(x) - r(x)'a(x)$ is positive, and hence $f(x)$ cannot vanish for any $x \in \Theta$ given $r(x) \geq 0$, i.e. $\mathscr{X} = \emptyset$.

Third, let us consider the case where $\hat{\gamma} = 0$. Let us prove that $\mathscr{X} \subseteq \mathscr{M}_1$. Let $x \in \mathscr{X}$, i.e. $f(x) = 0$ and $a(x) \geq 0$. Since M is positive semidefinite for definition of $\hat{\gamma}$, pre- and post-multiplying M by $b_{pol}(x,m)'$ and $b_{pol}(x,m)$, respectively, we get

$$\begin{aligned} 0 &\leq b_{pol}(x,m)'Mb_{pol}(x,m) \\ &= q(x)f(x)'f(x) - r(x)'a(x) \\ &= -r(x)'a(x). \end{aligned}$$

Since $r(x) \geq 0$ and $a(x) \geq 0$, this implies that

$$\begin{aligned} 0 &= b_{pol}(x,m)'Mb_{pol}(x,m) \\ &= r(x)'a(x) \\ &= r_i(x)a_i(x) \;\; \forall i = 1, \ldots, n_a. \end{aligned}$$

Also, since M is positive semidefinite, this implies that

$$b_{pol}(x,m) \in \ker(M)$$

i.e. $x \in \mathcal{M}_1$. Lastly, let us prove that $\mathcal{M}_1 \subseteq \mathcal{X}$. Let $x \in \mathcal{M}_1$, i.e. $b_{pol}(x,m) \in \ker(M)$, $x \in \Theta$ and $r_i(x)a_i(x) = 0$ for all $i = 1,\ldots,n_a$. It follows that

$$\begin{aligned} 0 &= b_{pol}(x,m)' M b_{pol}(x,m) \\ &= q(x)f(x)'f(x) - r(x)'a(x) \\ &= q(x)f(x)'f(x). \end{aligned}$$

Since $q(x) > 0$, one has that $f(x) = 0$, and hence $x \in \mathcal{X}$. □

Theorem 2.15 states that, for any chosen k, one can compute the sought set $\mathcal{X}$ by solving the SDP 2.69, and then by determining the set $\mathcal{M}_1$ in (2.76). This set can be found via trivial substitution from the set $\mathcal{M}$ in (2.77) (at least whenever $\mathcal{M}$ is finite), which can be computed as explained in Section 1.5.1.

Let us observe that, contrary to (2.60) which requires the solution of a sequence of SDPs, (2.69) amounts to solving an SDP only.

In (2.69)–(2.70), the integer k bounds the degrees of the polynomials $q(x)$ and $r_1(x),\ldots,r_{n_a}(x)$. Indeed, since $\partial p \leq 2m$, it follows that

$$\begin{aligned} \partial q &= 2k \\ \partial r_i &= 2m - 2\lceil \partial a_i/2 \rceil. \end{aligned} \tag{2.79}$$

Also, let us observe that Theorem 2.15 is still valid by redefining the polynomial $p(x)$ in (2.70) with

$$p(x) = q(x)f(x)'f(x) - zg(x) - r(x)'a(x) \tag{2.80}$$

where $g \in \mathcal{P}_n$ is any positive polynomial (such that $\partial g \leq 2m$).

Now we consider the specialization of Theorem 2.15 to the case where the sought set of solutions is given by

$$\tilde{\mathcal{X}} = \left\{ x \in \tilde{\Theta} : \ f(x) = 0 \right\} \tag{2.81}$$

where $f \in \mathcal{P}_n^{n_f}$ and $\tilde{\Theta}$ is described as

$$\tilde{\Theta} = \left\{ x \in \mathbb{R}^n : \ a(x) > 0 \right\}. \tag{2.82}$$

for $a \in \mathcal{P}_n^{n_a}$. Specifically, $\tilde{\Theta}$ is defined by strict inequalities while Θ is defined by non-strict inequalities.

Corollary 2.10. *Let $\hat{\gamma}$ be as in* (2.69)*, and suppose $\hat{\gamma} = 0$. Then, $\tilde{\mathcal{X}} = \mathcal{M}_3$ where*

$$\mathcal{M}_3 = \left\{ x \in \mathcal{M}_2 : \ x \in \tilde{\Theta} \right\} \tag{2.83}$$

and

$$\mathscr{M}_2 = \{x \in \mathscr{M}_1 : b_{pol}(x,m) \in \ker(M) \text{ and } b_{pol}(x,\partial r_i/2) \in \ker(R_i^*) \\ \forall i = 1,\dots,n_a\} \tag{2.84}$$

where m, M and R^* are as in Theorem 2.15.

Proof. Analogous to the proof of Theorem 2.15 by observing that, with $x \in \tilde{\Theta}$, one has that

$$r_i(x)a_i(x) = 0 \iff r_i(x) = 0$$

since $a(x) > 0$, and that

$$r_i(x) = 0 \iff b_{pol}(x,\partial r_i/2) \in \ker(R_i^*)$$

since R_i^* is a positive semidefinite SMR matrix of $r_i(x)$. □

As explained in Corollary 2.10, one can compute the sought set $\tilde{\mathscr{X}}$ by solving the SDP 2.69, and then by determining the set $\mathscr{M}_2$ which amounts to looking either for power vectors $b_{pol}(x,m)$ into $\ker(M)$ or power vectors $b_{pol}(x,\partial r_i/2)$ into $\ker(R_i^*)$. Clearly, computing $\mathscr{M}_2$ is not more complicated (and, possibly, simpler) than computing $\mathscr{M}$ in (2.77) since one can exploit any alternative $b_{pol}(x,\partial r_i/2) \in \ker(R_i^*)$, $i = 1,\dots,n_a$, for the condition $b_{pol}(x,m) \in \ker(M)$.

Example 2.12. Consider the computation of $\mathscr{X}$ in (2.66) with

$$f(x) = \begin{pmatrix} 1 - x_1 - x_1^3 + x_1x_2^2 \\ 1 + x_2 - x_2^3 + 3x_1^2 \end{pmatrix}.$$

Hence, $\Theta = \mathbb{R}^2$. By solving the SDP (2.69) with $k = 0$ we find $\hat{\gamma} = 0$. Then, Theorem 2.15 provides

$$\mathscr{M} = \{(3.104, 3.211)', (-2.751, 2.989)', (-0.865, 1.704)'\}$$

(in this case, $\dim(\ker(M)) = 3$). Hence, $\mathscr{X} = \mathscr{M}_1 = \mathscr{M}$ since $n_a = 0$. Figure 2.3 shows the curves $f_1(x) = 0$ and $f_2(x) = 0$, and the points in $\mathscr{X}$.

□

Example 2.13. Consider the computation of $\tilde{\mathscr{X}}$ in (2.81) with

$$\begin{aligned} f(x) &= \begin{pmatrix} 0.1x_1^4 + 0.2x_2^4 - 0.2x_1x_2^2 + 0.2x_2^3 - 2.5x_1^2 - 0.5x_2^2 + 3x_1 - 0.5x_2 + 2 \\ 0.2x_1^4 + 0.3x_2^4 + 0.5x_2^3 - 2.5x_2^2 - 3x_2 + 0.5 \end{pmatrix} \\ a(x) &= \begin{pmatrix} -3 + 4x_1 - x_1^2 \\ -3 + 4x_2 - x_2^2 \end{pmatrix}. \end{aligned}$$

Hence, $\tilde{\Theta} = (1,3)^2$. By solving the SDP (2.69) with $k = 0$ we find $\hat{\gamma} = 0$. Then, Corollary 2.10 provides

$$\mathscr{M}_2 = \{(2.287, 2.256)'\}$$

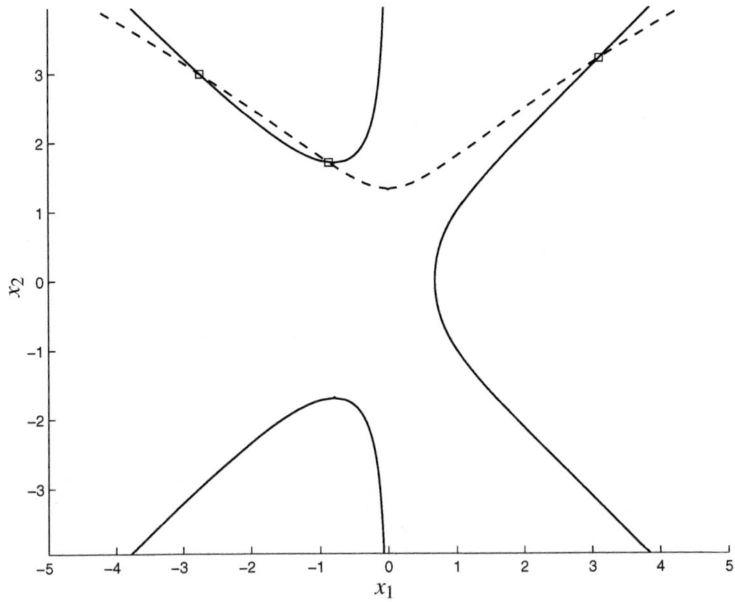

Fig. 2.3 Example 2.12. Curves $f_1(x) = 0$ (solid) and $f_2(x) = 0$ (dashed), and points in $\mathscr{X}$ ("□" markers).

(in this case, $\dim(\ker(M)) = 1$). We hence find that $\mathscr{M}_3 = \mathscr{M}_2$, and hence $\tilde{\mathscr{X}} = \mathscr{M}_3$. Figure 2.4 shows the curves $f_1(x) = 0$ and $f_2(x) = 0$, the region Θ, and the point in $\mathscr{X}$.

□

Example 2.14. Consider the computation of $\mathscr{X}$ in (2.66) with

$$f(x) = \begin{pmatrix} 1 - x_1^4 - x_2^4 - x_3^4 \\ 3 + 2x_1 - 4x_3 + x_1x_2 - 2x_1^4 + x_1^2x_2^2 - x_2^4 - 3x_3^4 \\ 1 + 2x_2 + x_1^3 - x_1^4 + x_1^2x_2x_3 + x_1x_2^2x_3 - x_2^4 - x_3^4 \end{pmatrix}.$$

Hence, $\Theta = \mathbb{R}^3$. By solving the SDP (2.69) with $k = 0$ we find $\hat{\gamma} = 0$. Then, Theorem 2.15 provides

$$\mathscr{M} = \{(-0.981, 0.506, -0.288)', (0.952, -0.367, 0.633)'\}$$

(in this case, $\dim(\ker(M)) = 2$). Hence, $\mathscr{X} = \mathscr{M}_1 = \mathscr{M}$ since $n_a = 0$. Figure 2.5 shows the surfaces $f_i(x) = 0$, $i = 1, 2, 3$, and the points in $\mathscr{X}$. □

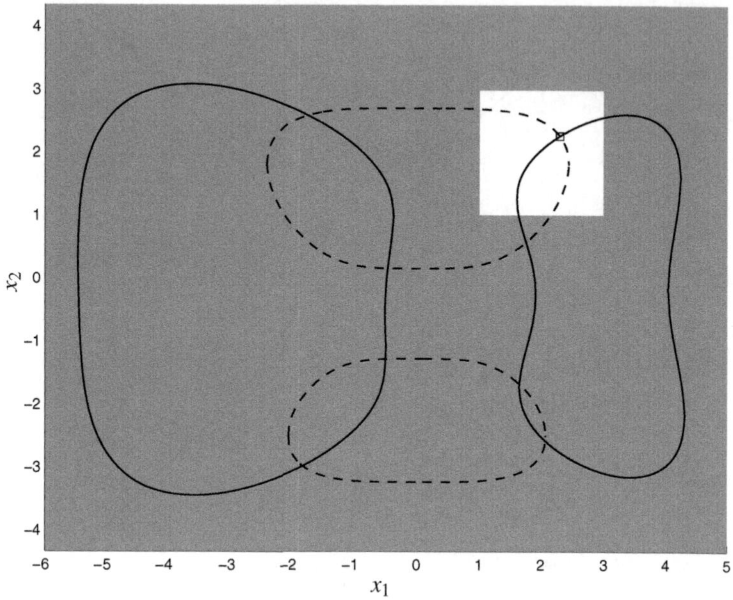

Fig. 2.4 Example 2.13. Curves $f_1(x) = 0$ (solid) and $f_2(x) = 0$ (dashed), set Θ (unfilled area), and point in $\mathscr{X}$ ("□" marker).

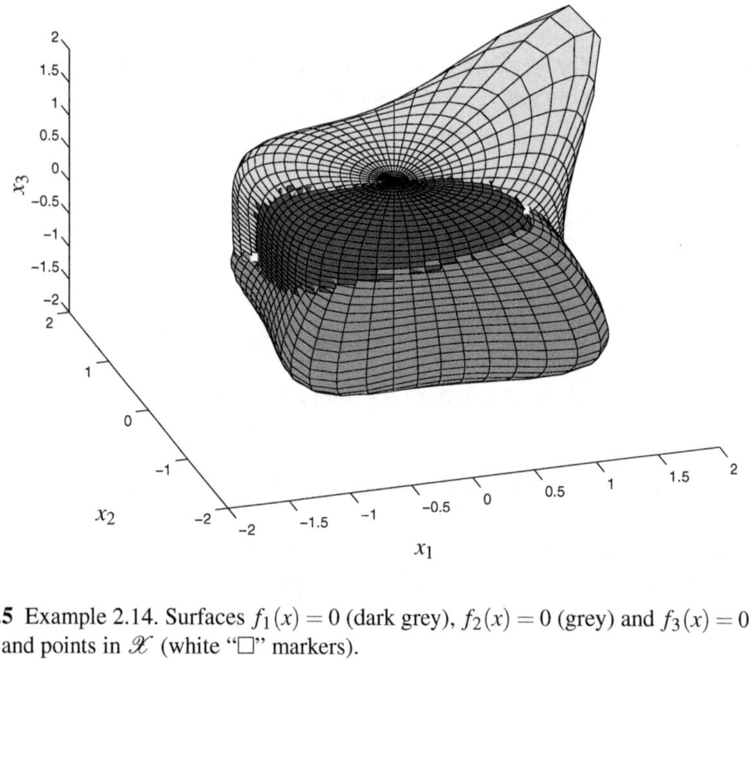

Fig. 2.5 Example 2.14. Surfaces $f_1(x) = 0$ (dark grey), $f_2(x) = 0$ (grey) and $f_3(x) = 0$ (light grey), and points in $\mathscr{X}$ (white "□" markers).

2.2.4 Positivity of Matrix Polynomials

Here we consider the problem of establishing whether

$$\left.\begin{array}{c} F(\theta) \geq 0 \\ \text{or} \\ F(\theta) > 0 \end{array}\right\} \ \forall \theta \in \Theta \tag{2.85}$$

where $F = F' \in \mathscr{P}_{n_\theta}^{n\times n}$ and $\Theta \subset \mathbb{R}^{n_\theta}$ is a semialgebraic set defined as

$$\Theta = \{\theta \in \mathbb{R}^{n_\theta} : \ a(\theta) \geq 0, \ b(\theta) = 0\} \tag{2.86}$$

for some $a \in \mathscr{P}_{n_\theta}^{n_a}$ and $b \in \mathscr{P}_{n_\theta}^{n_b}$.

This problem can be addressed by exploiting the Positivstellensatz techniques presented in Section 2.2.1 and the SMR of matrix polynomials. Specifically, let us define

$$G(\theta) = q(\theta)F(\theta) - \sum_{i=1}^{n_a} H_i(\theta)a_i(\theta) - \sum_{i=1}^{n_b} J_i(\theta)b_i(\theta) \tag{2.87}$$

for $q \in \mathscr{P}_{n_\theta}$ and $H_i = H_i', J_i = J_i' \in \mathscr{P}_{n_\theta}^{n\times n}$. For $k \in \mathbb{N}$ let us introduce the SDP

$$\hat{\gamma} = \sup_{Q,R,\mathbf{S},z,\alpha} \ z$$
$$\text{s.t.} \begin{cases} P(Q,R,\mathbf{S}) + L(\alpha) - zI \geq 0 \\ Q > 0 \\ R \geq 0 \\ \text{trace}(Q) = 1 \end{cases} \tag{2.88}$$

where

$$\begin{aligned} P(Q,R,\mathbf{S}) + L(\alpha) &= CSMR_{mat}(G) \\ q(\theta) &= b_{pol}(\theta,k)'Qb_{pol}(\theta,k) \\ R &= \text{diag}(R_1,\ldots,R_{n_a}) \\ R_i &= SMR_{mat}(H_i) \ \ \forall i = 1,\ldots,n_a \\ \mathbf{S} &= (\mathbf{S}_1,\ldots,\mathbf{S}_{n_b}) \\ \mathbf{S}_i &= COE_{mat}(J_i) \ \ \forall i = 1,\ldots,n_b \\ \partial G &\leq 2m \end{aligned} \tag{2.89}$$

where

$$m = \left\lceil \frac{\max\{\partial F, \partial a, \partial b\}}{2} \right\rceil + k. \tag{2.90}$$

Theorem 2.16. *Let $\hat{\gamma}$ be as in* (2.88). *The following statements hold.*

1. *If $\hat{\gamma} \geq 0$, then $F(\theta) \geq 0$ for all $\theta \in \Theta$.*
2. *If $\hat{\gamma} > 0$, then $F(\theta) > 0$ for all $\theta \in \Theta$.*

Proof. Let us consider the first statement. Suppose that there exists k such that $\hat{\gamma} \geq 0$. This means that there exist $q(\theta)$, $H_i(\theta)$ and $J_i(\theta)$ such that

$$\left.\begin{array}{l} G(\theta) \geq 0 \\ q(\theta) > 0 \\ H_i(\theta) \geq 0 \;\; \forall i = 1,\ldots,n_a \end{array}\right\} \forall \theta \in \mathbb{R}^{n_\theta}.$$

Let $\theta \in \Theta$. Since $a(\theta) \geq 0$ and $b(\theta) = 0$ it follows that

$$\begin{array}{l} 0 \leq G(\theta) \\ = q(\theta)F(\theta) - \sum_{i=1}^{n_a} H_i(\theta)a_i(\theta) - \sum_{i=1}^{n_b} J_i(\theta)b_i(\theta) \\ \leq q(\theta)F(\theta). \end{array}$$

Since $q(\theta) > 0$, this implies that $F(\theta) \geq 0$.

Next, let us consider the second statement. Suppose that there exists k such that $\hat{\gamma} > 0$. This means that there exist $q(\theta)$, $H_i(\theta)$ and $J_i(\theta)$ such that the conditions found in the previous case hold with $G(\theta) > 0$. Proceeding as in the previous case, one directly obtains that $F(\theta) > 0$. □

In (2.88)–(2.89), the integer k bounds the degrees of the matrix polynomials $q(\theta)$, $H_1(x),\ldots,H_{n_a}(x)$ and $J_1(x),\ldots,J_{n_b}(x)$. Indeed, since $\partial G \leq 2m$, it follows that

$$\begin{aligned} \partial q &= 2k \\ \partial H_i &= 2m - 2\lceil \partial a_i/2 \rceil \\ \partial J_i &= 2m - \partial b_i. \end{aligned} \tag{2.91}$$

Let us observe that Theorem 2.16 can be used also in the case of unconstrained θ, simply by choosing $n_a = n_b = 0$ in (2.86) (which yields $\Theta = \mathbb{R}^{n_\theta}$). In this case, $q(\theta)$ can be allowed to be nonnegative rather than positive, i.e. one can allow Q to be positive semidefinite in (2.88) analogously to Definition 2.1.

Also, let us observe that $\hat{\gamma}$ in (2.88) can be regarded as a generalized SOS index of matrix polynomials that considers both cases of unconstrained variable and constrained variable.

The following result provides a sufficient condition for establishing whether $F(\theta) \not\geq 0$ for some $\theta \in \Theta$.

Theorem 2.17. *Suppose that $\hat{\gamma}$ in (2.88) satisfies $\hat{\gamma} < 0$ for some $k \in \mathbb{N}$. Define*

$$\mathscr{M} = \left\{(x,\theta) \in \mathbb{R}_0^n \times \mathbb{R}^{n_\theta} : \; b_{pol}(\theta,m) \otimes x \in \ker(M) \;\; \text{and (1.192) holds}\right\} \tag{2.92}$$

where m is as in (2.90) and

$$M = P(Q^*,R^*,\mathbf{S}^*) + L(\alpha^*) - \hat{\gamma}I \tag{2.93}$$

where $P(Q^*,R^*,\mathbf{S}^*) + L(\alpha^*) - \hat{\gamma}I$ is the found optimal value of $P(Q,R,\mathbf{S}) + L(\alpha) - zI$ in (2.88). Define also

$$\mathscr{M}_1 = \left\{(x,\theta) \in \mathscr{M} : \; \theta \in \Theta \;\; \text{and} \;\; r(x,\theta)'a(\theta) = 0\right\}. \tag{2.94}$$

Then, there exists $\theta \in \Theta$ such that $F(\theta) \not\geq 0$ if $\mathscr{M}_1$ is nonempty. In particular,

$$F(\theta) \not\geq 0 \ \ \forall (x,\theta) \in \mathscr{M}_1. \tag{2.95}$$

Proof. Let $(x,\theta) \in \mathscr{M}_1$. We obtain:

$$\begin{aligned} 0 &= \Delta\left(M, b_{pol}(\theta,m), x\right) \\ &= q(\theta)x'F(\theta)x - r(x,\theta)'a(\theta) - s(x,\theta)'b(\theta) - \hat{\gamma}\|x \otimes b_{pol}(\theta,m)\|^2 \\ &= q(\theta)x'F(\theta)x - \hat{\gamma}\|b_{pol}(\theta,m) \otimes x\|^2. \end{aligned}$$

Since $q(\theta) > 0$, $\hat{\gamma} < 0$ and $\|b_{pol}(\theta,m) \otimes x\| > 0$, it follows that $x'F(\theta)x < 0$. Therefore, $F(\theta) \not\geq 0$. □

Theorem 2.17 provides a condition for establishing whether a matrix polynomial $F(\theta)$ is not positive semidefinite for some $\theta \in \Theta$. In particular, this happens if the set $\mathscr{M}_1$ in (2.94) is nonempty. This set can be found via trivial substitution from the set $\mathscr{M}$ in (2.92) (at least whenever $\mathscr{M}$ is finite), which can be computed as explained in Section 1.5.2 after (1.202).

Example 2.15. Consider problem (2.85) with

$$\begin{aligned} F(\theta) &= \begin{pmatrix} 1 & \theta_1 & \theta_2^2 & 0 \\ \star & 1 & \theta_2 & \theta_3^2 \\ \star & \star & 2+\theta_1^2 & \theta_3 \\ \star & \star & \star & 2+\theta_2 \end{pmatrix} \\ a(\theta) &= 1 - \theta_1^2 - \theta_2^2 - \theta_3^2 \\ b(\theta) &= \theta_1\theta_2 + \theta_2\theta_3 + \theta_1\theta_3 - \theta_1^2 - \theta_2^2 + \theta_3^2. \end{aligned}$$

With $k = 0$ the SDP (2.88) provides $\hat{\gamma} = 0.001$, and hence from Theorem 2.16 we conclude that $F(\theta) > 0$ for all $\theta \in \Theta$. □

Example 2.16. Consider problem (2.85) with

$$\begin{aligned} F(\theta) &= \begin{pmatrix} 3 - \theta_2\theta_3 + \theta_3^2 & \theta_2 - \theta_1\theta_3 & 3+\theta_2 \\ \star & 5 + \theta_1\theta_3 - \theta_2^2 & 1 + 2\theta_3 \\ \star & \star & 3 + \theta_1 + \theta_2^2 \end{pmatrix} \\ a(\theta) &= \begin{pmatrix} 1 - \theta_1^2 \\ 1 - \theta_2^2 \\ 1 - \theta_3^2 \end{pmatrix}. \end{aligned}$$

Hence, $\Theta = [-1,1]^3$. With $k = 0$ the SDP (2.88) provides $\hat{\gamma} = -0.474$. Let us use Theorem 2.17. We find that $\mathscr{M}$ in (2.92) for $k = 0$ is given by

$$\mathscr{M} = \left\{((-0.907, -0.190, 1.000)', (-0.517, 0.462, 0.287)')\right\}$$

(in this case, $\dim(\ker(M)) = 1$). For $(x,\theta) \in \mathscr{M}$ we have $a(\theta) \geq 0$ and $r(x,\theta)'a(\theta) = 0$, and hence $\mathscr{M}_1 = \mathscr{M}$. Therefore, $F(\theta) \not\geq 0$ since $\mathscr{M}_1 \neq \emptyset$. Indeed, $\mathrm{spc}(F(\theta)) = \{-0.741, 3.805, 7.220\}$ for $(x,\theta) \in \mathscr{M}_1$. □

2.3 Optimization over Special Sets

This section addresses the problems of establishing positivity or non-positivity of a polynomial over ellipsoids and over the simplex.

2.3.1 *Positivity over Ellipoids*

In this section we describe a method for investigating positivity of a polynomial over an ellipsoid. Let $f \in \mathscr{P}_n$ and consider the problem of establishing whether

$$f(x) > 0 \;\; \forall x \in \mathscr{B}(V,c) \tag{2.96}$$

where $\mathscr{B}(V,c)$ denotes the ellipsoid

$$\mathscr{B}(V,c) = \{x \in \mathbb{R}^n : \, x'Vx = c\} \tag{2.97}$$

for some $c \in \mathbb{R}$, $c \geq 0$, and $V \in \mathbb{S}^n$, $Q > 0$.

Without loss of generality, we assume that

$$\begin{cases} f(x_{any}) > 0 \\ \quad x_{any} = \sqrt{\dfrac{c}{V_{1,1}}}(1,0,\ldots,0)'. \end{cases} \tag{2.98}$$

Clearly, if this is not the case, $f(x)$ cannot be positive on $\mathscr{B}(V,c)$ since $x_{any} \in \mathscr{B}(V,c)$.

Let us introduce the following class of polynomials, which play a key role in investigating (2.96).

Definition 2.4 (Even Polynomial). A polynomial $f \in \mathscr{P}_n$ is an *even polynomial* of degree not greater than $2m \in \mathbb{N}$ if

$$f(x) = \sum_{i=0}^{m} u_{2i}(x) \tag{2.99}$$

where $u_{2i} \in \mathscr{P}_n$ is a homogeneous polynomial with $\partial u_{2i} = 2i$, $i = 0,\ldots,m$. □

If $f(x)$ is an even polynomial, then we define

$$g(x) = \sum_{i=0}^{m} u_{2i}(x) \left(\frac{x'Vx}{c}\right)^{m-i} \tag{2.100}$$

which is a homogeneous polynomial of degree $2m$. Otherwise, if $f(x)$ is not an even polynomial, we define $g(x)$ as

$$g(x) = \sum_{i=0}^{m} \bar{u}_{2i}(x) \left(\frac{x'Vx}{c}\right)^{m-i} \tag{2.101}$$

where $\bar{u}_{2i}(x)$, $i = 0, \ldots, m$, are the homogeneous polynomials of degree $2i$ satisfying

$$f(x)f(-x) = \sum_{i=0}^{m} \bar{u}_{2i}(x). \tag{2.102}$$

Observe, in fact, that $f(x)f(-x)$ is an even polynomial, and hence can be written according to (2.102). Let us define the following function:

$$\mathrm{HPE}(x,f,V,c) = \begin{cases} -\|x\|^2 & \text{if } f(x_{any}) \leq 0 \\ g(x) \text{ in (2.100)} & \text{if } f(x_{any}) > 0 \text{ and } f(x) \text{ is an even polynomial} \\ g(x) \text{ in (2.101)} & \text{otherwise.} \end{cases} \tag{2.103}$$

Theorem 2.18. *Define* $g(x) = \mathrm{HPE}(x,f,V,c)$. *Then,* (2.96) *holds if and only if* $g(x)$ *is positive definite.*

Proof. "$\Rightarrow$" Suppose that (2.96) holds. This implies that $f(x_{any}) > 0$. If $f(x)$ is an even polynomial, one has that $g(x) = f(x)$ for all $x \in \mathscr{B}(V,c)$, otherwise one has that $g(x) = f(x)f(-x)$ for all $x \in \mathscr{B}(V,c)$. In both cases, we have that $g(x) > 0$ for all $x \in \mathscr{B}(V,c)$. Since $g(x)$ is a homogeneous polynomial, this implies that $g(x)$ is positive definite.

"$\Leftarrow$" Suppose that $g(x)$ is positive definite. If $f(x)$ is an even polynomial, one has that $g(x) = f(x)$ for all $x \in \mathscr{B}(V,c)$, and hence (2.96) holds. If $f(x)$ is not an even polynomial, one has that $g(x) = f(x)f(-x)$ for all $x \in \mathscr{B}(V,c)$. Let us suppose for contradiction that (2.96) does not hold. This means that there exist $\tilde{x} \in \mathscr{B}(V,c)$ such that $f(\tilde{x}) \leq 0$. Since $f(x_{any}) > 0$, there exist $\bar{x} \in \mathscr{B}(V,c)$ such that $f(\bar{x}) = 0$. Consequently, $g(\bar{x}) = 0$, hence contradicting the assumption that $g(x)$ is positive definite. Therefore, (2.96) holds. □

The condition proposed in Theorem 2.18 relies on a suitable homogenization of $f(x)$ and establish that (2.96) is equivalent to the positive definiteness of the homogeneous polynomial $g(x)$. Notice that $g(x)$ is constructed by direct homogenization of $f(x)$ if this is an even polynomial, while it comes from homogenization of $f(x)f(-x)$ in the general case.

A possible way of checking the condition proposed in Theorem 2.18 consists of using the generalized SOS index of homogeneous polynomials, as explained in the following result.

Corollary 2.11. *Define* $g(x) = \mathrm{HPE}(x,f,V,c)$. *Then,* (2.96) *holds if and only if there exists* $k \in \mathbb{N}$ *such that* $\lambda_{hom}(g,k) > 0$.

Proof. Direct consequence of Theorems 2.18 and 2.5. □

The following theorem provides a sufficient condition for establishing whether $f(x)$ is negative for some $x \in \mathscr{B}(V,c)$.

Theorem 2.19. *Define* $g(x) = \mathrm{HPE}(x, f, V, c)$. *Suppose that* $\lambda_{hom}(g,k) \le 0$ *for some* $k \in \mathbb{N}$. *Define*

$$\mathscr{M} = \{x \in \mathbb{R}^n :\ b_{hom}(x,m) \in \ker(M) \ \ and\ (1.192)\ holds\} \tag{2.104}$$

where $m = \partial g/2 + k$ *and*

$$M = P(Q^*) + L(\alpha^*) - \lambda_{hom}(g,k)I \tag{2.105}$$

where $P(Q^*) + L(\alpha^*) - \lambda_{hom}(g,k)I$ *is the found optimal value of* $P(Q) + L(\alpha) - zI$ *in* (2.22) *for* $f(x)$ *replaced by* $g(x)$. *Define also*

$$\mathscr{M}_1 = \left\{ \pm x \sqrt{\frac{c}{x'Vx}},\ x \in \mathscr{M} \right\}. \tag{2.106}$$

Then,

$$\mathscr{M} \neq \emptyset \ \Rightarrow\ \exists x \in \mathscr{B}(V,c) :\ x \in \mathscr{M}_1 \ \ and\ \ f(x) \le 0. \tag{2.107}$$

Proof. Let $x \in \mathscr{M}$. It follows that

$$\begin{aligned} 0 &= b_{pol}(x,m)' M b_{pol}(x,m) \\ &= g(x) - \lambda_{hom}(g,k) \|b_{pol}(x,m)\|^2. \end{aligned}$$

Since $\lambda_{hom}(g,k) \le 0$ and $\|b_{hom}(x,m)\| > 0$, it follows that $g(x) \le 0$. This means that $g(x)$ is not positive definite, and hence (2.96) does not hold due to Theorem 2.18. Then, let us observe that, if $x \in \mathscr{M}$, one has that $x_-, x_+ \in \mathscr{M}_1$ where

$$x_- = -ax,\ \ x_+ = ax,\ \ a = \sqrt{\frac{c}{x'Vx}}.$$

Moreover,

$$\begin{aligned} g(x_-) &= g(x_+) \\ &= a^{\partial g} g(x) \\ &\le 0. \end{aligned}$$

These two points satisfy $x_-, x_+ \in \mathscr{B}(V,c)$. Moreover, if $f(x)$ is an even polynomial, one has that $f(x_+) = g(x_+)$, i.e. there exists $x_+ \in \mathscr{B}(V,c)$ such that $x_+ \in \mathscr{M}_1$ and $f(x_+) \le 0$. If $f(x)$ is not an even polynomial, one has that $f(x)f(-x) = g(x)$ for $x \in \{x_-, x_+\}$, i.e. there exists $x \in \{x_-, x_+\} \subset \mathscr{B}(V,c)$ such that $x \in \mathscr{M}_1$ and $f(x) \le 0$. □

Theorem 2.19 provides a condition for establishing whether a polynomial $f(x)$ is non-positive for some $x \in \mathscr{B}(V,c)$. In particular, this happens if the set $\mathscr{M}$ in (2.104) is nonempty. This set can be computed as explained in Section 1.5.2 after (1.193).

Example 2.17. Let us consider the problem of establish whether (2.96) holds with

$$\begin{aligned} f(x) &= x_1x_2^3 + 2x_1^2x_2 - x_2^2 + 6 \\ Q &= \begin{pmatrix} 1 & 1 \\ 1 & 2 \end{pmatrix} \\ c &= 1. \end{aligned}$$

First of all, let us observe that (2.98) holds, indeed

$$x_{any} = (1,0)', \quad f(x_{any}) = 6.$$

Define $g(x) = \text{HPE}(x, f, V, c)$. We have that

$$g(x) = 36y(x)^4 - 12x_2^2y(x)^3 + (12x_1x_2^3 + x_2^4)y(x)^2 - (4x_1^4x_2^2 + 2x_1x_2^5)y(x) + x_1^2x_2^8$$

where

$$y(x) = \frac{x'Vx}{c} = x_1^2 + 2x_1x_2 + 2x_2^2.$$

Hence:

$$\begin{aligned} g(x) = {} & 36x_1^8 + 288x_1^7x_2 + 1136x_1^6x_2^2 + 2812x_1^5x_2^3 + 4721x_1^4x_2^4 \\ & + 5474x_1^3x_2^5 + 4277x_1^2x_2^6 + 2068x_1x_2^7 + 484x_2^8. \end{aligned}$$

Then, we find that $\lambda_{hom}(g) = 9.968 \cdot 10^{-5}$, which implies from Corollary 2.11 that (2.96) holds. □

Example 2.18. Let us consider (2.96) with

$$\begin{aligned} f(x) &= -x_1^3 + x_2^3 + 2x_1x_2 + 1 \\ Q &= \begin{pmatrix} 3 & -1 \\ \star & 1 \end{pmatrix} \\ c &= 1. \end{aligned}$$

We have that (2.98) holds, indeed

$$x_{any} = (0.577, 0)', \quad f(x_{any}) = 0.808.$$

Then, we find that $\lambda_{hom}(g) = -0.006$. Let us use Theorem 2.19 with $k = 0$. We have that the sets $\mathscr{M}$ and $\mathscr{M}_1$ in (2.104) and (2.106) are given by

$$\begin{aligned} \mathscr{M} &= \{(0.125, 1.000)'\} \\ \mathscr{M}_1 &= \{\pm(0.140, 1.120)'\} \end{aligned}$$

(in this case, $\dim(\ker(M)) = 1$). Since $\mathscr{M}$ is not empty, it follows that $f(x)$ is negative for some $x \in \mathscr{B}(V, c)$. Indeed, $x = -(0.140, 1.120)' \in \mathscr{M}_1$ satisfies $x \in \mathscr{B}(V, c)$ and $f(x) = -0.089$. □

2.3.2 *Positivity over the Simplex*

Let us introduce the following set.

Definition 2.5 (Simplex). The set defined as

$$\Upsilon_n = \{x \in \mathbb{R}^n : \ |x| = 1, \ x_i \geq 0\} \tag{2.108}$$

is called *simplex*. □

The problem considered in this section is formulated as follows: given $f \in \mathscr{P}_n$, establish whether

$$\left.\begin{array}{c} f(x) \geq 0 \\ \text{or} \\ f(x) > 0 \end{array}\right\} \ \forall x \in \Upsilon_n. \tag{2.109}$$

In order to address (2.109), let us express $f(x)$ as

$$f(x) = \sum_{i=0}^{m} u_i(x) \tag{2.110}$$

where $u_i \in \mathscr{P}_n$, $i = 0,\ldots,m$, are homogeneous polynomials with $\partial u_i = i$. Let us define

$$g(x) = \sum_{i=0}^{m} u_i(x)|x|^{m-i}. \tag{2.111}$$

It is straightforward to verify that $g(x)$ is a homogeneous polynomial of degree m. Moreover, from (2.111) it directly follows that

$$g(x) = f(x) \ \ \forall x \in \Upsilon_n. \tag{2.112}$$

Let us introduce the notation

$$\text{sq}(x) = (x_1^2,\ldots,x_n^2)' \tag{2.113}$$

and

$$\text{sqr}(x) = (\sqrt{x_1},\ldots,\sqrt{x_n})', \tag{2.114}$$

and let us define

$$h(x) = g(\text{sq}(x)). \tag{2.115}$$

Observe that $h(x)$ is a homogeneous polynomial of degree $2m$.

Theorem 2.20. *Consider $f \in \mathscr{P}_n$ and define $h(x)$ as in* (2.115)*. The following statements hold.*

1. *$f(x) \geq 0$ for all $x \in \Upsilon_n$ if and only if $h(x)$ is positive semidefinite.*
2. *$f(x) > 0$ for all $x \in \Upsilon_n$ if and only if $h(x)$ is positive definite.*

Proof. Let us consider the first statement. Suppose that $f(x) \geq 0$ for all $x \in \Upsilon_n$. It follows that $g(x) \geq 0$ for all $x \in \Upsilon_n$. Suppose for contradiction that $h(x)$ is not positive semidefinite. This implies that there exists $\tilde{x}$ such that $h(\tilde{x}) < 0$. Define

$$\begin{aligned} \bar{x} &= \text{sq}(\tilde{x})/a \\ a &= (1,\ldots,1)'\text{sq}(\tilde{x}). \end{aligned}$$

It follows that $\bar{x} \in \Upsilon_n$. Moreover,

$$\text{sqr}(\bar{x}) = \tilde{x}/\sqrt{a}.$$

Hence:

$$\begin{aligned} g(\bar{x}) &= h(\tilde{x}/\sqrt{a}) \\ &= h(\tilde{x})a^{-\partial h/2} \\ &< 0 \end{aligned}$$

hence contradicting that $g(x) \geq 0$ for all $x \in \Upsilon_n$. Therefore, $h(x)$ is positive semidefinite.

Then, suppose that $h(x)$ is positive semidefinite. This means that $g(\text{sq}(x)) \geq 0$ for all x. Suppose for contradiction that there exists $\tilde{x} \in \Upsilon_n$ such that $f(\tilde{x}) < 0$. It follows that $g(\tilde{x}) < 0$. Define

$$\bar{x} = \text{sqr}(\tilde{x}).$$

It follows that

$$\begin{aligned} g(\text{sq}(\bar{x})) &= g(\tilde{x}) \\ &< 0 \end{aligned}$$

hence contradicting that $g(\text{sq}(x)) \geq 0$ for all x. Therefore, $f(x) \geq 0$ for all $x \in \Upsilon_n$.

Similarly, one can prove the second statement. □

As explained in the following result, the conditions proposed in Theorem 2.20 can be checked by using the generalized SOS index of homogeneous polynomials.

Corollary 2.12. *Consider $f \in \mathscr{P}_n$ and define $h(x)$ as in* (2.115). *The following statements hold.*

1. $f(x) \geq 0$ *for all* $x \in \Upsilon_n$ *if and only if there exists* $k \in \mathbb{N}$ *such that* $\lambda_{hom}(h,k) \geq 0$.
2. $f(x) > 0$ *for all* $x \in \Upsilon_n$ *if and only if there exists* $k \in \mathbb{N}$ *such that* $\lambda_{hom}(h,k) > 0$.

Proof. Direct consequence of Theorems 2.20 and 2.5. □

A sufficient condition for establishing whether $f(x)$ is negative for some $x \in \Upsilon_n$ can be obtained as explained hereafter.

Theorem 2.21. *Consider $f \in \mathscr{P}_n$ and define $h(x)$ as in* (2.115). *Suppose that $\lambda_{hom}(h,k) < 0$ for some $k \in \mathbb{N}$. Define*

$$\mathscr{M} = \{x \in \mathbb{R}^n : \ b_{hom}(x,m) \in \ker(M) \ \textit{and (1.192) holds}\} \tag{2.116}$$

where $m = \partial f + k$ *and*

$$M = P(Q^*) + L(\alpha^*) - \lambda_{hom}(h,k)I \tag{2.117}$$

where $P(Q^*) + L(\alpha^*) - \lambda_{hom}(h,k)I$ *is the found optimal value of* $P(Q) + L(\alpha) - zI$ *in* (2.22) *for* $f(x)$ *replaced by* $h(x)$. *Define also*

$$\mathscr{M}_1 = \left\{ \frac{y}{|y|},\ y = \mathrm{sq}(x),\ x \in \mathscr{M} \right\}. \tag{2.118}$$

Then,

$$\mathscr{M} \neq \emptyset \ \Rightarrow\ \exists x \in \Upsilon_n : x \in \mathscr{M}_1 \ \text{and}\ f(x) < 0. \tag{2.119}$$

Proof. Let $x \in \mathscr{M}$. It follows that

$$\begin{aligned} 0 &= b_{pol}(x,m)' M b_{pol}(x,m) \\ &= h(x) - \lambda_{hom}(h,k)\|b_{hom}(x,m)\|^2. \end{aligned}$$

Since $\lambda_{hom}(h,k) < 0$ and $\|b_{hom}(x,m)\| > 0$, it follows that $h(x) < 0$. This means that $h(x)$ is not positive definite, and hence (2.109) does not hold due to Theorem 2.20. Then, let us observe that, if $x \in \mathscr{M}$, one has that $\tilde{x} \in \mathscr{M}_1$ where

$$\tilde{x} = \frac{y}{a},\ y = \mathrm{sq}(x),\ a = |y|.$$

Moreover,

$$\mathrm{sqr}(\tilde{x}) = x/\sqrt{a}$$

and

$$\begin{aligned} g(\tilde{x}) &= h(x/\sqrt{a}) \\ &= h(x)a^{-\partial h/2} \\ &< 0. \end{aligned}$$

The point $\tilde{x}$ satisfies $\tilde{x} \in \Upsilon_n$. Moreover, one has that $f(\tilde{x}) = g(\tilde{x})$, i.e. there exists $\tilde{x} \in \Upsilon_n$ such that $\tilde{x} \in \mathscr{M}_1$ and $f(\tilde{x}) < 0$. □

Theorem 2.21 provides a condition for establishing whether a polynomial $f(x)$ is negative for some $x \in \Upsilon_n$. In particular, this happens if the set $\mathscr{M}$ in (2.116) is nonempty. This set can be computed as explained in Section 1.5.2 after (1.193).

Example 2.19. Let us consider (2.109) with $n = 3$ and

$$f(x) = x_1^3 - 8x_1x_2^2 + x_2^3 + x_2x_3^2 + x_3 + 1.$$

We find that $\lambda_{hom}(h) = 0.041$, and hence $f(x) > 0$ for all $x \in \Upsilon_3$. □

Example 2.20. Let us consider (2.109) with $n = 4$ and

$$f(x) = -3x_1^2x_2 - 6x_2^2x_3 - 9x_3^2x_4 - x_1x_2 + x_3x_4 + x_4^2 + 1.$$

We find that $\lambda_{hom}(h) = -0.008$. Let us use Theorem 2.21 with $k = 0$. We have that the sets $\mathscr{M}$ and $\mathscr{M}_1$ in (2.116)–(2.118) are given by

$$\begin{aligned}\mathscr{M} &= \{(0.000, 0.000, 1.000, 0.629)', (0.000, 0.000, 1.000, -0.629)'\} \\ \mathscr{M}_1 &= \{(0.000, 0.000, 0.716, 0.284)'\}\end{aligned}$$

(in this case, $\dim(\ker(M)) = 2$). Since $\mathscr{M}$ is not empty, it follows that $f(x)$ is negative for some $x \in \Upsilon_4$. Indeed, $x = (0.000, 0.000, 0.716, 0.284)' \in \mathscr{M}_1$ satisfies $x \in \Upsilon_4$ and $f(x) = -0.026$. □

2.4 Rank Constrained SMR Matrices of SOS Polynomials

As we have seen in this chapter and as we will see in the following ones, looking for power vectors in linear subspaces plays a key role in numerous problems, such as establishing non-positivity of a polynomial, establishing tightness of a found lower bound of a rational function, solving systems of polynomial equations, and establishing tightness of a found estimate of the DA.

Power vectors can be searched for in linear subspaces with the procedure proposed in Section 1.5. Although this procedure works well in typical cases, there can be circumstances where the procedure either cannot be applied or does not provide satisfactory results.

Specifically, this procedure requires that the dimension of the linear subspace be not greater than a certain threshold, but unfortunately there can be situations where this is not the case. Also, even if this dimension is smaller than the allowed threshold and the procedure can be applied, its numerical accuracy may worsen as the dimension increases.

The idea to cope with these problems is to replace, if possible, the linear subspace under consideration with an equivalent having a smaller dimension.

In details, the linear subspace under consideration in the above mentioned problems is the null space of a positive semidefinite SMR matrix obtained in an LMI problem. As we have understood in Chapter 1, the SMR matrix of a polynomial is not unique in general. And so is the positive semidefinite SMR matrix of a polynomial (which, clearly, has to be SOS).

For instance, consider the SOS polynomial

$$f(x) = (1 - x_1^2 - x_2^2)^2 + (1 - x_2)^2.$$

This polynomial admits the positive semidefinite SMR matrices

$$F_1 = \begin{pmatrix} 2 & 0 & -1 & -1 & 0 & -1 \\ \star & 0 & 0 & 0 & 0 & 0 \\ \star & \star & 1 & 0 & 0 & 0 \\ \star & \star & \star & 1 & 0 & 1 \\ \star & \star & \star & \star & 0 & 0 \\ \star & \star & \star & \star & \star & 1 \end{pmatrix}, \quad F_2 = \begin{pmatrix} 2.000 & 0.000 & -1.000 & -1.196 & 0.000 & -1.000 \\ \star & 0.392 & 0.000 & 0.000 & -0.392 & 0.000 \\ \star & \star & 1.000 & 0.392 & 0.000 & 0.000 \\ \star & \star & \star & 1.000 & 0.000 & 0.804 \\ \star & \star & \star & \star & 0.392 & 0.000 \\ \star & \star & \star & \star & \star & 1.000 \end{pmatrix}$$

which are built with respect to $b_{pol}(x,2)$ since $\partial f = 4$. Indeed, $b_{pol}(x,2)'F_i b_{pol}(x,2) = f(x)$ and $F_i \geq 0$ for all $i = 1,2$. However, F_1 and F_2 have different null space dimension, in fact

$$\begin{aligned} \dim(\ker(F_1)) &= 4 \\ \dim(\ker(F_2)) &= 2. \end{aligned}$$

Hence, it is natural to ask the following questions: is it possible to look for a positive semidefinite SMR matrix of a SOS polynomial with the smallest null space dimension?

By denoting the polynomial with $f \in \mathscr{P}_n$ and defining $F + L(\alpha) = CSMR_{pol}(f)$, this problem can be formulated as

$$\begin{aligned} &\inf_{\alpha} \dim(\ker(F + L(\alpha))) \\ &\text{s.t. } F + L(\alpha) \geq 0. \end{aligned} \tag{2.120}$$

Equivalently, we can rewrite the above problem as

$$\begin{aligned} &\sup_{\alpha} \operatorname{rank}(F + L(\alpha)) \\ &\text{s.t. } F + L(\alpha) \geq 0 \end{aligned} \tag{2.121}$$

i.e. the problem amounts to finding a positive semidefinite SMR matrix with the largest rank.

Interestingly, the above problem can be solved with LMIs, in particular with a finite sequence of LMI feasibility tests. Indeed, the previous problems can be rewritten as

$$\sup \left\{ m = 0, \ldots, d_{pol}(n, \partial f) : \text{(2.123) holds for some } \alpha \right\} \tag{2.122}$$

where

$$\begin{cases} \operatorname{rank}(F + L(\alpha)) \geq m \\ F + L(\alpha) \geq 0. \end{cases} \tag{2.123}$$

As explained in Appendix B, the generic condition $\operatorname{rank}(V) \geq m$ with $V \geq 0$ can be written via LMIs in the entries of V and other additional variables. Consequently, (2.123) can be written via LMIs only.

2.5 Notes and References

The generalized SOS indexes and their use for investigating positivity and non-positivity of polynomials described in Sections 2.1.1 and 2.1.2 extend the ideas proposed in [33] by exploiting results on the gap between positive polynomials and SOS polynomials.

The minimization of rational functions via SOS programming and the tightness test for the obtained lower bound derived in Sections 2.1.3 and 2.2.2 extend the use of determining lower bounds of polynomials via SOS programming introduced e.g. in [105, 83] and the investigation of optimality of such bounds introduced e.g. in [36]. Alternative approaches were proposed e.g. via the theory of moments [68].

The application of SOS programming for solving systems of polynomial equations and inequalities described in Section 2.2.3 was proposed in [28, 29] for the case of systems of polynomial equations, and in [22] for the case of systems of polynomial equations and inequalities.

The investigation of positivity and non-positivity of matrix polynomials over semialgebraic sets derived in Section 2.2.4 extend the use of the Positivstellensatz exploited in Section 2.2.2 for computing the minimum of rational functions to the case of matrix polynomials. Related works include the methods in [102] and [56] which consider the case of semialgebraic sets defined by matrix polynomial inequalities, and are based on SOS programming and the theory of moments, respectively.

The technique for investigating positivity and non-positivity of polynomials over an ellipsoid, described in Section 2.3.1, was proposed in [36, 31]. The technique for establishing positivity and non-positivity of polynomials over the simplex, described in Section 2.3.2, was introduced in [30, 32].

Lastly, the search for positive semidefinite SMR matrices of a SOS polynomial with the largest rank was proposed in [22].

SOS programming has been used for optimization over polynomials since long time, see e.g. [105, 92, 40, 87, 106, 75, 83]. See also recent contributions about sum of squares with rational coefficients [46, 64]. Optimization over polynomials is largely exploited in control systems, see e.g. the tutorials [63, 82], the book [53], the special issue [34], and the survey [24]. A large number of applications can be found in robust control [58, 56, 101, 33], time-delay systems [85], hybrid systems [89], systems biology [47, 26], and nonlinear systems as it will become clear from the second part of this book.

There are several software tools which allow one to solve optimization problems with SOS polynomials, see e.g. [109, 90, 54, 72]. In particular, the techniques presented in this chapter are available in the freely downloadable MATLAB toolbox SMRSOFT [27].

Part II: Domain of Attraction

Chapter 3
Dynamical Systems Background

This chapter introduces some basic preliminaries about stability analysis of dynamical systems, in particular continuous-time, time-invariant, nonlinear systems. First, the concept of equilibrium point is introduced, hence providing the notion of local asymptotical stability and its characterization based on LFs. Then, the definition of DA of an equilibrium point is provided, which is the set of initial conditions from which the state converges to such an equilibrium point. It is recalled how inner estimates can be obtained by looking for sublevel sets of LFs included in the region where the temporal derivatives are negative, moreover basic properties of such estimates are reviewed. The chapter hence proceeds by considering uncertain nonlinear systems, in particular nonlinear systems with coefficients depending on a time-invariant uncertain vector constrained in a given set of interest. The concept of common equilibrium point is introduced, hence providing the notion of robust local asymptotical stability and its characterization based on common and parameter-dependent LFs. Then, the definition of RDA of a common equilibrium point is provided, which is the set of initial conditions from which the state converges to such an equilibrium point for all admissible uncertainties. It is recalled how inner estimates can be obtained by looking for either sublevel sets of common LFs or parameter-dependent sublevel sets of parameter-dependent LFs.

3.1 Equilibrium Points of Nonlinear Systems

Let us consider a dynamical system of the form

$$\begin{cases} \dot{x}(t) = f(x(t)) \\ x(0) = x_{init} \end{cases} \tag{3.1}$$

where $x(t) \in \mathbb{R}^n$ is the state, $t \in \mathbb{R}$ is the time, $x_{init} \in \mathbb{R}^n$ is the initial condition, $f : \mathbb{R}^n \to \mathbb{R}^n$ defines the evolution of the state, and $\dot{x}(t)$ denotes the temporal derivative of the state, i.e.

G. Chesi: Domain of Attraction, LNCIS 415, pp. 89–103, 2011.
springerlink.com

$$\dot{x}(t) = \frac{dx(t)}{dt}. \tag{3.2}$$

Such a system is commonly said a autonomous, continuous-time, time-invariant, nonlinear system. For such a system we define the equilibrium points as follows.

Definition 3.1. Let $x_{eq} \in \mathbb{R}^n$ be such that

$$f(x_{eq}) = 0_n. \tag{3.3}$$

Then, x_{eq} is said an *equilibrium point* of (3.1). □

Hence, an equilibrium point is a zero of the function $f(x)$, and corresponds to a constant solution of the differential equation in (3.1). We denote the set of equilibrium points of (3.1) as

$$\mathscr{X} = \{x \in \mathbb{R}^n : f(x) = 0_n\}. \tag{3.4}$$

The set $\mathscr{X}$ consists of the intersections of the hyper-surfaces $\{x \in \mathbb{R}^n : f_i(x) = 0\}$, $i = 1, \ldots, n$. Except special cases, $\mathscr{X}$ is nonempty and finite. Special cases include linear systems (i.e., systems where $f(x)$ is linear): in such a case, $\mathscr{X}$ is a linear subspace.

When studying the properties of an equilibrium point, it can be convenient to assume that such an equilibrium point is the origin. Let us observe that this can be done without loss of generality. In fact, let us define the new variable

$$\tilde{x}(t) = x(t) - x_{eq}. \tag{3.5}$$

It follows that (3.1) can be rewritten with respect to $\tilde{x}$ as

$$\begin{cases} \dot{\tilde{x}}(t) = \tilde{f}(\tilde{x}(t)) \\ \tilde{x}(0) = \tilde{x}_{init} \end{cases} \tag{3.6}$$

where

$$\begin{aligned} \tilde{f}(\tilde{x}) &= f(\tilde{x} + x_{eq}) \\ \tilde{x}_{init} &= x_{init} - x_{eq}. \end{aligned} \tag{3.7}$$

Moreover, the equilibrium point x_{eq} for (3.1) is shifted into the origin for (3.6) since $\tilde{x}_{eq} = x_{eq} - x_{eq} = 0_n$.

Therefore, in the sequel we will consider that the system under investigation is given by

$$\begin{cases} \dot{x}(t) = f(x(t)) \\ x(0) = x_{init} \\ f(0_n) = 0_n \end{cases} \tag{3.8}$$

for which the origin is an equilibrium point.

3.2 Stability

Let us indicate the solution of (3.8) at time $t \geq 0$ with $\varphi(t, x_{init})$, i.e.

$$\begin{cases} \dot{\varphi}(t, x_{init}) = f(\varphi(t, x_{init})) \\ \varphi(0, x_{init}) = x_{init}. \end{cases} \tag{3.9}$$

The following definition introduces the concept of locally asymptotically stable equilibrium point.

Definition 3.2. Suppose that

$$\forall \varepsilon > 0 \, \exists \delta > 0 : \ \|x_{init}\| < \delta \Rightarrow \|\varphi(t, x_{init})\| < \varepsilon \ \ \forall t \geq 0 \tag{3.10}$$

and

$$\lim_{t \to \infty} \varphi(t, x_{init}) = 0_n \ \ \forall x_{init} : \ \|x_{init}\| < \delta. \tag{3.11}$$

Then, the origin is said a *locally asymptotically stable equilibrium point* for (3.8). □

A locally asymptotically stable equilibrium point is hence characterized by having a nonempty neighborhood from which the trajectories remain arbitrarily close to the equilibrium point and eventually converge to it. In the case that (3.10) holds but (3.11) does not, the equilibrium point is said locally stable.

Locally asymptotically stable equilibrium points can be also globally asymptotically stable according to the following definition.

Definition 3.3. Suppose that

$$\lim_{t \to \infty} \varphi(t, x_{init}) = 0_n \ \ \forall x_{init} \in \mathbb{R}^n. \tag{3.12}$$

Then, the origin is said a *globally asymptotically stable equilibrium point* for (3.8). □

Locally asymptotically stable equilibrium points can be characterized through Lyapunov stability theory as explained in the following result.

Theorem 3.1 ([65]). *Let $v : \mathbb{R}^n \to \mathbb{R}$ satisfy the following conditions:*

$$\begin{array}{c} v(x) \text{ is positive definite} \\ v(x) \text{ is continuously differentiable} \\ \dot{v}(x) \text{ is locally negative definite, i.e.} \\ \exists \varepsilon > 0 : \ \dot{v}(x) < 0 \, \forall x \in \mathbb{R}^n_0, \ \|x\| < \varepsilon \end{array} \tag{3.13}$$

where

$$\dot{v}(x) = \frac{dv(x)}{dt} = \nabla v(x)' f(x). \tag{3.14}$$

Then, the origin is a locally asymptotically stable equilibrium point for (3.8), *and* $v(x)$ *is said a* LF *for the origin.*

Hence, a LF is a nonnegative and continuously differentiable function which vanishes only at the equilibrium point and decreases along the trajectories of the system in a neighborhood of the equilibrium point. Clearly, such a function does not exist if the origin is not a locally asymptotically stable equilibrium point.

Lyapunov stability theory can be used also to characterize global asymptotical stability as explained hereafter.

Theorem 3.2 ([65]). *Let $v : \mathbb{R}^n \to \mathbb{R}$ satisfy the following conditions:*

$$\begin{array}{c} v(x) \textit{ is positive definite} \\ v(x) \textit{ is continuously differentiable} \\ v(x) \textit{ is radially unbounded, i.e.} \\ \|x\| \to \infty \ \Rightarrow \ v(x) \to \infty \\ \dot{v}(x) \textit{ is negative definite.} \end{array} \tag{3.15}$$

Then, the origin is a globally asymptotically stable equilibrium point for (3.8), *and $v(x)$ is said a* global LF *for such an equilibrium point.*

Local stability can be investigated also through the linear approximation of (3.8) at the origin. Indeed, let us introduce the following definition.

Definition 3.4. Suppose that $f(x)$ is continuously differentiable at the origin, and define

$$\begin{cases} \dot{\hat{x}}(t) = A\hat{x}(t) \\ \hat{x}(0) = \hat{x}_{init} \\ A = \mathrm{lp}(f) \end{cases} \tag{3.16}$$

where

$$\mathrm{lp}(f) = \left. \frac{df(x)}{dx} \right|_{x=0_n}. \tag{3.17}$$

Then, (3.16) is said the *linearized system* of (3.8) at the origin. □

The following result provides some conditions for investigating local stability of the origin of (3.8) based on A. To this end, let us introduce first the following definition.

Definition 3.5. Let $A \in \mathbb{R}^{n\times n}$, and suppose that all the eigenvalues of A have negative real part, i.e.

$$\mathrm{Re}(\lambda) < 0 \ \ \forall \lambda \in \mathrm{spc}(A) \tag{3.18}$$

where

$$\mathrm{spc}(A) = \{\lambda \in \mathbb{C} : \ \det(\lambda I - A) = 0\}. \tag{3.19}$$

Then, A is said a *Hurwitz matrix*. □

Theorem 3.3 ([65]). *Suppose that $f(x)$ is continuously differentiable at the origin, and define $A = \mathrm{lp}(f)$. Then:*

1. *the origin is a locally stable equilibrium point if A is Hurwitz;*
2. *the origin is not a locally stable equilibrium point if A has an eigenvalue with positive real part.*

Whenever A is Hurwitz, the origin admits a quadratic LF. Indeed, let us introduce the following preliminary result.

Theorem 3.4 ([65]). *Consider the* Lyapunov equation

$$VA + A'V + W = 0 \tag{3.20}$$

where $V, W \in \mathbb{S}^n$ and $A \in \mathbb{R}^{n\times n}$. If A is Hurwitz and $W > 0$, there exists a unique solution V for (3.20)*, and this solution satisfies $V > 0$.*

Therefore, whenever $A = \mathrm{lp}(f)$ is Hurwitz, the function

$$v(x) = x'Vx \tag{3.21}$$

obtained with V solution of (3.20) for any $W > 0$, is a LF for the origin of (3.8). Let us observe that (3.20) is linear in V, which means that determining V satisfying (3.20) amounts to solving a system of linear equations.

3.3 DA

We are now ready to introduce the definition of DA of an equilibrium point.

Definition 3.6. Define the set

$$\mathscr{D} = \left\{ x_{init} \in \mathbb{R}^n : \lim_{t\to\infty} \varphi(t, x_{init}) = 0_n \right\}. \tag{3.22}$$

Then, $\mathscr{D}$ is said the *DA* of the origin for (3.8). □

Hence, the DA of an equilibrium point (the origin in this case) is the set of initial conditions from which the trajectory of the system converges to the equilibrium point itself. Clearly:

$$0_n \in \mathscr{D} \tag{3.23}$$

and from Definition 3.2 it follows directly that

$$\begin{array}{c} \exists \delta > 0 : \; \{x \in \mathbb{R}^n : \; \|x\| \le \delta\} \subseteq \mathscr{D} \\ \Updownarrow \\ 0_n \text{ is a locally asymptotically stable equilibrium point.} \end{array} \tag{3.24}$$

It is well-known that the DA is a complicated set, and hence it is usually approximated with estimates of simple shape. To this end, let us introduce the following definition.

Definition 3.7. For $v : \mathbb{R}^n \to \mathbb{R}$ and $c \in \mathbb{R}$, let us define the set

$$\mathscr{V}(c) = \{x \in \mathbb{R}^n : \; v(x) \le c\}. \tag{3.25}$$

Then, $\mathscr{V}(c)$ is said a *sublevel set* of $v(x)$. □

Hence, a sublevel set of $v(x)$ is the set of points for which $v(x)$ takes values less than or equal to a certain constant. The sublevel sets of a LF own particular properties, as explained in the following result.

Theorem 3.5. *Let $v(x)$ be a LF for the origin of (3.8), and suppose that $v(x)$ is radially unbounded. Then, the following properties hold:*

1. $\mathscr{V}(c) = \emptyset$ *for all* $c < 0$*;*
2. $\mathscr{V}(0) = \{0_n\}$*;*
3. $0_n \in \mathscr{V}(c)$ *for all* $c \ge 0$*;*
4. $\mathscr{V}(c)$ *is compact for all* c*;*
5. $\mathscr{V}(c) \to \mathbb{R}^n$ *as* $c \to \infty$.

Proof. The properties follow directly from the fact that $v(x)$ is positive definite and continuous on $\mathbb{R}^n$ (according to the definitions of LFs given in this book). □

The following result explains how a LF for the origin of (3.8) can be used to compute estimates of the DA.

Theorem 3.6 ([65]). *Let the origin be a locally asymptotically stable equilibrium point of* (3.8), *and let $v(x)$ be a LF for it, with $v(x)$ radially unbounded. Let us suppose that*

$$\mathscr{V} \subseteq \mathscr{Z} \tag{3.26}$$

for some $c \in \mathbb{R}$, where

$$\mathscr{Z} = \{x \in \mathbb{R}^n : \; \dot{v}(x) < 0\} \cup \{0_n\}. \tag{3.27}$$

Then,

$$\mathscr{V}(c) \subseteq \mathscr{D} \tag{3.28}$$

i.e. $\mathscr{V}(c)$ is an inner estimate of $\mathscr{D}$.

Theorem 3.6 states that estimates of $\mathscr{D}$ are given by any set $\mathscr{V}(c)$ included in the region where the temporal derivative of $v(x)$ is negative union the origin. Let us also observe that the estimates provided by Theorem 3.6 are inner estimates of $\mathscr{D}$: this guarantees that, for any initial condition chosen inside such estimates, the trajectory of (3.8) converges to the origin.

3.4 Controlled Systems

In the case of non-autonomous systems, equilibrium points are defined relatively to a given input. Specifically, consider a dynamical system of the form

$$\begin{cases} \dot{x}(t) = f(x(t), u(t)) \\ y(t) = g(x(t), u(t)) \\ x(0) = x_{init} \end{cases} \tag{3.29}$$

where $x(t) \in \mathbb{R}^n$ is the state, $t \in \mathbb{R}$ is the time, $u(t) \in \mathbb{R}^{n_u}$ is the input, $y(t) \in \mathbb{R}^{n_y}$ is the output, $x_{init} \in \mathbb{R}^n$ is the initial condition, $f : \mathbb{R}^n \times \mathbb{R}^{n_u} \to \mathbb{R}^n$ defines the evolution of the state, and $g : \mathbb{R}^n \times \mathbb{R}^{n_u} \to \mathbb{R}^{n_y}$ defines the output. For such a system equilibrium points are defined as follows.

Definition 3.8. Given $\bar{u} \in \mathbb{R}^{n_u}$, let $x_{eq} \in \mathbb{R}^n$ be such that

$$f(x_{eq}, \bar{u}) = 0_n. \tag{3.30}$$

Then, x_{eq} is said an *equilibrium point* of (3.29) with input $u(t) = \bar{u}$. □

Similarly to the case of autonomous systems, it can be convenient to assume that the equilibrium point of interest is the origin, and the corresponding input and output are the null signals. This can be done without loss of generality by defining the new variables

$$\begin{aligned} \tilde{x}(t) &= x(t) - x_{eq} \\ \tilde{y}(t) &= y(t) - y_{eq} \\ \tilde{u}(t) &= u(t) - \bar{u} \\ y_{eq} &= g(x_{eq}, \bar{u}). \end{aligned} \tag{3.31}$$

In fact, it follows that (3.29) can be rewritten as

$$\begin{cases} \dot{\tilde{x}}(t) = \tilde{f}(\tilde{x}(t), \tilde{u}(t)) \\ \tilde{y}(t) = \tilde{g}(\tilde{x}(t), \tilde{u}(t)) \\ \tilde{x}(0) = \tilde{x}_{init} \end{cases} \tag{3.32}$$

where

$$\begin{aligned} \tilde{f}(\tilde{x}, \tilde{u}) &= f(\tilde{x} + x_{eq}, \tilde{u} + \bar{u}) \\ \tilde{g}(\tilde{x}, \tilde{u}) &= g(\tilde{x} + x_{eq}, \tilde{u} + \bar{u}) - y_{eq} \\ \tilde{x}_{init} &= x_{init} - x_{eq}. \end{aligned} \tag{3.33}$$

In controlled systems it is natural to address the design of controllers enlarging the DA of an equilibrium point of interest. For such a problem, the controller should not change the equilibrium point. Specifically, let us suppose according to the previous discussion that the origin is the equilibrium point of interest with $\bar{u} = 0_{n_u}$ and $y_{eq} = 0_{n_y}$, and let us denote with $u = k(y)$ the feedback control input. It follows that $k(y)$ should satisfy

$$\begin{aligned} f(0_n, k(y)) &= 0_n \\ y &= g(0_n, k(y)). \end{aligned} \tag{3.34}$$

3.5 Common Equilibrium Points of Uncertain Nonlinear Systems

Let us consider an uncertain dynamical system of the form

$$\begin{cases} \dot{x}(t) = f(x(t),\theta) \\ x(0) = x_{init} \\ \theta \in \Theta \end{cases} \tag{3.35}$$

where $x(t) \in \mathbb{R}^n$ is the state, $t \in \mathbb{R}$ is the time, $x_{init} \in \mathbb{R}^n$ is the initial condition, $\theta \in \mathbb{R}^{n_\theta}$ is the uncertainty, $\Theta \subseteq \mathbb{R}^{n_\theta}$ is the admissible region for the uncertainty, and $f : \mathbb{R}^n \times \mathbb{R}^{n_\theta} \to \mathbb{R}^n$ defines the evolution of the state. Such a system is commonly said a autonomous, continuous-time, time-invariant, uncertain, nonlinear system.

Let us observe that the equilibrium points of (3.35) can be function of the uncertainty. In the sequel we will focus on equilibrium points that do not depend on the uncertainty. Such equilibrium points are defined as follows.

Definition 3.9. Let $x_{eq} \in \mathbb{R}^n$ be such that

$$f(x_{eq},\theta) = 0_n \;\; \forall \theta \in \Theta. \tag{3.36}$$

Then, x_{eq} is said a *common equilibrium point* of (3.35). □

Hence, a common equilibrium point is a zero in x of the function $f(x,\theta)$ for all admissible values of θ, and corresponds to a constant solution of the differential equation in (3.35). We denote the set of common equilibrium points of (3.35) as

$$\mathscr{X} = \{x \in \mathbb{R}^n : \; f(x,\theta) = 0_n \;\; \forall \theta \in \Theta\}. \tag{3.37}$$

The set $\mathscr{X}$ consists of the intersections of the hyper-surfaces $\{x \in \mathbb{R}^n : \; f_i(x,\theta) \; \forall \theta \in \Theta\} = 0$, $i = 1,\ldots,n$. Except special cases, $\mathscr{X}$ is nonempty and finite.

As in the case of uncertainty-free systems previously considered, it can be convenient to assume that the equilibrium point of interest is the origin. Let us observe that this can be done without loss of generality also in the case of uncertain systems. In fact, let us define the new variable

$$\tilde{x}(t) = x(t) - x_{eq}. \tag{3.38}$$

It follows that (3.35) can be rewritten with respect to $\tilde{x}$ as

$$\begin{cases} \dot{\tilde{x}}(t) = \tilde{f}(\tilde{x}(t),\theta) \\ \tilde{x}(0) = \tilde{x}_{init} \\ \theta \in \Theta \end{cases} \tag{3.39}$$

where

$$\begin{aligned} \tilde{f}(\tilde{x},\theta) &= f(\tilde{x} + x_{eq},\theta) \\ \tilde{x}_{init} &= x_{init} - x_{eq}. \end{aligned} \tag{3.40}$$

Moreover, the equilibrium point x_{eq} for (3.35) is shifted into the origin for (3.39) since $\tilde{x}_{eq} = x_{eq} - x_{eq} = 0_n$.

Therefore, in the sequel we will consider that the system under investigation is given by

$$\begin{cases} \dot{x}(t) = f(x(t), \theta) \\ x(0) = x_{init} \\ \theta \in \Theta \\ f(0_n, \theta) = 0_n \ \forall \theta \in \Theta \end{cases} \tag{3.41}$$

for which the origin is a common equilibrium point.

3.6 Robust Stability

Let us indicate the solution of (3.41) at time $t \geq 0$ with $\varphi(t, x_{init}, \theta)$, i.e.

$$\begin{cases} \dot{\varphi}(t, x_{init}, \theta) = f(\varphi(t, x_{init}, \theta), \theta) \\ \varphi(0, x_{init}, \theta) = x_{init}. \end{cases} \tag{3.42}$$

The following definition introduces the concept of robustly locally asymptotically stable common equilibrium point.

Definition 3.10. Suppose that

$$\forall \varepsilon > 0 \, \exists \delta > 0 : \ \|x_{init}\| < \delta \ \Rightarrow \ \|\varphi(t, x_{init}, \theta)\| < \varepsilon \ \forall t \geq 0 \, \forall \theta \in \Theta \tag{3.43}$$

and

$$\lim_{t \to \infty} \varphi(t, x_{init}, \theta) = 0_n \ \forall x_{init} : \ \|x_{init}\| < \delta \ \forall \theta \in \Theta. \tag{3.44}$$

Then, the origin is said a *robustly locally asymptotically stable common equilibrium point* for (3.41). □

A robustly locally asymptotically stable common equilibrium point is hence characterized by having a nonempty neighborhood from which the trajectories remain arbitrarily close to the equilibrium point and eventually converge to it for all admissible uncertainties. In the case that (3.43) holds but (3.44) does not, the common equilibrium point is said robustly locally stable.

Robustly locally asymptotically stable common equilibrium points can be also robustly globally asymptotically stable according to the following definition.

Definition 3.11. Suppose that

$$\lim_{t \to \infty} \varphi(t, x_{init}, \theta) = 0_n \ \forall x_{init} \in \mathbb{R}^n \ \forall \theta \in \Theta. \tag{3.45}$$

Then, the origin is said a *robustly globally asymptotically stable common equilibrium point* for (3.41). □

Robustly locally asymptotically stable common equilibrium points can be characterized through LFs as explained in the following result.

Theorem 3.7. *Let* $v : \mathbb{R}^n \to \mathbb{R}$ *satisfy the following conditions:*

$$\begin{array}{c} v(x) \text{ is positive definite} \\ v(x) \text{ is continuously differentiable} \\ \dot{v}(x,\theta) \text{ is locally negative definite for all } \theta \in \Theta, \text{ i.e.} \\ \exists \varepsilon > 0 : \ \dot{v}(x,\theta) < 0 \ \ \forall x \in \mathbb{R}_0^n, \ \|x\| < \varepsilon \ \ \forall \theta \in \Theta \end{array} \tag{3.46}$$

where

$$\dot{v}(x,\theta) = \frac{dv(x)}{dt} = \nabla v(x)' f(x,\theta). \tag{3.47}$$

Then, the origin is a robustly locally asymptotically stable common equilibrium point for (3.41)*, and* $v(x)$ *is said a* common LF *for such an equilibrium point.*

Proof. Direct consequence of Theorem 3.1 and Definition 3.10. □

Hence, a common LF is a nonnegative and continuously differentiable function which vanishes only at the common equilibrium point and decreases along the trajectories of the system in a neighborhood of this point for all admissible uncertainties. Clearly, such a function does not exist if the origin is not a robustly locally asymptotically stable common equilibrium point.

LFs can be used also to characterize robust global asymptotical stability as explained hereafter.

Theorem 3.8. *Let* $v : \mathbb{R}^n \to \mathbb{R}$ *satisfy the following conditions:*

$$\begin{array}{c} v(x) \text{ is positive definite} \\ v(x) \text{ is continuously differentiable} \\ v(x) \text{ is radially unbounded} \\ \dot{v}(x,\theta) \text{ is negative definite for all } \theta \in \Theta. \end{array} \tag{3.48}$$

Then, the origin is a robustly globally asymptotically stable common equilibrium point for (3.41)*, and* $v(x)$ *is said a* global common LF *for such a common equilibrium point.*

Proof. Direct consequence of Theorem 3.2 and Definition 3.11. □

Common LFs may be conservative in order to prove robust asymptotic stability of a common equilibrium point. In order to cope with this problem, one can use LFs that depend on the uncertainty as explained hereafter.

Theorem 3.9. *Let* $v : \mathbb{R}^n \times \mathbb{R}^{n_\theta} \to \mathbb{R}$ *satisfy the following conditions:*

$$\begin{array}{c} v(x,\theta) \text{ is positive definite for all } \theta \in \Theta \\ v(x,\theta) \text{ is continuously differentiable for all } \theta \in \Theta \\ \dot{v}(x,\theta) \text{ is locally negative definite for all } \theta \in \Theta \end{array} \tag{3.49}$$

where

$$\dot{v}(x,\theta) = \frac{dv(x,\theta)}{dt} = \nabla_x v(x,\theta)' f(x,\theta). \tag{3.50}$$

Then, the origin is a robustly locally asymptotically stable common equilibrium point for (3.41)*, and* $v(x,\theta)$ *is said a* parameter-dependent LF *for such an equilibrium point.*

Proof. Direct consequence of Theorem 3.1 and Definition 3.10. □

Theorem 3.10. *Let* $v : \mathbb{R}^n \times \mathbb{R}^{n_\theta} \to \mathbb{R}$ *satisfy the following conditions:*

$$\begin{array}{c} v(x,\theta) \text{ is positive definite for all } \theta \in \Theta \\ v(x,\theta) \text{ is continuously differentiable for all } \theta \in \Theta \\ v(x,\theta) \text{ is radially unbounded for all } \theta \in \Theta \\ \dot{v}(x,\theta) \text{ is negative definite for all } \theta \in \Theta. \end{array} \tag{3.51}$$

Then, the origin is a robustly globally asymptotically stable common equilibrium point for (3.41)*, and* $v(x,\theta)$ *is said a* global parameter-dependent LF *for such an equilibrium point.*

Proof. Direct consequence of Theorem 3.2 and Definition 3.11. □

Robust local stability can be investigated also through the linear approximation of (3.41) at the origin. Indeed, let us introduce the following definition.

Definition 3.12. Suppose that $f(x,\theta)$ is continuously differentiable at the origin for all $\theta \in \Theta$, and define

$$\begin{cases} \dot{\hat{x}}(t) = A(\theta)\hat{x}(t) \\ \hat{x}(0) = \hat{x}_{init} \\ \theta \in \Theta \\ A(\theta) = \mathrm{lp}(f(\cdot,\theta)) \end{cases} \tag{3.52}$$

where

$$\mathrm{lp}(f(\cdot,\theta)) = \left.\frac{df(x,\theta)}{dx}\right|_{x=0_n}. \tag{3.53}$$

Then, (3.52) is said the *linearized system* of (3.41) at the origin. □

The following result provides some conditions for investigating robust local stability of the origin of (3.41) based on $\mathrm{lp}(f,\theta)$.

Theorem 3.11. *Suppose that* $f(x,\theta)$ *is continuously differentiable in x at the origin, and define* $A(\theta) = \mathrm{lp}(f(\cdot,\theta))$*. Then:*

1. *the origin is a robustly locally stable common equilibrium point if* $A(\theta)$ *is Hurwitz for all* $\theta \in \Theta$*;*
2. *the origin is not a robustly locally stable equilibrium point if* $A(\theta)$ *has an eigenvalue with positive real part for some* $\theta \in \Theta$*.*

Proof. Direct consequence of Theorem 3.3 and Definition 3.10. □

Whenever $A(\theta)$ is Hurwitz for all $\theta \in \Theta$, the origin admits a quadratic parameter-dependent LF. Indeed, let us introduce the following preliminary result.

Theorem 3.12. *Consider the* parametrized Lyapunov equation

$$V(\theta)A(\theta) + A(\theta)'V(\theta) + W(\theta) = 0 \tag{3.54}$$

where $V, W : \mathbb{R}^{n_\theta} \to \mathbb{S}^n$ *and* $A : \mathbb{R}^{n_\theta} \to \mathbb{R}^{n \times n}$*. If* $A(\theta)$ *is Hurwitz and* $W(\theta) > 0$ *for all* $\theta \in \Theta$*, there exists a unique solution* $V(\theta)$ *for* (3.54)*, and this solution satisfies* $V(\theta) > 0$ *for all* $\theta \in \Theta$*.*

Proof. Direct consequence of Theorem 3.4. □

Therefore, whenever $A(\theta) = \mathrm{lp}(f(\cdot, \theta))$ is Hurwitz for all $\theta \in \Theta$, the function

$$v(x, \theta) = x'V(\theta)x \tag{3.55}$$

obtained with $V(\theta)$ solution of (3.54) for any $W(\theta) > 0$ for all $\theta \in \Theta$, is a parameter-dependent LF for the origin of (3.41).

3.7 RDA

We are now ready to introduce the definition of RDA of a common equilibrium point.

Definition 3.13. Define the set

$$\mathscr{R} = \left\{ x_{init} \in \mathbb{R}^n : \lim_{t \to \infty} \varphi(t, x_{init}, \theta) = 0_n \ \ \forall \theta \in \Theta \right\}. \tag{3.56}$$

Then, $\mathscr{R}$ is said the *RDA* of the origin for (3.41). □

Hence, the RDA of a common equilibrium point (the origin in this case) is the set of initial conditions from which the trajectory of the system converges to the equilibrium point itself for all admissible uncertainties. Clearly,

$$0_n \in \mathscr{R} \tag{3.57}$$

and from Definition 3.10 it follows directly that

$$\begin{array}{c} \exists \delta > 0 : \ \{x \in \mathbb{R}^n : \ \|x\| \leq \delta\} \subseteq \mathscr{R} \\ \Updownarrow \\ 0_n \text{ is a robustly locally asymptotically stable common equilibrium point.} \end{array} \tag{3.58}$$

Obviously, the RDA is given by the intersection of all the DAs of the origin obtained for any admissible value of the uncertainty, i.e.

$$\mathscr{R} = \bigcap_{\theta \in \Theta} \mathscr{D}(\theta) \tag{3.59}$$

where $\mathscr{D}(\theta)$ is the DA of the origin defined in (3.22) and parametrized by θ.

The following result explains how a common LF for the origin of (3.41) can be used to compute estimates of the RDA.

Theorem 3.13. *Let the origin be a robustly locally asymptotically stable equilibrium point of* (3.41)*, and let $v(x)$ be a common LF for it, with $v(x)$ radially unbounded. Let us suppose that*

$$\mathscr{V}(c) \subseteq \mathscr{Z}(\theta) \ \ \forall \theta \in \Theta \tag{3.60}$$

for some $c \in \mathbb{R}$, where

$$\mathscr{Z}(\theta) = \{x \in \mathbb{R}^n : \ \dot{v}(x,\theta) < 0\} \cup \{0_n\}. \tag{3.61}$$

Then,

$$\mathscr{V}(c) \subseteq \mathscr{R} \tag{3.62}$$

i.e. $\mathscr{V}(c)$ is an inner estimate of $\mathscr{R}$.

Proof. Direct consequence of Theorem 3.6 and Definition 3.13. □

Theorem 3.13 states that estimates of $\mathscr{R}$ are given by any set $\mathscr{V}(c)$ included in the region where the temporal derivative of $v(x)$ is negative for all admissible uncertainties union the origin.

The RDA can be estimated also by using parameter-dependent LFs. To this end, let us introduce the following definition.

Definition 3.14. For $v : \mathbb{R}^n \times \mathbb{R}^{n_\theta} \to \mathbb{R}$ and $c : \mathbb{R}^{n_\theta} \to \mathbb{R}$, let us define the set

$$\mathscr{V}(c(\theta),\theta) = \{x \in \mathbb{R}^n : \ v(x,\theta) \leq c(\theta)\}. \tag{3.63}$$

Then, $\mathscr{V}(c(\theta),\theta)$ is said a *parameter-dependent sublevel set* of $v(x,\theta)$. □

The following result explains how a parameter-dependent LF for the origin of (3.41) can be used to compute estimates of the RDA.

Theorem 3.14. *Let the origin be a robustly locally asymptotically stable equilibrium point of* (3.41)*, and let $v(x,\theta)$ be a parameter-dependent LF for it, with $v(x,\theta)$ radially unbounded for all $\theta \in \Theta$. Let us suppose that*

$$\mathscr{V}(c(\theta),\theta) \subseteq \mathscr{Z}(\theta) \ \ \forall \theta \in \Theta \tag{3.64}$$

for some $c : \mathbb{R}^q \to \mathbb{R}$. Then,

$$\mathscr{V}(c(\theta),\theta) \subseteq \mathscr{R} \ \ \forall \theta \in \Theta \tag{3.65}$$

i.e. $\mathscr{V}(c(\theta),\theta)$ is an inner estimate of $\mathscr{R}$ for any admissible θ.

Proof. Direct consequence of Theorem 3.6 and Definition 3.13. □

Theorem 3.14 states that estimates of $\mathscr{R}$ are given by any set $\mathscr{V}(c(\theta),\theta)$ included in the region where the temporal derivative of $v(x,\theta)$ is negative for all admissible θ union the origin.

Let us observe that θ has been assumed time-invariant in (3.35). As it is known, common LFs prove stability for both cases of time-invariant and time-varying uncertainty. This means that the estimates of the RDA provided by common LFs are also estimates of the RDA defined for time-varying uncertainties, i.e. the set

$$\tilde{\mathscr{R}} = \left\{ x_{init} \in \mathbb{R}^n : \lim_{t\rightarrow\infty} \varphi(t, x_{init}, \theta(\cdot)) = 0_n \;\; \forall \theta(\cdot) : \; \theta(t) \in \Theta \;\; \forall t \geq 0 \right\}. \tag{3.66}$$

Clearly,

$$\tilde{\mathscr{R}} \subseteq \mathscr{R}. \tag{3.67}$$

3.8 Robustly Controlled Systems

In the case of non-autonomous uncertain systems, common equilibrium points are defined relatively to a given input. Specifically, consider a dynamical system of the form

$$\left\{ \begin{aligned} \dot{x}(t) &= f(x(t),\theta,u(t)) \\ y(t) &= g(x(t),\theta,u(t)) \\ x(0) &= x_{init} \\ \theta &\in \Theta \end{aligned} \right. \tag{3.68}$$

where $x(t) \in \mathbb{R}^n$ is the state, $t \in \mathbb{R}$ is the time, $u(t) \in \mathbb{R}^{n_u}$ is the input, $y(t) \in \mathbb{R}^{n_y}$ is the output, $x_{init} \in \mathbb{R}^n$ is the initial condition, $\theta \in \mathbb{R}^{n_\theta}$ is the uncertainty, $\Theta \subseteq \mathbb{R}^{n_\theta}$ is the admissible region for the uncertainty, $f : \mathbb{R}^n \times \mathbb{R}^{n_\theta} \times \mathbb{R}^{n_u} \rightarrow \mathbb{R}^n$ defines the evolution of the state, and $g : \mathbb{R}^n \times \mathbb{R}^{n_\theta} \times \mathbb{R}^{n_u} \rightarrow \mathbb{R}^{n_y}$ defines the output. For such a system common equilibrium points are defined as follows.

Definition 3.15. Given $\bar{u} \in \mathbb{R}^{n_u}$, let $x_{eq} \in \mathbb{R}^n$ be such that

$$f(x_{eq},\theta,\bar{u}) = 0_n \;\; \forall \theta \in \Theta. \tag{3.69}$$

Then, x_{eq} is said a *common equilibrium point* of (3.68) with input $u(t) = \bar{u}$. □

Similarly to the case of autonomous uncertain systems, it can be convenient to assume that the common equilibrium point of interest is the origin, and the corresponding input and output are the null signals. This can be done without loss of generality by defining the new variables in (3.31) by replacing $y_{eq} = g(x_{eq},\bar{u})$ with $y_{eq}(\theta) = g(x_{eq},\theta,\bar{u})$. In fact, it follows that (3.68) can be rewritten as

$$\left\{ \begin{aligned} \dot{\tilde{x}}(t) &= \tilde{f}(\tilde{x}(t),\theta,\tilde{u}(t)) \\ \tilde{y}(t) &= \tilde{g}(\tilde{x}(t),\theta,\tilde{u}(t)) \\ \tilde{x}(0) &= \tilde{x}_{init} \\ \theta &\in \Theta \end{aligned} \right. \tag{3.70}$$

where

$$\begin{aligned}\tilde{f}(\tilde{x},\theta,\tilde{u}) &= f(\tilde{x}+x_{eq},\theta,\tilde{u}+\bar{u})\\ \tilde{g}(\tilde{x},\theta,\tilde{u}) &= g(\tilde{x}+x_{eq},\theta,\tilde{u}+\bar{u})-y_{eq}(\theta)\\ \tilde{x}_{init} &= x_{init}-x_{eq}.\end{aligned} \tag{3.71}$$

In robustly controlled systems it is natural to address the design of controllers enlarging the RDA of an equilibrium point of interest. For such a problem, the controller should not change the equilibrium point. Specifically, let us suppose according to the previous discussion that the origin is the equilibrium point of interest with $\bar{u} = 0_{n_u}$, and let us consider the case where

$$y_{eq}(\theta) = 0_{n_y}. \tag{3.72}$$

Let us denote with $u = k(y)$ the feedback control input. It follows that $k(y)$ should satisfy

$$\left.\begin{aligned} f(0_n,\theta,k(y)) &= 0_n\\ y &= g(0_n,\theta,k(y))\end{aligned}\right\} \ \forall\theta\in\Theta. \tag{3.73}$$

3.9 Notes and References

The background concepts summarized in this chapter have been explained and studied in several classical works, in particular [65] for general stability properties of nonlinear dynamical systems, and [73, 48, 37, 38, 86, 1] for the characterization of the DA.

The estimation of the DA using LFs has been proposed since long time. In particular, pioneering contributions were proposed by Zubov who introduced the well-known Zubov equation [120], see also the various applications and re-elaborations of this method such as [74, 88, 14], and by La Salle who introduced the well-known La Salle theorem [98]. Other contributions include the use of maximal LFs [119, 97], piecewise LFs [78, 77], and logical compositions of LFs [3]. See also the references mentioned in Chapter 4.

Several non-Lyapunov techniques for estimating the DA have been introduced in the literature, in particular the trajectory reverse method described in [48], and techniques for determining reachable sets such as [2].

The DA for controlled systems has been studied from differential viewpoints, for instance in systems with high-gain feedback [67], power systems [39], stabilization problems [104], saturated systems [59] and quantized systems [107]. See also the references mentioned in Chapters 4–6.

The reader is referred to [5, 49, 6, 33] and references therein for general robust stability properties of uncertain dynamical systems. The estimation of the RDA in the case of uncertain systems was considered in [13, 43]. See also the references mentioned in Chapters 5–6.

Chapter 4
DA in Polynomial Systems

This chapter addresses the estimation of the DA of equilibrium points of polynomial systems, i.e. dynamical systems whose dynamics is described by a polynomial function of the state, by using polynomial LFs and SOS programming. It is shown that a condition for establishing whether a sublevel set of a LF is an estimate of the DA can be obtained in terms of an LMI feasibility test by exploiting SOS polynomials and their representation through the SMR. Then, the problem of computing the LEDA provided by a LF is addressed, showing that in general a candidate of such an estimate can be found by solving a GEVP, which is a convex optimization problem. A sufficient and necessary condition for establishing whether the found candidate is tight is hence provided by looking for power vectors in linear subspaces. Moreover, the case of quadratic LFs is considered, showing that the estimation of the DA in such a case can be performed by establishing positivity of homogeneous polynomials obtained without the introduction of unknown multipliers. Then, the chapter addresses the problem of determining optimal estimates of the DA by using variable LFs. Three strategies are hence described, one that consists of maximizing either the volume of the estimate or an approximation of it, one that consists of enlarging the estimate by maximizing the size of a set with fixed shape included in the estimate, and one that attempts to establish global asymptotical stability of the equilibrium point. Lastly, the use of the previously described techniques for designing polynomial static output controllers to enlarge the DA, is discussed.

4.1 Polynomial Systems

With the term polynomial system we refer to a dynamical system where the temporal derivative of the state is a polynomial function of the state. In this chapter we address the problem of estimating and controlling the DA of an equilibrium point of polynomial systems. As explained in Section 3.3, we can assume without loss of generality that the equilibrium point under consideration is the origin. Hence, let us start by considering the polynomial system described by

G. Chesi: Domain of Attraction, LNCIS 415, pp. 105–149, 2011.
springerlink.com

$$\begin{cases} \dot{x}(t) = f(x(t)) \\ x(0) = x_{init} \\ f(0_n) = 0_n \end{cases} \tag{4.1}$$

where $x(t) \in \mathbb{R}^n$ is the state, $t \in \mathbb{R}$ is the time, $\dot{x}(t)$ is the temporal derivative of the state, $x_{init} \in \mathbb{R}^n$ is the initial condition, and $f \in \mathscr{P}_n^n$. In the sequel we will drop the dependence on the time for ease of notation.

System (4.1) has at least one equilibrium point (the origin). Depending on $f(x)$, (4.1) may have other equilibrium points, whose number can be either finite or infinite. Whenever finite, this number (denoted by $\mathrm{card}(\mathscr{X})$ where $\mathscr{X}$ is the set of equilibrium points of (4.1)) satisfies

$$\mathrm{card}(\mathscr{X}) \leq \partial f^n.$$

Example 4.1. Let us consider the electric circuit in Figure 4.1. By denoting with y_1 the voltage v of the resistive element and with y_2 its temporal derivative, the circuit can be described by

$$\begin{cases} \dot{y}_1 = y_2 \\ \dot{y}_2 = -y_1 - \sqrt{\dfrac{L}{C}} \left.\dfrac{dh(v)}{dv}\right|_{v=y_1} y_2 \end{cases}$$

where L is the inductance, C is the capacitance, and $h(v)$ is the current i in the resistive element. Let us assume $L/C = 1$. The choice

$$h(v) = -v + \frac{v^3}{3}. \tag{4.2}$$

corresponds to a negative resistive element, and the circuit is known as Van der Pol oscillator [65]. Indeed, in such a case the circuit presents a stable limit cycle surrounding the origin, which is the unique equilibrium point, and which is unstable. This limit cycle can be estimated by reverting the dynamics of the system, which has the effect of turning the origin into a locally asymptotically stable equilibrium point: the region delimited by the limit cycle is the DA of the origin in the new system. Specifically, the reverted system can be obtained by choosing $x_1 = y_1$ and $x_2 = -y_2$, which provides

$$\begin{cases} \dot{x}_1 = x_2 \\ \dot{x}_2 = -x_1 - x_2(1 - x_1^2). \end{cases} \tag{4.3}$$

□

Example 4.2. Let us consider the predator-prey system

$$\begin{cases} \dot{y}_1 = -3y_1 + 4y_1^2 - 0.5y_1y_2 - y_1^3 \\ \dot{y}_2 = -2.1y_2 + y_1y_2 \end{cases} \tag{4.4}$$

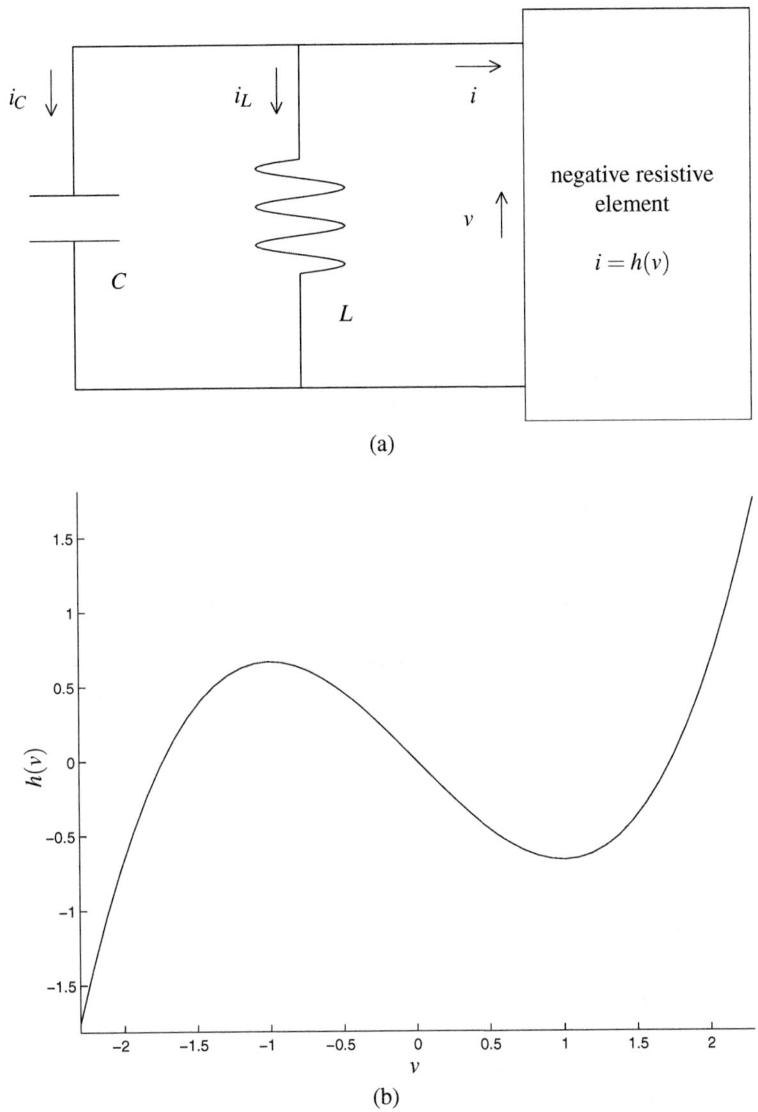

Fig. 4.1 Example 4.1 (Van der Pol circuit). (a) An electric circuit with an inductor, a capacitor, and a negative resistive element. (b) Voltage-current characteristic of the negative resistive element with the choice (4.2).

where y_1 and y_2 are the prey and predator populations [48]. This system has four equilibrium points as follows:

$$y_{eq1} = \begin{pmatrix} 0 \\ 0 \end{pmatrix}, \quad y_{eq2} = \begin{pmatrix} 2.1 \\ -1.98 \end{pmatrix}, \quad y_{eq3} = \begin{pmatrix} 1 \\ 0 \end{pmatrix}, \quad y_{eq4} = \begin{pmatrix} 3 \\ 0 \end{pmatrix}. \tag{4.5}$$

The first two equilibrium points are locally asymptotically stable, while the others are unstable. □

4.2 Estimates via LFs

In this section we describe a method for computing estimates of the DA based on polynomial LFs and SOS programming.

4.2.1 Establishing Estimates

The following result provides a condition for establishing whether a sublevel set is an inner estimate of the DA by testing the positivity of some functions.

Theorem 4.1. *Let $v : \mathbb{R}^n \to \mathbb{R}$ be continuously differentiable, positive definite and radially unbounded, and let $c \in \mathbb{R}$ be positive. Suppose that there exists $q : \mathbb{R}^n \to \mathbb{R}^2$ such that*

$$\left.\begin{array}{l} p(x) > 0 \\ q(x) > 0 \end{array}\right\} \quad \forall x \in \mathbb{R}_0^n \tag{4.6}$$

where

$$p(x) = -q(x)' \begin{pmatrix} \dot{v}(x) \\ c - v(x) \end{pmatrix} \tag{4.7}$$

and

$$\dot{v}(x) = \nabla v(x)' f(x). \tag{4.8}$$

Then, $v(x)$ is a LF for the origin of (4.1), *and $\mathcal{V}(c) \subseteq \mathcal{D}$.*

Proof. Suppose that (4.6) holds, and let $x \in \mathcal{V}(c) \setminus \{0_n\}$. Then, from the first inequality it follows that

$$\begin{aligned} 0 &< -q_1(x)\dot{v}(x) - q_2(x)(c - v(x)) \\ &\leq -q_1(x)\dot{v}(x) \end{aligned}$$

since $q_2(x) > 0$ from the second inequality in (4.6), and since $v(x) \leq c$. Moreover, since $q_1(x) > 0$ from the second inequality in (4.6), this implies that $\dot{v}(x) < 0$. Hence, it follows that (3.26) holds. Since $c > 0$, this implies that (3.13) holds, i.e. $v(x)$ is a Lyapunov for the origin. From Theorem 3.6 we conclude that $\mathcal{V}(c) \subseteq \mathcal{D}$. □

Theorem 4.1 provides a condition for establishing whether $\mathscr{V}(c)$ is included in $\mathscr{D}$ through the introduction of the auxiliary function $q(x)$. This condition does not require a priori knowledge of the fact whether $v(x)$ is LF for the origin of (4.1): indeed, it is easy to see that (4.6) cannot be satisfied for any positive c if $v(x)$ is not a LF for the origin of (4.1).

In the sequel we will consider the case where $q(x)$ and $v(x)$ are polynomials. In such a case, the condition of Theorem 4.1 can be checked through the SOS index. Now, since $\dot{v}(0_n) = 0$, one has that

$$(4.6) \text{ holds } \Rightarrow \; p(x) \text{ and } q_2(x) \text{ are locally quadratic polynomials.}$$

This means that, a necessary condition for the SOS index of these polynomials to be positive, is that such a SOS index is defined with respect to the power vector $b_{lin}(x,\cdot)$. The following result summarizes this discussion.

Corollary 4.1. *Let $v \in \mathscr{P}_n$ be positive definite, and let $c \in \mathbb{R}$ be positive. Suppose that there exist $q \in \mathscr{P}_n^2$ such that*

$$\begin{cases} \lambda_{qua}(p) > 0 \\ \lambda_{pol}(q_1) > 0 \\ \lambda_{qua}(q_2) > 0 \end{cases} \tag{4.9}$$

where $p(x)$ is as in (4.7)*. Then, $v(x)$ is a LF for the origin of* (4.1)*, and $\mathscr{V}(c) \subseteq \mathscr{D}$.*

Proof. Suppose that (4.9) holds. From Section 1.3 this implies that $p(x)$ and q_2 are positive definite, and that $q_1(x)$ is positive. Hence, (4.6) holds. Moreover, $v(x)$ is polynomial and positive definite, and hence it is also radially unbounded. Therefore, from Theorem 4.1 we conclude the proof. □

Corollary 4.1 provides a condition for establishing whether $\mathscr{V}(c)$ is an inner estimate of $\mathscr{D}$ through the SOS index. This condition can be checked through an LMI feasibility test. Indeed, with $q \in \mathscr{P}_n^2$ and $p(x)$ is as in (4.7) let us define

$$\begin{aligned} P(Q)+L(\alpha) &= CSMR_{qua}(p) \\ Q &= \mathrm{diag}(Q_1,Q_2) \\ Q_1 &= SMR_{pol}(q_1) \\ Q_2 &= SMR_{qua}(q_2). \end{aligned} \tag{4.10}$$

Corollary 4.2. *Let $v \in \mathscr{P}_n$ be positive definite, and let $c \in \mathbb{R}$ be positive. Define the quantities in* (4.10)*. Suppose that there exist Q and α such that the following LMIs hold:*

$$\begin{cases} P(Q)+L(\alpha) > 0 \\ Q > 0 \\ \mathrm{trace}(Q_1) = 1. \end{cases} \tag{4.11}$$

Then, $v(x)$ is a LF for the origin of (4.1)*, and $\mathscr{V}(c) \subseteq \mathscr{D}$.*

Proof. Suppose that (4.11) holds. This implies that (4.9) holds, and from Corollary 4.1 we conclude the proof. □

Corollary 4.2 shows how (4.9) can be converted into an LMI feasibility test. Let us observe that the constraint $\mathrm{trace}(Q_1) = 1$ normalizes the variables involved in the test: in fact,

$$\text{(4.11) holds for some } Q \text{ and } \alpha \iff \text{(4.11) holds for } \rho Q \text{ and } \rho\alpha \text{ for all } \rho > 0.$$

Let us consider the selection of the degrees of $q_1(x)$ and $q_2(x)$. A possibility is to choose them in order to maximize the degrees of freedom in (4.11) for a fixed degree of $p(x)$. This is equivalent to require that the degrees of $q_1(x)\dot{v}(x)$ and $q_2(x)(c - v(x))$, rounded to the smallest following even integers, are equal, i.e.

$$\left\lceil \frac{\partial q_1 \dot{v}}{2} \right\rceil = \left\lceil \frac{\partial q_2 (c-v)}{2} \right\rceil.$$

This can be achieved as follows:

$$\boxed{\begin{array}{l} \bullet \text{ choose an even degree for } q_1(x) \text{ denoted by } 2k; \\ \bullet \text{ set the degree of } q_2(x) \text{ as } \partial q_2 = 2\left\lceil \dfrac{\partial f - 1}{2} \right\rceil + 2k; \\ \bullet \text{ hence, the degree of } p(x) \text{ is given by } \partial p = \partial v + \partial q_2. \end{array}} \tag{4.12}$$

The number of scalar variables in the LMIs (4.11) is given by

$$\omega_1 = l_{qua}(n, \partial p) + t_{pol}(n, 2k) + t_{qua}(n, \partial q_2) - 1. \tag{4.13}$$

Tables 4.1–4.2 show this number in some cases according to the choice (4.12). In the sequel we will assume that the degrees of $q_1(x)$ and $q_2(x)$ are chosen according to (4.12) unless specified otherwise.

Table 4.1 Number of scalar variables in the LMIs (4.11) for $k = 0$.

$(n=2)$ $\partial f \setminus \partial v$	2	4	6
2 or 3	6	23	66
4 or 5	35	78	162

$(n=3)$ $\partial f \setminus \partial v$	2	4	6
2 or 3	20	116	440
4 or 5	155	479	1303

Table 4.2 Number of scalar variables in the LMIs (4.11) for $k = 1$.

$(n=2)$ $\partial f \setminus \partial v$	2	4	6
2 or 3	40	83	167
4 or 5	113	197	340

$(n=3)$ $\partial f \setminus \partial v$	2	4	6
2 or 3	164	488	1312
4 or 5	633	1457	3234

4.2.2 Choice of the LF

Let us observe that a necessary condition for (4.11) to hold is that

$$\begin{cases} V_2 > 0 \\ V_2A + A'V_2 < 0 \\ A = \mathrm{lp}(f) \end{cases} \tag{4.14}$$

where $\mathrm{lp}(f)$ is the state matrix of the linearized system in (3.17), and $V_2 = V_2'$ defines the quadratic part of $v(x)$, i.e.

$$v_2(x) = x'V_2x \tag{4.15}$$

where $v(x)$ is expressed as

$$v(x) = \sum_{i=2}^{\partial v} v_i(x) \tag{4.16}$$

where $v_i(x)$ is a homogeneous polynomial of degree i. Condition (4.14) holds if and only if the linearized system is asymptotically stable (i.e., A is Hurwitz) and $v_2(x)$ is a LF for it. Hence, the latter is a requirement for (4.11) to hold. This requirement can be removed as explained in Section 7.4.

Based on this discussion, a simple way to choose a LF $v(x)$ for estimating the DA is as follows:

• choose $W > 0$ and find V_2 that satisfies the Lyapunov equation $V_2A + A'V_2 + W = 0$; • define $v_2(x) = x'V_2x$; • define $v(x)$ as in (4.16) where $v_3(x), v_4(x), \ldots$ are arbitrarily chosen under the constraint that $v(x)$ is positive definite.	(4.17)

Let us observe that one can establish whether $v(x)$ is positive definite through the generalized SOS index according to Corollary 2.2, specifically

$$\exists k \in \mathbb{N} : \lambda_{qua}(v,k) > 0 \Rightarrow v(x) \text{ is positive definite.} \tag{4.18}$$

Example 4.3. For the Van der Pol circuit (4.3), let us consider the problem of establishing whether $\mathscr{V}(c)$ is a subset of the DA of the origin with the choice

$$\begin{aligned} v(x) &= x_1^2 + x_1x_2 + x_2^2 \\ c &= 0.5. \end{aligned}$$

For this system we have

$$A = \begin{pmatrix} 0 & 1 \\ -1 & -1 \end{pmatrix}$$

and hence A is Hurwitz. Moreover, $v(x)$ is a LF for the origin. In fact, (4.14) holds with

$$V_2 = \begin{pmatrix} 1 & 0.5 \\ \star & 1 \end{pmatrix}.$$

Let us use Corollary 4.2 choosing $k = 0$ in (4.12). Since $\partial v = 2$ and $\partial f = 3$, we have

$$\partial q_2 = 2, \ \ \partial p = 4.$$

Let us select the power vectors

$$b_{lin}(x, \partial p/2) = (x_1, x_2, x_1^2, x_1 x_2, x_2^2)', \ \ b_{pol}(x, \partial q_1/2) = 1, \ \ b_{lin}(x, \partial q_2/2) = x.$$

The SMR matrices in (4.11) are given by

$$P(Q) = \begin{pmatrix} 1-0.5g_1 & 0.5-0.5g_2 & 0 & 0 & 0 \\ \star & 1-0.5g_3 & 0 & 0 & 0 \\ \star & \star & g_1 & -0.5+0.5g_1+g_2 & -1+0.5g_1+g_2+0.5g_3 \\ \star & \star & \star & 0 & g_2+0.5g_3 \\ \star & \star & \star & \star & g_3 \end{pmatrix}$$

$$Q_1 = 1, \ \ Q_2 = \begin{pmatrix} g_1 & g_2 \\ \star & g_3 \end{pmatrix}, \ \ L(\alpha) = \begin{pmatrix} 0 & 0 & 0 & -\alpha_1 & -\alpha_2 \\ \star & 0 & \alpha_1 & \alpha_2 & 0 \\ \star & \star & 0 & 0 & -\alpha_3 \\ \star & \star & \star & 2\alpha_3 & 0 \\ \star & \star & \star & \star & 0 \end{pmatrix}.$$

Hence, the LMIs in (4.11) have 6 scalar variables. We find that these LMIs hold with

$$Q_2 = \begin{pmatrix} 0.565 & -0.004 \\ \star & 0.877 \end{pmatrix}, \ \ \alpha = (0.000, 0.000, 0.222)'.$$

Therefore, $\mathscr{V}(c) \subseteq \mathscr{D}$. □

4.2.3 LEDA

The condition provided in Corollary 4.2 is built for a chosen scalar c, which determines the extension of the estimate $\mathscr{V}(c)$. Clearly, it would be helpful if one could search for the largest c for which the LMIs in (4.11) are feasible. In fact, this would allow one to obtain the largest estimate of the DA that can be obtained through the LF $v(x)$. Such an estimate is defined as follows.

Definition 4.1. Let $v(x)$ be a LF for the origin of (4.1), and define the quantity

$$\begin{aligned} \gamma = & \sup_c \ c \\ & \text{s.t.} \ \mathscr{V}(c) \subseteq \mathscr{X} \end{aligned} \tag{4.19}$$

where $\mathscr{X}$ is given by (3.27). Then, $\mathscr{V}(\gamma)$ is said the *LEDA* of the origin provided by the LF $v(x)$. □

The LEDA of a LF $v(x)$ is the largest sublevel set of $v(x)$ included in the region where the temporal derivative of $v(x)$ is negative, and therefore is the largest estimate of the DA that can be obtained through $v(x)$. Indeed, supposing that γ is bounded, one has that

$$\begin{cases} \mathscr{V}(c) \subseteq \mathscr{D} \ \forall c < \gamma \\ \exists x: \ v(x) = \gamma \text{ and } \dot{v}(x) = 0. \end{cases}$$

From Corollary 4.2 one can define a natural lower bound of γ as

$$\begin{aligned} \hat{\gamma} = & \sup_{c,Q,\alpha} \ c \\ & \text{s.t.} \begin{cases} P(c,Q) + L(\alpha) > 0 \\ Q > 0 \\ \text{trace}(Q_1) = 1 \end{cases} \end{aligned} \tag{4.20}$$

where $P(c,Q)$ is the SMR matrix of $p(x)$ in (4.10) where the dependence on c has been made explicit. Let us observe that the computation of this lower bound is not straightforward because the first constraint in (4.20) is a BMI (due to the product of c with Q), which may lead to nonconvex optimization problems.

A way to cope with this problem is to fix Q_2, since in such a case the constraints in (4.20) are LMIs. However, this solution may require to increase the degree of Q_1, and consequently the degree of $p(x)$, which may significantly increase the computational burden.

Another way to compute $\hat{\gamma}$ where Q_2 does not need to be fixed is via a one-parameter sweep on c in Corollary 4.2. This search can be conducted in several ways, for instance via a bisection algorithm in order to speed up the convergence.

A further way to obtain $\hat{\gamma}$ where Q_2 does not need to be fixed is via a quasi-convex optimization problem. Indeed, let us define the polynomial

$$p_1(x) = -q(x)' \begin{pmatrix} \dot{v}(x) \\ -v(x) \end{pmatrix} \tag{4.21}$$

and its SMR matrix as

$$P_1(Q) + L(\alpha) = CSMR_{qua}(p_1) \tag{4.22}$$

where Q is as in (4.10). Let us define

$$V = SMR_{qua}(v) \tag{4.23}$$

and let us introduce the following preliminary result.

Theorem 4.2. *Let $v(x)$ and V be related as in* (4.23)*, and let $q_2(x)$ and Q_2 be related as in* (4.10)*. For $\mu \in \mathbb{R}$ define*

$$p_2(x) - q_2(x)(1 + \mu v(x)) \tag{4.24}$$

and

$$P_2 = T_1'(\mathrm{diag}(1,\mu V)\otimes Q_2)T_1 \tag{4.25}$$

where T_1 *is the matrix satisfying*

$$\begin{pmatrix} 1 \\ b_{lin}(x,\partial v/2) \end{pmatrix} \otimes b_{lin}(x,\partial q_2/2) = T_1 b_{lin}(x,\partial p_2/2). \tag{4.26}$$

Then, P_2 *satisfies*

$$P_2 = SMR_{qua}(p_2). \tag{4.27}$$

Moreover,

$$\mu V > 0 \text{ and } Q_2 > 0 \ \Rightarrow\ P_2 > 0. \tag{4.28}$$

Proof. Let us pre- and post-multiply P_2 by $b_{lin}(x,\partial p_2/2)'$ and $b_{lin}(x,\partial p_2/2)$, respectively. We get:

$$\begin{aligned}
&b_{lin}(x,\partial p_2/2)'P_2 b_{lin}(x,\partial p_2/2)\\
&= b_{lin}(x,\partial p_2/2)'T_1'(\mathrm{diag}(1,\mu V)\otimes Q_2)T_1 b_{lin}(x,\partial p_2/2)\\
&= \left(\begin{pmatrix} 1 \\ b_{lin}(x,\partial v/2) \end{pmatrix} \otimes b_{lin}(x,\partial q_2/2)\right)'(\mathrm{diag}(1,\mu V)\otimes Q_2)\\
&\quad \cdot \begin{pmatrix} 1 \\ b_{lin}(x,\partial v/2) \end{pmatrix} \otimes b_{lin}(x,\partial q_2/2)\\
&= (1+\mu v(x))q_2(x)\\
&= p_2(x)
\end{aligned}$$

i.e. (4.27) holds. Then, let us observe that $\mu V > 0$ and $Q_2 > 0$ implies that

$$\mathrm{diag}(1,\mu V)\otimes Q_2 > 0.$$

Moreover, T_1 is a full column matrix, and hence it follows that (4.28) holds. □

Theorem 4.2 provides an explicit expression for an SMR matrix of $p_2(x)$. As explained in the theorem, this matrix is guaranteed to be positive definite provided that μV and Q_2 are positive definite. The following result shows how $\hat{\gamma}$ can be obtained via a quasi-convex optimization problem.

Theorem 4.3. *Let* $v \in \mathscr{P}_n$ *be positive definite, and let* $\mu > 0$*. Define the quantities in* (4.22)–(4.23)*. Assume that* $V > 0$ *and let* $P_2(Q_2)$ *be the SMR matrix of* $p_2(x)$ *in* (4.25) *where the dependence on* Q_2 *has been made explicit. Then,* $\hat{\gamma}$ *in* (4.20) *is given by*

$$\hat{\gamma} = -\frac{z^*}{1+\mu z^*} \tag{4.29}$$

where z^* is the solution of the GEVP

$$
z^* = \inf_{Q,\alpha,z} z \quad \text{s.t.} \begin{cases} zP_2(Q_2) + P_1(Q) + L(\alpha) > 0 \\ Q > 0 \\ 1+\mu z > 0 \\ \text{trace}(Q_1) = 1. \end{cases} \tag{4.30}
$$

Proof. First of all, observe that $v(x)$ is positive definite since $V > 0$, and that $p(x)$ and $p_2(x)$ have the same degree due to (4.12). Now, suppose that the constraints in (4.30) hold. Let us pre- and post-multiply the first inequality in (4.30) by $b_{lin}(x,\partial p/2)'$ and $b_{lin}(x,\partial p/2)$, respectively, where $x \neq 0_n$. We get:

$$
\begin{aligned}
0 &< b_{lin}(x,\partial p/2)'\left(zP_2(Q_2)+P_1(Q)+L(\alpha)\right) b_{lin}(x,\partial p/2) \\
&= zp_2(x) + p_1(x) \\
&= zq_2(x)(1+\mu v(x)) - q(x)' \begin{pmatrix} \dot{v}(x) \\ -v(x) \end{pmatrix} \\
&= -q_1(x)\dot{v}(x) - q_2(x)(-z - z\mu v(x) - v(x)) \\
&= -q_1(x)\dot{v}(x) - (1+\mu z) q_2(x) \left(\frac{-z}{1+\mu z} - v(x) \right).
\end{aligned}
$$

Hence, the first inequality in (4.20) coincides with the first inequality in (4.30) whenever $q_2(x)$ and c are replaced by

$$
\begin{aligned}
q_2(x) &\to q_2(x)(1+\mu z) \\
c &\to \frac{-z}{1+\mu z}.
\end{aligned}
$$

Since $1+\mu z$ is positive, it follows that the constraints in (4.20) are equivalent to those in (4.30), and hence (4.29) holds. □

Theorem 4.3 states that the LEDA provided by the LF $v(x)$ can be found by solving (4.30) provided that $v(x)$ admits a positive definite SMR matrix V. The optimization problem (4.30) is a GEVP, and belongs to the class of quasi-convex optimization problems. See Appendix A for details about GEVPs.

Let us observe that the condition $V > 0$ in Theorem 4.3 implies that $v(x)$ is SOS and positive definite. If this condition does not hold, then $\hat{\gamma}$ cannot be calculated as in (4.29) in general because the matrix $P_2(Q_2)$ could not be positive definite under the condition $Q_2 > 0$.

Example 4.4. For the Van der Pol circuit (4.3), let us consider the problem of determining $\hat{\gamma}$ in (4.20) with $v(x) = x_1^2 + x_1x_2 + x_2^2$. Let us use Theorem 4.3 with $k = 0$ in (4.12). The SMR matrices in (4.30) and the matrix T_1 in (4.25) are given by those in Example 4.3 and

$$T_1 = \begin{pmatrix} 1\,0\,0\,0\,0 \\ 0\,1\,0\,0\,0 \\ 0\,0\,1\,0\,0 \\ 0\,0\,0\,1\,0 \\ 0\,0\,0\,1\,0 \\ 0\,0\,0\,0\,1 \end{pmatrix}, \; P_2(Q_2) = \begin{pmatrix} g_1 & g_2 & 0 & 0 & 0 \\ \star & g_3 & 0 & 0 & 0 \\ \star & \star & \mu g_1 & \mu(0.5g_1+g_2) & 0.5\mu g_2 \\ \star & \star & \star & \mu(g_1+g_2+g_3) & \mu(g_2+0.5g_3) \\ \star & \star & \star & \star & \mu g_3 \end{pmatrix}$$

$$P_1(Q) = \begin{pmatrix} 1 & 0.5 & 0 & 0 & 0 \\ \star & 1 & 0 & 0 & 0 \\ \star & \star & g_1 & -0.5+0.5g_1+g_2 & -1+0.5g_1+g_2+0.5g_3 \\ \star & \star & \star & 0 & g_2+0.5g_3 \\ \star & \star & \star & \star & g_3 \end{pmatrix}$$

$$Q_1 = 1, \; Q_2 = \begin{pmatrix} g_1 & g_2 \\ \star & g_3 \end{pmatrix}.$$

Hence, the GEVP (4.30) has 7 scalar variables. We simply choose $\mu = 1$, and we find the optimal values

$$a^* = -0.478, \; Q_2 = \begin{pmatrix} 1.030 & 0.162 \\ \star & 1.356 \end{pmatrix}, \; \alpha = (0.000, 0.000, 0.811)'.$$

From a^* we obtain $\hat{\gamma} = 0.915$. Figure 4.2a shows the estimate $\mathscr{V}(\hat{\gamma})$ and the curve $\dot{v}(x) = 0$. As we can see from the figure, the estimate $\mathscr{V}(\hat{\gamma})$ is tangent to the curve $\dot{v}(x) = 0$, hence implying that $\hat{\gamma}$ is tight, i.e. $\hat{\gamma} = \gamma$.

We repeat the procedure for the LF

$$v(x) = x_1^2 + x_1x_2 + x_2^2 + x_1^4 + x_1^2x_2^2 + x_2^4$$

hence finding that $\hat{\gamma} = 0.493$. This lower bound is not tight, in fact by repeating the computation with $k = 1$ in (4.12) we find the new lower bound $\hat{\gamma} = 3.047$. Figure 4.2b shows the found results. As we can see from the figure, this new lower bound is tight. □

Example 4.5. For the predator-prey system in (4.4), let us consider the problem of estimating the DA of the two locally asymptotically stable equilibrium points, in particular determining $\hat{\gamma}$ in (4.20) with

$$v(x) = 4x_1^2 + x_1x_2 + 2x_2^2$$

where

$$x = y - y_{eq}, \; y_{eq} \in \{y_{eq1}, y_{eq2}\}.$$

The GEVP (4.30) has 7 scalar variables. With $k = 0$ in (4.12) we find $\hat{\gamma} = 3.376$ for y_{eq1} and $\hat{\gamma} = 0.022$ for y_{eq2}. Figure 4.3 shows the estimates $\mathscr{V}(\hat{\gamma})$ and the curves $\dot{v}(x) = 0$. As we can see from the figures, the estimates $\mathscr{V}(\hat{\gamma})$ are tangent to the curves $\dot{v}(x) = 0$, hence implying that the found lower bounds $\hat{\gamma}$ are tight. □

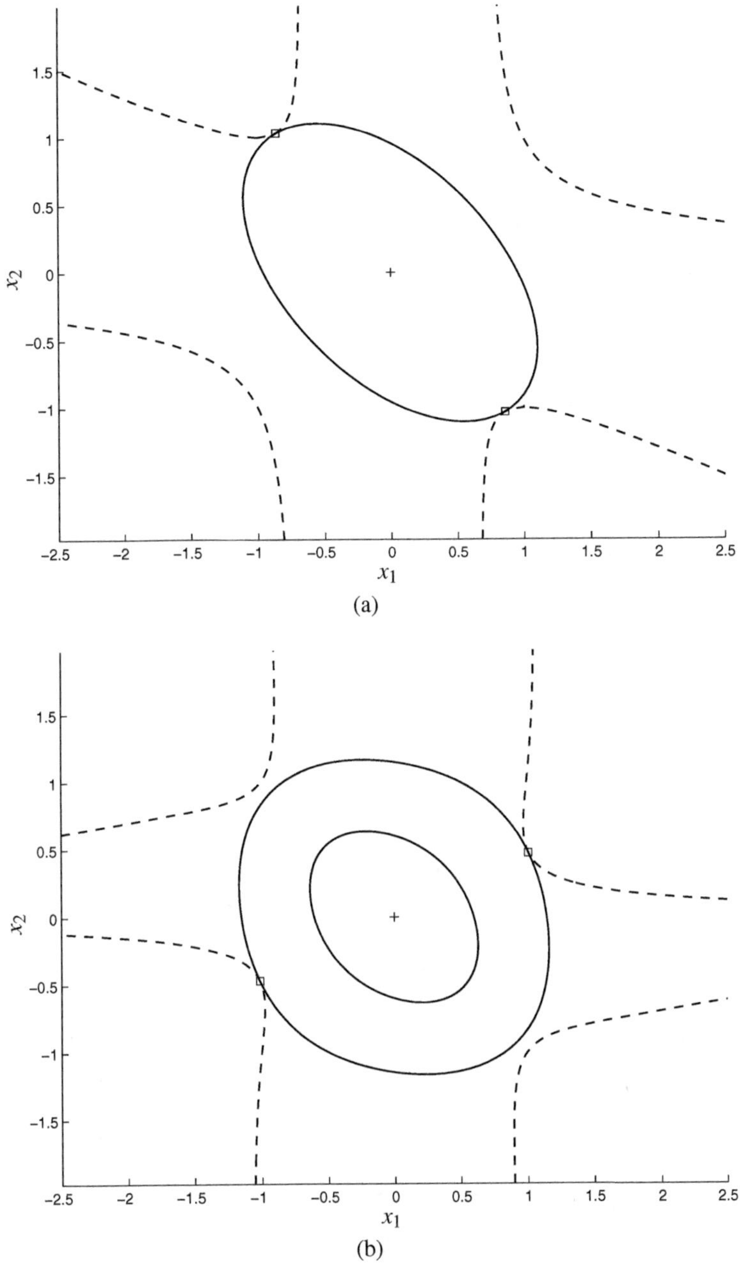

Fig. 4.2 Example 4.4 (Van der Pol circuit). Boundary of $\mathscr{V}(\hat{\gamma})$ (solid) and curve $\dot{v}(x) = 0$ (dashed) for $v(x) = x_1^2 + x_1x_2 + x_2^2$ (a) and $v(x) = x_1^2 + x_2^2 + x_1^4 + x_1^2x_2^2 + x_2^4$ (b). The "+" denotes the equilibrium point. In (b), the small estimate corresponds to $k = 0$, while the large estimate corresponds to $k = 1$.

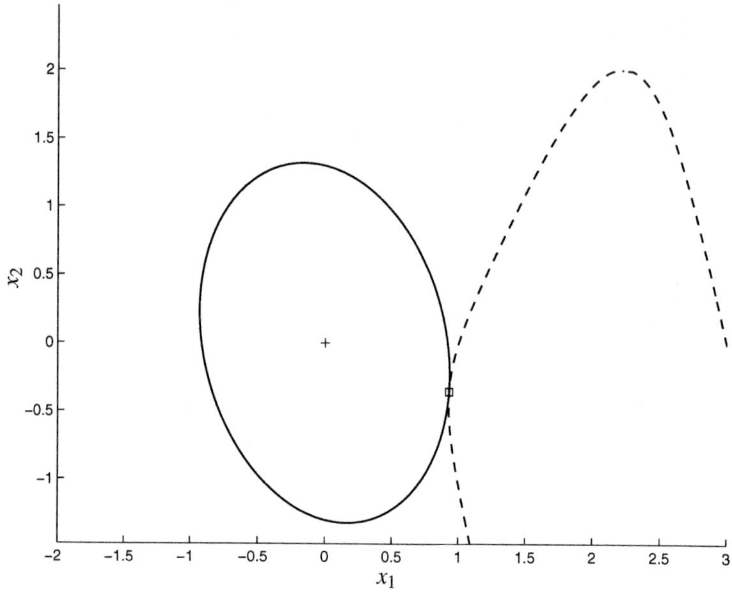

(a)

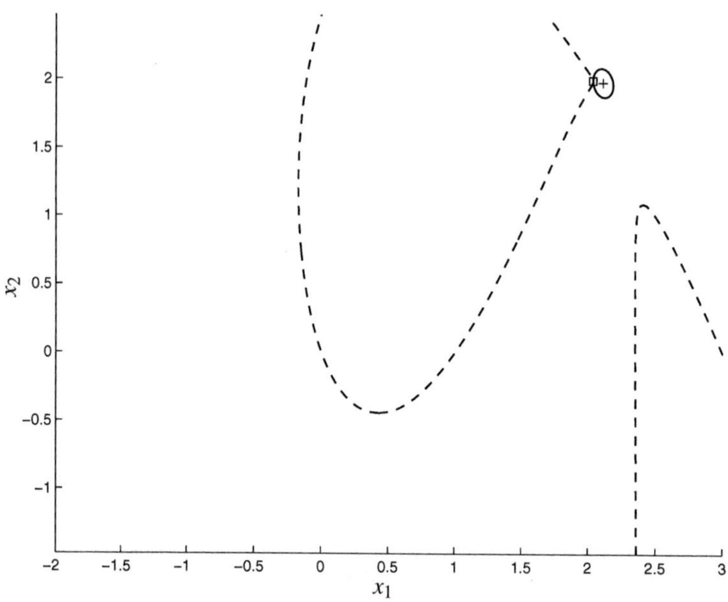

(b)

Fig. 4.3 Example 4.5 (predator-prey system). Boundary of $\mathscr{V}(\hat{\gamma})$ (solid) and curve $\dot{v}(x) = 0$ (dashed) for y_{eq1} (a) and y_{eq2} (b). The "+" denote the equilibrium points.

Example 4.6. Let us consider the system

$$\begin{cases} \dot{y}_1 = y_1 - y_1^3 + 0.5y_3 + y_2^2 \\ \dot{y}_2 = -y_2 - y_2^3 + 0.5y_3^2 - y_1^2 \\ \dot{y}_3 = y_1 + 2y_2 - y_3^3 + y_1^2 - y_2^2. \end{cases} \tag{4.31}$$

This system has three equilibrium points as follows:

$$y_{eq1} = \begin{pmatrix} 1.362 \\ -0.829 \\ 0.956 \end{pmatrix}, \; y_{eq2} = \begin{pmatrix} -1.110 \\ -0.524 \\ -1.063 \end{pmatrix}, \; y_{eq3} = \begin{pmatrix} 0 \\ 0 \\ 0 \end{pmatrix}. \tag{4.32}$$

The first two equilibrium points are locally asymptotically stable, while the other is unstable.

Let us consider the problem of estimating the DA of the first two equilibrium points, in particular determining $\hat{\gamma}$ in (4.20) with the choice

$$v(x) = x_1^2 + x_2^2 + x_3^2$$

where

$$x = y - y_{eq}, \; y_{eq} \in \{y_{eq1}, y_{eq2}\}.$$

The GEVP (4.30) has 21 scalar variables. With $k = 0$ in (4.12) we find $\hat{\gamma} = 0.291$ for y_{eq1} and $\hat{\gamma} = 1.803$ for y_{eq2}. Figure 4.4 shows the estimates $\mathscr{V}(\hat{\gamma})$ and the surfaces $\dot{v}(x) = 0$. □

4.2.4 Estimate Tightness

Once that the lower bound $\hat{\gamma}$ has been found, a natural question concerns its tightness: is $\hat{\gamma}$ equal to the sought γ in (4.19)? As we have seen in the previous examples, one can answer to this question by visual inspection if $n = 2$, however this becomes difficult or even impossible for larger values of n. The following result provides a necessary and sufficient condition for answering to this question.

Theorem 4.4. *Let γ and $\hat{\gamma}$ be as in* (4.19) *and* (4.20)*, respectively. Suppose that $0 < \gamma < \infty$, and define*

$$\mathscr{M} = \{x \in \mathbb{R}^n : \; b_{lin}(x, \partial p/2) \in \ker(M)\} \tag{4.33}$$

where

$$M = P(\hat{\gamma}, Q^*) + L(\alpha^*) \tag{4.34}$$

and $P(\hat{\gamma}, Q^*) + L(\alpha^*)$ is the found optimal value of $P(c, Q) + L(\alpha)$ in (4.20) *(or, equivalently, the found optimal value of $zP_2(Q_2) + P_1(Q) + L(\alpha)$ in* (4.30)*). Define also*

$$\mathscr{M}_1 = \{x \in \mathscr{M} : \; v(x) = \hat{\gamma} \text{ and } \dot{v}(x) = 0\}. \tag{4.35}$$

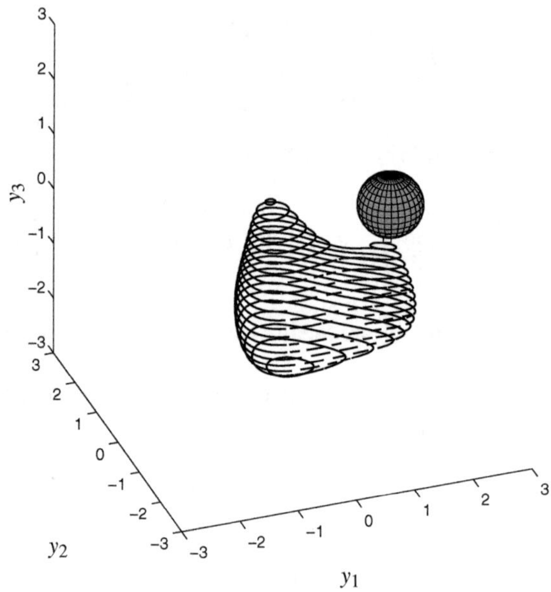

(a)

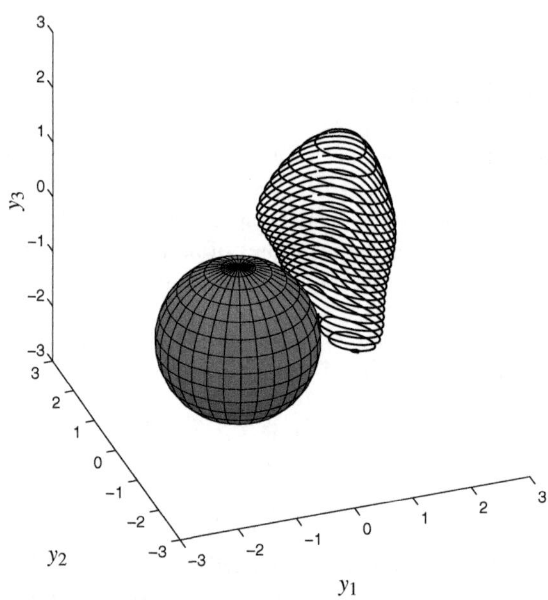

(b)

Fig. 4.4 Example 4.6. Boundary of $\mathscr{V}(\hat{\gamma})$ (solid) and surface $\dot{v}(x) = 0$ (dashed) for y_{eq1} (a) and y_{eq2} (b).

Then,

$$\hat{\gamma} = \gamma \iff \mathscr{M}_1 \neq \emptyset. \tag{4.36}$$

Proof. "$\Rightarrow$" Suppose that $\hat{\gamma} = \gamma$, and let x^* be the tangent point between the surface $\dot{v}(x) = 0$ and the sublevel set $\mathscr{V}(\gamma)$, i.e.

$$\begin{aligned} \dot{v}(x^*) &= 0 \\ v(x^*) &= \gamma. \end{aligned}$$

Pre- and post-multiplying M by $b_{lin}(x^*, \partial p/2)'$ and $b_{lin}(x^*, \partial p/2)$, respectively, we get

$$\begin{aligned} 0 &\leq b_{lin}(x^*, \partial p/2)' M b_{lin}(x^*, \partial p/2) \\ &= -q_1^*(x^*)\dot{v}(x^*) - q_2^*(x^*)(\gamma - v(x^*)) \\ &= 0 \end{aligned}$$

since M is positive semidefinite for definition of $\hat{\gamma}$, $\dot{v}(x^*) = 0$ and $\gamma - v(x^*) = 0$. This implies that

$$b_{lin}(x^*, \partial p/2) \in \ker(M)$$

i.e. $x^* \in \mathscr{M}$. Hence, $x^* \in \mathscr{M}_1$, i.e. $\mathscr{M}_1 \neq \emptyset$.

"$\Leftarrow$" Suppose that $\mathscr{M}_1 \neq \emptyset$ and let x be a point of $\mathscr{M}_1$. It follows that $v(x) = \hat{\gamma}$ and $\dot{v}(x) = 0$. Since $\hat{\gamma}$ is a lower bound of γ, this implies that x is a tangent point between the surface $\dot{v}(x) = 0$ and the sublevel set $\mathscr{V}(\hat{\gamma})$. Hence, $\hat{\gamma} = \gamma$. □

Theorem 4.4 provides a condition for establishing whether the lower bound provided by (4.20) is tight. In particular, this happens if the set $\mathscr{M}_1$ in (4.35) is nonempty. This set can be found via trivial substitution from the set $\mathscr{M}$ in (4.33) (at least whenever $\mathscr{M}$ is finite), which can be computed as explained in Section 1.5.2. In the following, Example 4.7 shows a situation where $\mathscr{M}$ is infinite.

Example 4.7. For the Van der Pol circuit (4.3), let us consider the problem of establishing tightness of $\hat{\gamma}$ found in Example 4.4 with $v(x) = x_1^2 + x_1x_2 + x_2^2$. Let us use Theorem 4.4. We have that

$$M = \begin{pmatrix} 0.208 & 0.173 & -0.000 & 0.000 & -0.000 \\ \star & 0.144 & -0.000 & 0.000 & -0.000 \\ \star & \star & 0.319 & -0.007 & -0.227 \\ \star & \star & \star & 0.328 & 0.278 \\ \star & \star & \star & \star & 0.389 \end{pmatrix}$$

and hence

$$\ker(M) = \left\langle \begin{pmatrix} 0.000 \\ -0.000 \\ 0.470 \\ 0.565 \\ 0.678 \end{pmatrix}, \begin{pmatrix} 0.640 \\ -0.769 \\ -0.000 \\ 0.000 \\ -0.000 \end{pmatrix} \right\rangle.$$

Since $b_{lin}(x, \partial p/2) = (x_1, x_2, x_1^2, x_1x_2, x_2^2)'$, it follows that $\mathscr{M}$ is not finite in this case, in particular

$$\mathscr{M} = \{\rho(0.640, -0.769)' \ \forall \rho \in \mathbb{R}\}.$$

By imposing that $x \in \mathscr{M}$ satisfies $v(x) = \hat{\gamma}$ we find $x = \pm(-0.859, 1.031)'$, which also satisfies $\dot{v}(x) = 0$. Hence,

$$\mathscr{M}_1 = \{\pm(-0.859, 1.031)'\}$$

and from (4.36) we conclude that $\hat{\gamma} = \gamma$. Figure 4.2a show the points in $\mathscr{M}_1$ ("□" markers).

We repeat the procedure for the case $v(x) = x_1^2 + x_1x_2 + x_2^2 + x_1^4 + x_1^2x_2^2 + x_2^4$ hence finding again that $\hat{\gamma}$ is tight, in particular

$$\mathscr{M}_1 = \{\pm(1.012, 0.471)'\}.$$

Figure 4.2b show the points in $\mathscr{M}_1$ ("□" markers). □

Example 4.8. For the predator-prey system (4.4), let us consider the problem of establishing tightness of $\hat{\gamma}$ found in Example 4.5 for y_{eq1}. Let us use Theorem 4.4. We find that

$$\mathscr{M} = \mathscr{M}_1 = \{(0.929, -0.354)'\}$$

and from (4.36) we conclude that $\hat{\gamma} = \gamma$. Figure 4.3a shows the point in $\mathscr{M}_1$ ("□" marker).

We repeat the procedure for y_{eq2} hence finding again that $\hat{\gamma}$ is tight, in particular

$$\mathscr{M} = \mathscr{M}_1 = \{(-0.075, 0.015)'\}$$

(the point in $\mathscr{M}$ and $\mathscr{M}_1$ is expressed for convenience in the coordinates system of (4.4)). Figure 4.3b shows the point in $\mathscr{M}_1$ ("□" marker). □

Example 4.9. For the system (4.31), let us consider the problem of establishing tightness of $\hat{\gamma}$ found for y_{eq1} in Example 4.6. Let us use Theorem 4.4. We find that

$$\mathscr{M} = \mathscr{M}_1 = \{(-0.124, -0.205, -0.483)'\}$$

(the point in $\mathscr{M}$ and $\mathscr{M}_1$ is expressed for convenience in the coordinates system of (4.31)). From (4.36) we conclude that $\hat{\gamma} = \gamma$. Figure 4.4a shows the point in $\mathscr{M}_1$ ("□" marker).

We repeat the procedure for y_{eq2} hence finding again that $\hat{\gamma}$ is tight, in particular

$$\mathscr{M}_1 = \{(1.336, 0.027, 0.134)'\}$$

(the point in $\mathscr{M}$ and $\mathscr{M}_1$ is expressed for convenience in the coordinates system of (4.31)). Figure 4.3b shows the point in $\mathscr{M}_1$ ("□" marker). □

4.3 Estimates via Quadratic LFs

In the case that of quadratic LFs, another technique can be used to estimate the DA as explained in this section.

4.3.1 Establishing Estimates

The following result provides a reformulation of the condition (3.28) via a one-parameter sequence of positivity tests.

Theorem 4.5. *Let $v : \mathbb{R}^n \to \mathbb{R}$ be positive definite, continuously differentiable and radially unbounded, and let $c \in \mathbb{R}$ be positive. Suppose that*

$$\tilde{v}(\tilde{c}) > 0 \ \ \forall \tilde{c} \in (0, c] \tag{4.37}$$

where

$$\tilde{v}(\tilde{c}) = \min_{x \in \partial \mathscr{V}(\tilde{c})} -\dot{v}(x). \tag{4.38}$$

Then, $v(x)$ is a LF for the origin of (4.1), *and $\mathscr{V}(c) \subseteq \mathscr{D}$.*

Proof. Suppose that (4.37) holds. Then, it follows that

$$-\dot{v}(x) > 0 \ \ \forall x \in \mathscr{V}(c) \setminus \{0_n\}$$

and hence (3.26) holds. Since $c > 0$, this implies that (3.13) holds, i.e. $v(x)$ is a Lyapunov for the origin. From Theorem 3.6 we conclude that $\mathscr{V}(c) \subseteq \mathscr{D}$. □

The condition proposed in Theorem 4.5 can be investigated via a one-parameter sequence of LMI feasibility tests. Indeed, let us express $v(x)$ as

$$\begin{array}{r} v(x) = x'Vx \\ V > 0. \end{array} \tag{4.39}$$

Corollary 4.3. *Let $v(x)$ be as in* (4.39), *and let $c \in \mathbb{R}$ be positive and $k \in \mathbb{N}$ be nonnegative. Suppose that*

$$\lambda_{hom}(w(\cdot, \tilde{c}), k) > 0 \ \ \forall \tilde{c} \in (0, c] \tag{4.40}$$

where $w(x, \tilde{c})$ is the homogeneous polynomial given by

$$w(x, \tilde{c}) = \mathrm{HPE}(x, -\dot{v}(x), V, \tilde{c}) \tag{4.41}$$

and $\mathrm{HPE}(x, -\dot{v}(x), V, \tilde{c})$ *is defined in Section 2.3.1. Then, $v(x)$ is a LF for the origin of* (4.1), *and $\mathscr{V}(c) \subseteq \mathscr{D}$.*

Proof. Suppose that (4.40) holds. Then, (4.37) holds, and from Theorem 4.5 we conclude the proof. □

According to Corollary 4.3, a sufficient condition for establishing whether $\mathscr{V}(c)$ is included in $\mathscr{D}$ can be obtained via a one-parameter sequence of evaluations of the generalized SOS index of the homogeneous polynomial $w(x,\tilde{c})$, which amounts to solving a convex optimization problem in the form of an LMI feasibility test.

Let us observe that $w(x,\tilde{c})$ is defined in two different ways depending on the structure of $f(x)$. Specifically, let us introduce the following definition.

Definition 4.2. The polynomial function $f(x)$ (and the system (4.1)) is said *odd* if $f(x)$ can be expressed as

$$f(x) = \sum_{i=0}^{j} h_{2i+1}(x) \tag{4.42}$$

where $h_{2i+1} \in \mathscr{P}_n^n$, $i = 0, \ldots, j$, is a vector of homogeneous polynomials of degree $2i+1$. □

Indeed, if (4.1) is odd, then $-\dot{v}(x)$ is an even polynomial, and hence $w(x,\tilde{c})$ is either $-\|x\|^2$ or as in (2.100). If (4.1) is not odd, then $-\dot{v}(x)$ is not an even polynomial, and hence $w(x,\tilde{c})$ is either $-\|x\|^2$ or as in (2.101).

It is interesting to observe that, in some situations, this sequence of tests can be reduced to one test only. Specifically, let us introduce the following definition.

Definition 4.3. The polynomial function $f(x)$ (and the system (4.1)) is said *two-forms* if $f(x)$ can be expressed as

$$f(x) = Ax + h(x) \tag{4.43}$$

where $A \in \mathbb{R}^{n\times n}$ and $h \in \mathscr{P}_n^n$ is a vector of homogeneous polynomials of degree ∂f. □

Indeed, if (4.1) is two-forms, the function $w(x,c)$ can be expressed as

$$w(x,c) = c^{-\delta} w_1(x) + w_2(x) \tag{4.44}$$

where, for ∂f even, the quantities $w_1(x)$, $w_2(x)$ and δ are given by

$$\begin{aligned} w_1(x) &= (x'(VA + A'V)x)^2 (x'Vx)^\delta \\ w_2(x) &= -(2x'Vh(x))^2 \\ \delta &= \partial f - 1, \end{aligned} \tag{4.45}$$

while, for ∂f odd, they are given by

$$\begin{aligned} w_1(x) &= x'(VA + A'V)x(x'Vx)^\delta \\ w_2(x) &= 2x'Vh(x) \\ \delta &= (\partial f - 1)/2. \end{aligned} \tag{4.46}$$

The next result explains how the condition provided in Corollary 4.3 boils down to one evaluation only of the generalized SOS index in the case that (4.1) is two-forms.

Theorem 4.6. *Let $v(x)$ be as in (4.39), and let $c \in \mathbb{R}$ be positive and $k \in \mathbb{N}$ be non-negative. Suppose that (4.1) is two-forms, and suppose that $v(x)$ is a LF for the system linearized at the origin. Define $w(x,c)$ as in (4.44), and suppose that*

$$\lambda_{hom}(w(\cdot,c),k) > 0. \tag{4.47}$$

Then, $\mathscr{V}(c) \subseteq \mathscr{D}$.

Proof. Suppose that (4.47) holds. The quantity z^* in the definition of the generalized SOS index $\lambda_{hom}(w(\cdot,c),k)$ in (2.22) is given by

$$z^* = \max_{Q,z,\alpha} z$$
$$\text{s.t.} \begin{cases} c^{-\delta}P_1(Q) + P_2(Q) + L(\alpha) - zI \geq 0 \\ Q \geq 0 \\ \text{trace}(Q) = 1 \end{cases}$$

where $P_1(Q)$ and $P_2(Q)$ are SMR matrices of $q(x)w_1(x)$ and $q(x)w_2(x)$, respectively. Since $v(x)$ is a LF for the system linearized at the origin, it follows that $w_1(x)$ is positive definite, and furthermore $w_1(x)$ admits a positive definite SMR matrix. This implies that $P_1(Q)$ can be chosen as $\tilde{P}_1(Q)$ which satisfies

$$Q \geq 0 \;\Rightarrow\; \tilde{P}_1(Q) \geq 0.$$

Now, let us observe that, since $\delta > 0$,

$$\begin{aligned} 0 &< c^{-\delta}P_1(Q) + P_2(Q) + L(\alpha) \\ &= c^{-\delta}\tilde{P}_1(Q) + P_2(Q) + L(\alpha + c^{-\delta}\tilde{\alpha}) \\ &= \tilde{c}^{-\delta}\tilde{P}_1(Q) + P_2(Q) + L(\alpha + c^{-\delta}\tilde{\alpha}) + (c^{-\delta} - \tilde{c}^{-\delta})\tilde{P}_1(Q) \\ &< \tilde{c}^{-\delta}\tilde{P}_1(Q) + P_2(Q) + L(\alpha + c^{-\delta}\tilde{\alpha}) \\ &= \tilde{c}^{-\delta}P_1(Q) + P_2(Q) + L\left(\alpha + \left(c^{-\delta} - \tilde{c}^{-\delta}\right)\tilde{\alpha}\right) \end{aligned}$$

for all $\tilde{c} \in (0,c]$, where $\tilde{\alpha}$ is such that

$$P_1(Q) = \tilde{P}_1(Q) + L(\tilde{\alpha}).$$

This implies that (4.40) holds, and from Corollary 4.3 we conclude that the proof. □

Theorem 4.6 states that one can establish whether $\mathscr{V}(c)$ is an inner set of $\mathscr{D}$ via one LMI feasibility test only provided that (4.1) is two-forms and that $v(x)$ is a LF for the system linearized at the origin.

Computing $\lambda_{hom}(w(\cdot,c),k)$ requires to solve the SDP (2.22) where the number of scalar variables is given by

$$\omega_2 = l_{hom}(n,2m) + t_{hom}(n,2k) - 1 \tag{4.48}$$

where

$$m = \left\lceil \frac{\partial f + 1}{2} \right\rceil + k. \tag{4.49}$$

Table 4.3 shows this number in some cases.

Table 4.3 Number of scalar variables in the SDP (2.22) for $k = 0$.

(non-odd $f(x)$)

$\partial f \setminus n$	2	3	4
2	3	27	126
3	6	75	465
4	10	165	1310
5	15	315	3115

(odd $f(x)$)

$\partial f \setminus n$	2	3	4
3	1	6	20
5	3	27	126

4.3.2 LEDA

Here we investigate the determination of the LEDA provided by a quadratic LF $v(x)$, i.e. the computation of γ in (4.19) for such a $v(x)$. From Corollary 4.3 one can define a natural lower bound of γ as

$$\begin{aligned} \hat{\gamma} = \sup_{c} \; & c \\ \text{s.t. } & \lambda_{hom}(w(\cdot,\tilde{c}),k) > 0 \;\; \forall \tilde{c} \in (0,c]. \end{aligned} \tag{4.50}$$

In the case that (4.1) is two-forms, the computation of $\hat{\gamma}$ can be simplified by exploiting Theorem 4.6. Indeed, it directly follows that

$$\begin{aligned} \hat{\gamma} = \sup_{c} \; & c \\ \text{s.t. } & \lambda_{hom}(w(\cdot,c),k) > 0. \end{aligned} \tag{4.51}$$

For $k = 0$, the lower bound $\hat{\gamma}$ can be found via an SDP only. In fact, let us introduce SMR matrices of the homogeneous polynomials $w_1(x)$ and $w_2(x)$ in (4.44) according to

$$\begin{aligned} W_1 &= SMR_{hom}(w_1) \\ W_2 + L(\alpha) &= CSMR_{hom}(w_2). \end{aligned} \tag{4.52}$$

Theorem 4.7. *Let $v(x)$ be as in* (4.39). *Suppose that* (4.1) *is two-forms, and that $v(x)$ is a LF for the system linearized at the origin. Define the matrices in* (4.52). *Then, $\hat{\gamma}$ in* (4.50) *for $k = 0$ is given by*

$$\hat{\gamma} = (z^*)^{-1/\delta} \tag{4.53}$$

where

$$\begin{aligned} z^* &= \inf_{z,\alpha} z \\ &\quad s.t. \;\; zW_1 + W_2 + L(\alpha) > 0. \end{aligned} \tag{4.54}$$

Proof. Let us observe that $\lambda_{hom}(w(\cdot,c))$ is given by

$$\begin{aligned} \lambda_{hom}(w(\cdot,c)) &= \max_{t,\alpha} \; t \\ &\quad \text{s.t.} \;\; c^{-\delta} W_1 + W_2 + L(\alpha) - tI \geq 0. \end{aligned}$$

Hence, the constraint in (4.51) holds for $k = 0$ if and only if the constraint in (4.54) holds with $c = z^{-1/\delta}$. Therefore, (4.53) holds. □

Example 4.10. For the Van der Pol circuit (4.3), let us consider the problem of estimating the DA of the origin with $v(x) = x_1^2 + x_1x_2 + x_2^2$. In this case, (4.1) is two-forms, and hence we can use Theorem 4.7. We have that

$$W_1 = \begin{pmatrix} 1.000 & 1.000 & 1.500 \\ \star & 0.000 & 1.000 \\ \star & \star & 1.000 \end{pmatrix}, \; W_2 = \begin{pmatrix} 0.000 & -0.500 & -1.000 \\ \star & 0.000 & 0.000 \\ \star & \star & 0.000 \end{pmatrix}$$
$$L(\alpha) = \begin{pmatrix} 0 & 0 & -\alpha_1 \\ \star & 2\alpha_1 & 0 \\ \star & \star & 0 \end{pmatrix}.$$

Hence, the SDP (4.54) contains 2 scalar variables. We find the optimal values

$$z^* = 1.093, \;\; \alpha_1 = 0.903$$

and from z^* we obtain $\hat{\gamma} = 0.915$ in agreement with Example 4.4. □

Example 4.11. For the predator-prey system (4.4), let us consider the problem of estimating the DA of the two locally asymptotically stable equilibrium points in Example 4.2 with $v(x) = 4x_1^2 + x_1x_2 + 2x_2^2$. To this end, let us find $\hat{\gamma}$ in (4.50) with $k = 0$. The computation of $\lambda_{hom}(w(\cdot,\tilde{c}))$ amounts to solving an SDP with 7 scalar variables. The lower bound $\hat{\gamma}$ corresponds to the smallest value of $\tilde{c}$ for which $\lambda_{hom}(w(\cdot,\tilde{c}))$ vanishes. This is given by $\hat{\gamma} = 3.376$ for y_{eq1} and $\hat{\gamma} = 0.022$ for y_{eq2} in agreement with Example 4.5. Figure 4.5 shows $\lambda_{hom}(w(\cdot,\tilde{c}))$. □

Example 4.12. For the system (4.31), let us consider the problem of estimating the DA of the two locally asymptotically stable equilibrium points in Example 4.6 with $v(x) = x_1^2 + x_2^2 + x_3^2$. To this end, let us find $\hat{\gamma}$ in (4.50) with $k = 0$. The computation of $\lambda_{hom}(w(\cdot,\tilde{c}))$ amounts to solving an SDP with 76 scalar variables. We find $\hat{\gamma} = 0.291$ for y_{eq1} and $\hat{\gamma} = 1.803$ for y_{eq2} in agreement with Example 4.6. Figure 4.6 shows $\lambda_{hom}(w(\cdot,\tilde{c}))$. □

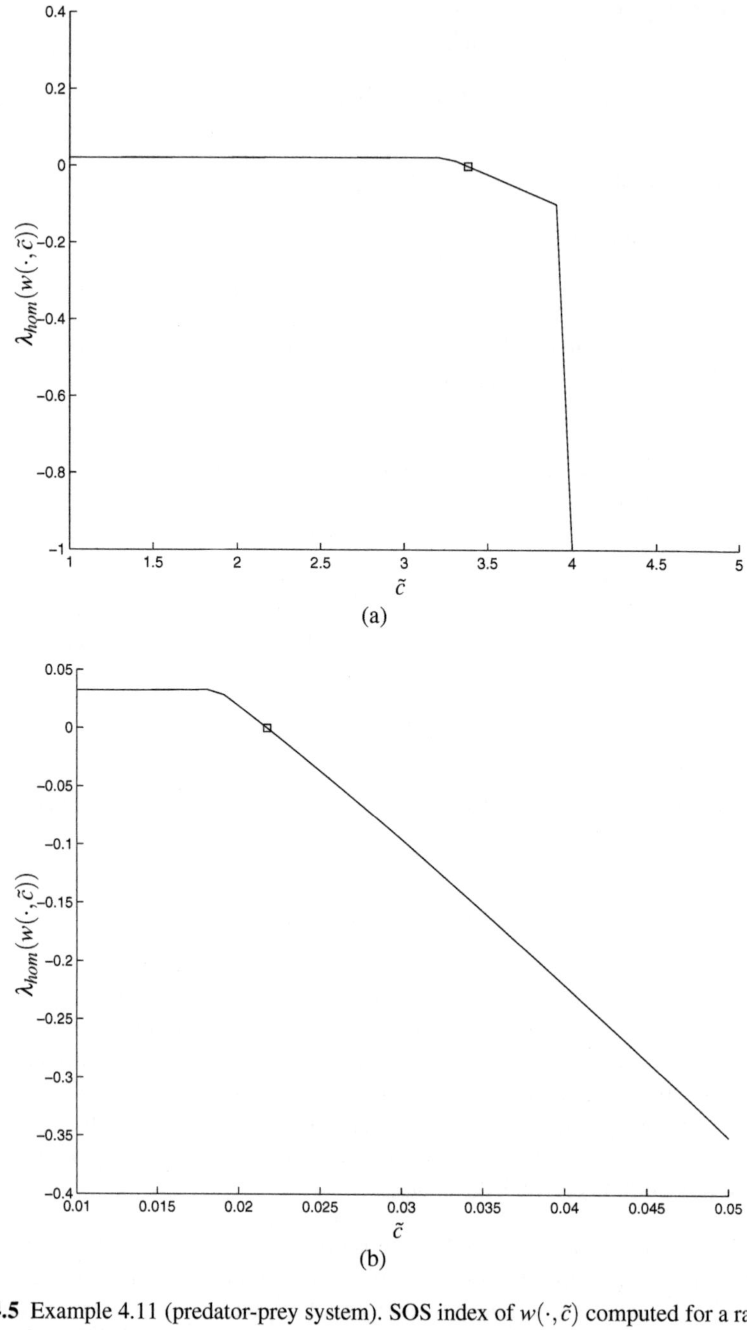

Fig. 4.5 Example 4.11 (predator-prey system). SOS index of $w(\cdot,\tilde{c})$ computed for a range of values of $\tilde{c}$ (solid) for y_{eq1} (a) and y_{eq2} (b). The "□" marker shows the lower bound $\hat{\gamma}$.

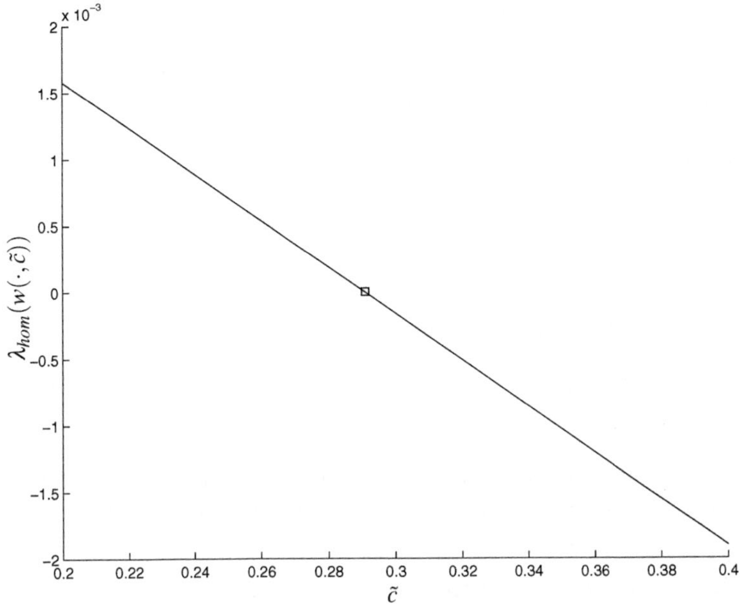

(a)

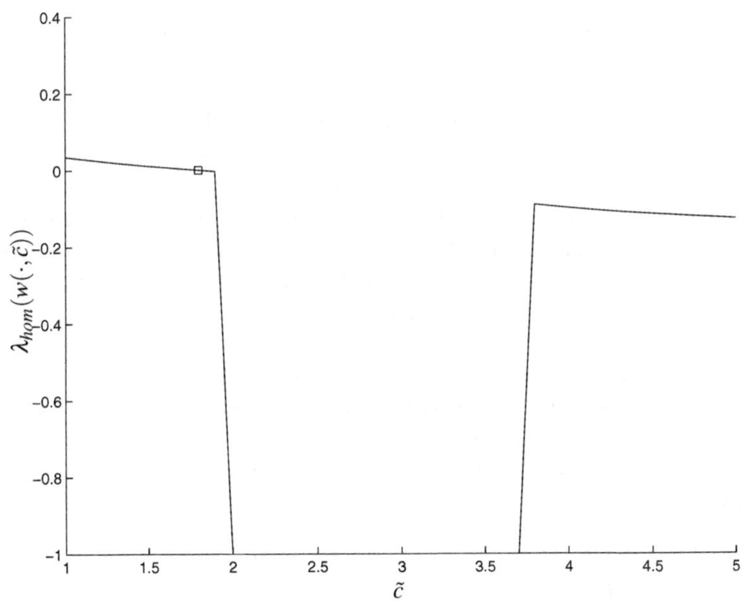

(b)

Fig. 4.6 Example 4.12. SOS index of $w(\cdot,\tilde{c})$ computed for a range of values of $\tilde{c}$ (solid) for y_{eq1} (a) and y_{eq2} (b). The "□" marker shows the lower bound $\hat{\gamma}$.

4.3.3 Estimate Tightness

The following result explains that the conservatism of the lower bound $\hat{\gamma}$ is monotonically non-increasing with the integer k.

Corollary 4.4. *Let $\hat{\gamma}$ be as in* (4.50). *Then:*

$$k_1 \leq k_2 \Rightarrow \hat{\gamma}|_{k=k_1} \leq \hat{\gamma}|_{k=k_2}. \tag{4.55}$$

Proof. Direct consequence of the definition of $\hat{\gamma}$ and Corollary 2.5. □

Depending on n and ∂f, it is possible to guarantee that the lower bound $\hat{\gamma}$ is tight as described in the following result.

Theorem 4.8. *Let γ and $\hat{\gamma}$ be as in* (4.19) *and* (4.50)*, respectively. Suppose that $0 < \gamma < \infty$, and suppose that one of the following conditions is satisfied:*

1. $n \leq 2$;
2. $n = 3$, $\partial f = 3$ and $f(x)$ does not contain monomials of degree 2.

Then, $\hat{\gamma} = \gamma$ for an arbitrarily chosen $k \in \mathbb{N}$.

Proof. We prove the theorem for $k = 0$, which implies that the theorem holds for any $k \in \mathbb{N}$ from Theorem 2.1.

First, suppose that $n \leq 2$. Then, for each fixed c we have that $w(x,c)$ is a homogeneous polynomial in either one or two scalar variables. Since a homogeneous polynomial in n variables is equivalent to a polynomial in $n-1$ variables, it follows from Theorem 1.10 that $w(x,c)$ is positive definite if and only if $\lambda_{hom}(w(\cdot,c)) > 0$.

Then, suppose that $n = 3$, $\partial f = 3$ and $f(x)$ does not contain monomials of degree 2. Then, for each fixed c we have that $w(x,c)$ is a homogeneous polynomial in three scalar variables of degree 4 since (4.43) holds, and again from Theorem 1.10 we have that $w(x,c)$ is positive definite if and only if $\lambda_{hom}(w(\cdot,c)) > 0$. □

Theorem 4.8 provides an a priori condition for non-conservatism. Let us observe that this condition is independent on the integer k used to obtain $\hat{\gamma}$ in (4.50), and hence one can simply use $k = 0$. The following result explains that one can also exploit a posteriori conditions for establishing whether the found lower bound is tight or not.

Theorem 4.9. *Let γ and $\hat{\gamma}$ be as in* (4.19) *and* (4.50)*, respectively. Suppose that $0 < \gamma < \infty$, and define*

$$\mathscr{M} = \{x \in \mathbb{R}^n : \ b_{hom}(x,m) \in \ker(M) \ \text{ and } (1.192) \text{ holds}\} \tag{4.56}$$

where $m = \partial w/2 + k$ and either

$$M = P(Q^*) + L(\alpha^*) \tag{4.57}$$

where $P(Q^*)+L(\alpha^*)$ *is the found optimal value of* $P(Q)+L(\alpha)$ *in* (2.24) *when computing* $\lambda_{hom}(w(\cdot,\hat{\gamma}),k)$, *or*

$$M = z^* W_1 + W_2 + L(\alpha^*) \tag{4.58}$$

where $z^* W_1 + W_2 + L(\alpha^*)$ *is the found optimal value of* $zW_1 + W_2 + L(\alpha)$ *in* (4.54). *Define also*

$$\mathscr{M}_1 = \left\{ \pm x \sqrt{\frac{\hat{\gamma}}{v(x)}},\ x \in \mathscr{M} \right\} \tag{4.59}$$

and

$$\mathscr{M}_2 = \{ x \in \mathscr{M}_1 : \dot{v}(x) = 0 \}. \tag{4.60}$$

Then,

$$\hat{\gamma} = \gamma \iff \mathscr{M}_2 \neq \emptyset. \tag{4.61}$$

Proof. "$\Rightarrow$" Suppose that $\hat{\gamma} = \gamma$, and let x^* be the tangent point between the surface $\dot{v}(x) = 0$ and the sublevel set $\mathscr{V}(\gamma)$, i.e.

$$\begin{aligned} \dot{v}(x^*) &= 0 \\ v(x^*) &= \gamma. \end{aligned}$$

Pre- and post-multiplying M by $b_{hom}(x^*,m)'$ and $b_{hom}(x^*,m)$, respectively, we get for the case (4.57)

$$\begin{aligned} 0 &\le b_{hom}(x^*,m)' M b_{hom}(x^*,m) \\ &= q^*(x^*) w(x^*,\hat{\gamma}) \\ &= -q^*(x^*)\dot{v}(x^*) \\ &= 0 \end{aligned}$$

and for the case (4.58)

$$\begin{aligned} 0 &\le b_{hom}(x^*,m)' M b_{hom}(x^*,m) \\ &= w(x^*,\hat{\gamma}) \\ &= -\dot{v}(x^*) \\ &= 0 \end{aligned}$$

i.e., in both cases one has that $0 = b_{hom}(x^*,m)' M b_{hom}(x^*,m)$. Moreover, M is positive semidefinite for definition of $\hat{\gamma}$. This implies that

$$b_{hom}(x^*,m) \in \ker(M)$$

i.e. $ax^* \in \mathscr{M}$ for some $a \in \mathbb{R}_0$. Then, $x^* \in \mathscr{M}_1$, and $x^* \in \mathscr{M}_2$, i.e. $\mathscr{M}_2 \neq \emptyset$.

"$\Leftarrow$" Suppose that $\mathscr{M}_2 \neq \emptyset$ and let x be a point of $\mathscr{M}_2$. It follows that $\dot{v}(x) = 0$. Moreover, since $x \in \mathscr{M}_1$, it follows that $v(x) = \hat{\gamma}$. Since $\hat{\gamma}$ is a lower bound of γ, this implies that x is a tangent point between the surface $\dot{v}(x) = 0$ and the sublevel set $\mathscr{V}(\hat{\gamma})$. Hence, $\hat{\gamma} = \gamma$. □

Theorem 4.9 provides a condition for establishing whether the lower bound provided either by (2.24) or by (4.54) is tight. In particular, this happens if the set $\mathscr{M}_2$ in (4.60) is nonempty. This set can be found via trivial substitution from the set $\mathscr{M}$ in (4.56) (at least whenever $\mathscr{M}$ is finite), which can be computed as explained in Section 1.5.2 after (1.193).

Example 4.13. For the Van der Pol circuit (4.3), let us consider the lower bound $\hat{\gamma}$ found in Example 4.10. Let us exploit Theorem 4.9. We have that

$$\ker(M) = \langle(-0.470, 0.565, -0.678)'\rangle.$$

Since $b_{lin}(x, m_0) = (x_1^2, x_1x_2, x_2^2)'$, it follows that

$$\mathscr{M} = \left\{(-0.833, 1.000)'\right\}.$$

Moreover,

$$\mathscr{M}_2 = \mathscr{M}_1 = \left\{\pm(-0.859, 1.031)'\right\}$$

and hence we conclude that $\hat{\gamma} = \gamma$. These results are in agreement with those found in Examples 4.4 and 4.7. □

Example 4.14. For the predator-prey system (4.4), let us consider the lower bound $\hat{\gamma}$ found in Example 4.11 for y_{eq1}. Let us exploit Theorem 4.9. We have that

$$\mathscr{M} = \left\{(1.000, -0.381)'\right\}.$$

Moreover,

$$\mathscr{M}_1 = \left\{\pm(0.929, -0.354)'\right\}$$

and

$$\mathscr{M}_2 = \left\{(0.929, -0.354)'\right\}.$$

Hence, we conclude that $\hat{\gamma} = \gamma$. Similarly, we conclude that the lower bound for y_{eq2} is tight (here we have $\mathscr{M}_2 = \{(-0.075, 0.015)'\}$). These results are in agreement with those found in Examples 4.5 and 4.8. □

Example 4.15. Let us consider $\hat{\gamma}$ in Example 4.12. Let us exploit Theorem 4.9. We have that

$$\mathscr{M} = \left\{(0.258, 0.424, 1.000)'\right\}.$$

Moreover,

$$\mathscr{M}_1 = \left\{\pm(0.125, 0.205, 0.483)'\right\}$$

and

$$\mathscr{M}_2 = \left\{(-0.125, -0.205, -0.483)'\right\}.$$

Hence, we conclude that $\hat{\gamma} = \gamma$. Similarly, we conclude that the lower bound for y_{eq2} is tight (here we have $\mathscr{M}_2 = \{(1.336, 0.027, 0.134)'\}$). These results are in agreement with those found in Examples 4.6 and 4.9. □

4.4 Optimal Estimates

In the previous sections we have considered the case of a fixed LF. Clearly, it would be useful to let this function vary in some given class in order to reduce the conservatism of the obtained estimates of the DA. Hence, we consider in this section the problem of obtaining estimates that are optimal according to specified criteria among the estimates provided by polynomial LFs of a chosen degree.

Before proceeding, let us observe that some of the LMIs in the optimization problems derived in the previous sections become BMIs if the LF is allowed to vary. This leads to optimization problems that are possibly nonconvex. Such problems can be solved locally in several ways, either by searching in the variables space through gradient-like methods, or by a sequence of LMI problems where the variables are alternatively frozen. In the sequel we will discuss some of these strategies.

4.4.1 *Maximizing the Volume of the Estimate*

The first criterion that we consider consists of maximizing the volume of the estimate, i.e.

$$\begin{array}{ll} \sup\limits_{v} & \mathrm{vol}(\mathscr{V}(\gamma)) \\ \text{s.t.} & \begin{cases} v(x) > 0 \ \ \forall x \in \mathbb{R}_0^n \\ v \in \mathscr{P}_n \end{cases} \end{array} \tag{4.62}$$

where $\mathrm{vol}(\cdot)$ denotes the volume, and γ defines the LEDA provided by $v(x)$ according to (4.19).

One way to address this problem is:

1. parametrizing the LF $v(x)$;
2. for any fixed LF, estimating γ with the lower bound $\hat{\gamma}$ provided by (4.20);
3. evaluating the volume of the found estimate, i.e. $\mathrm{vol}(\mathscr{V}(\hat{\gamma}))$;
4. updating the LF through a gradient-like method.

Concerning the parametrization of the LFs, a simple choice consists of adopting an SMR matrix of $v(x)$ as parameter, in particular the matrix $V = V'$ in

$$v(x) = b_{lin}(x, \partial v/2)' V b_{lin}(x, \partial v/2). \tag{4.63}$$

However, V contains redundant entries since is not unique in general. Hence, let us parametrize $v(x)$ through

$$\mathbf{v} = COE_{qua}(v). \tag{4.64}$$

This vector $\mathbf{v}$ contains the coefficients of the monomials of $v(x)$ with degree not smaller than 2. Let us observe that this parametrization is still redundant for our problem, since there exist different vectors $\mathbf{v}$ that provide the same estimate. Indeed,

$$\mathscr{V}(c)\big|_{\mathbf{v}} = \mathscr{V}(zc)\big|_{z\mathbf{v}} \ \ \forall c, z > 0. \tag{4.65}$$

We can cope with this issue by simply normalizing $\mathbf{v}$, for example according to $\|\mathbf{v}\| = 1$. Hence, the problem (4.62) is replaced by

$$\begin{aligned} &\sup_{\mathbf{v},\alpha_v} \operatorname{vol}(\mathscr{V}(\hat{\gamma})) \\ &\text{s.t.} \begin{cases} V(\mathbf{v}) + L(\alpha_v) > 0 \\ \|\mathbf{v}\| = 1 \end{cases} \end{aligned} \tag{4.66}$$

where

$$V(\mathbf{v}) + L(\alpha_v) = CSMR_{qua}(v). \tag{4.67}$$

Concerning the evaluation of the volume of the found estimate, it should be observed that this step may be difficult. Specifically, consider $c > 0$. In the case that $v(x)$ is a quadratic form (i.e., $\partial v = 2$), then $\mathscr{V}(c)$ is an ellipsoid and its volume can trivially be calculated with the explicit formula

$$\operatorname{vol}(\mathscr{V}(c)) = k_0 \sqrt{\frac{c^n}{\det(V)}} \tag{4.68}$$

where k_0 is a positive scalar depending only on n. However, in the case that $v(x)$ is non-quadratic (i.e., $\partial v > 2$), the computation of $\operatorname{vol}(\mathscr{V}(c))$ is nontrivial. Hence, it can be useful to approximate $\operatorname{vol}(\mathscr{V}(c))$ with tractable expressions, e.g.

$$\operatorname{vol}(\mathscr{V}(c)) \approx \left(\frac{c^{k_1}}{\det(V)}\right)^{k_2} \tag{4.69}$$

or

$$\operatorname{vol}(\mathscr{V}(c)) \approx \left(\frac{c}{\operatorname{trace}(V)}\right)^{k_3} \tag{4.70}$$

where V is an SMR matrix of $v(x)$ according to (4.63) with $V > 0$, and $k_1, k_2, k_3 \in \mathbb{R}$ are positive.

Another way to address problem (4.62) consists of alternatively freezing some of the variables in the BMIs obtained by letting $v(x)$ vary, hence solving a sequence of LMI problems. Specifically, let us consider the approximation of the volume of this estimate as in (4.70). Problem (4.62) is replaced by

$$\begin{aligned} &\inf_{c,Q,V,\alpha} \frac{\operatorname{trace}(V)}{c} \\ &\text{s.t.} \begin{cases} P(c,Q,V) + L(\alpha) > 0 \\ Q > 0 \\ V > 0 \\ \operatorname{trace}(Q_1) = 1 \end{cases} \end{aligned} \tag{4.71}$$

where $P(c,Q,V)$ is the SMR matrix of $p(x)$ in (4.10) where the dependence on V and c has been made explicit. We have that the first inequality of problem (4.71) is a BMI due to the products of Q with V and c. Hence, we proceed by alternatively solving two subproblems obtained from (4.71). In particular, for a fixed V with $V > 0$, we solve

$$\begin{aligned} &\sup_{c,Q,\alpha} \ c \\ &\text{s.t.} \ \begin{cases} P(c,Q,V) + L(\alpha) > 0 \\ Q > 0 \\ \text{trace}(Q_1) = 1 \end{cases} \end{aligned} \tag{4.72}$$

which is equivalent to solving the GEVP (4.30). The current estimate of the DA is $\mathscr{V}(c)$ where c is updated with its optimal value in (4.72). Then, we also update Q with its optimal value in (4.72), and for such a fixed value of Q we solve

$$\begin{aligned} &\inf_{V,\alpha} \ \text{trace}(V) \\ &\text{s.t.} \ \begin{cases} P(c,Q,V) + L(\alpha) > 0 \\ V > 0 \end{cases} \end{aligned} \tag{4.73}$$

which is an SDP. The current estimate of the DA is $\mathscr{V}(c)$ where $v(x)$ is updated with the optimal value of V in (4.73) according to (4.63). Finally, the procedure is repeated using in (4.72) the found V. The cost $\text{trace}(V)/c$ is monotonically non-increasing in the procedure, which is terminated when the cost decrease is lesser than a chosen threshold.

Example 4.16. Let us consider problem (4.62) in order to obtain optimal estimates of the DA of:

1. the origin of Van der Pol circuit (4.3);
2. the two locally asymptotically stable equilibrium points of the predator-prey system (4.4);
3. the two locally asymptotically stable equilibrium points of (4.31).

Hence, we address problem (4.71) by solving iteratively (4.72)–(4.73) in the case of polynomial LFs of degree 2, 4 and 6. The found LFs, that we denote with $v_2(x)$, $v_4(x)$ and $v_6(x)$, are obtained by using the following initializations:

1. in the case of $v_2(x)$, the LF used in Examples 4.4–4.6;
2. in the case of $v_4(x)$ and $v_6(x)$, the LFs $v_2(x)$ and $v_4(x)$, respectively.

Table 4.4 and Figures 4.7–4.8 show the obtained results. In Figure 4.7a, the dotted line is the boundary of the DA, which has been found with the trajectory reverse method [48] and which represents an unstable limit cycle. □

4.4.2 Enlarging the Estimate with Fixed Shape Sets

The second criterion that we consider for obtaining less conservative estimates of the DA consists of maximizing the size of a set of fixed shape included in the estimate.

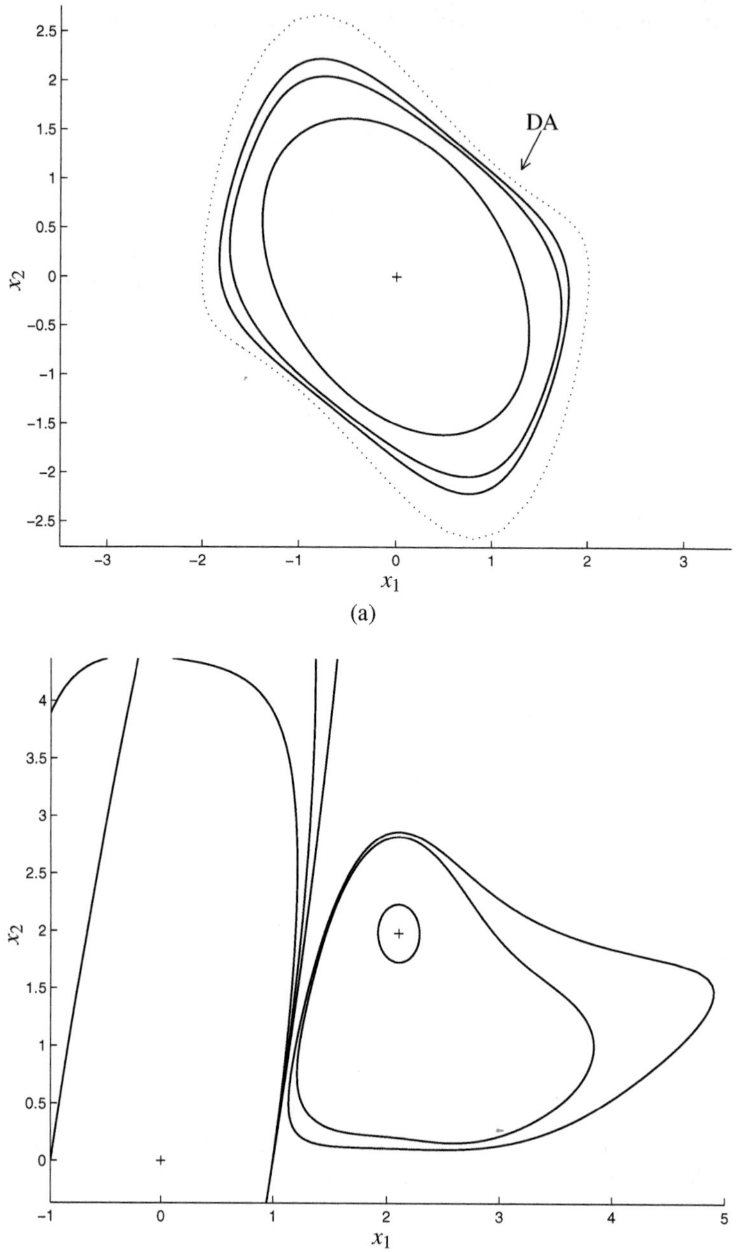

Fig. 4.7 Example 4.16. Boundaries of the estimates of the DA found with variable LFs of degree 2 (smallest curve), 4 and 6 (biggest curve) by maximizing an approximation of the volume for the Van der Pol circuit (4.3) (a) and the predator-prey system (4.4) (b). The "+" denote the equilibrium points. In (a), the dotted line is the boundary of the DA.

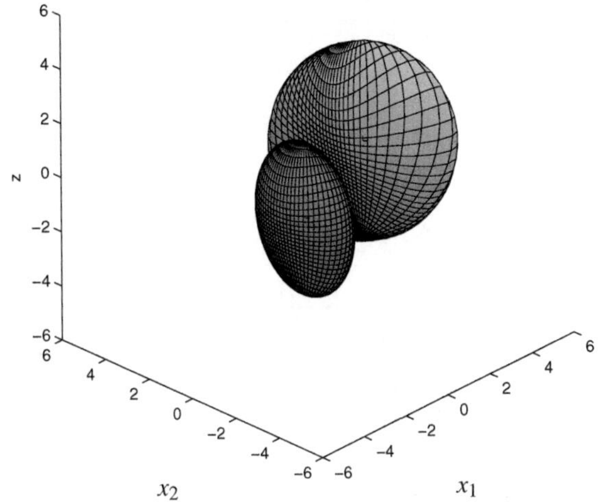

(a)

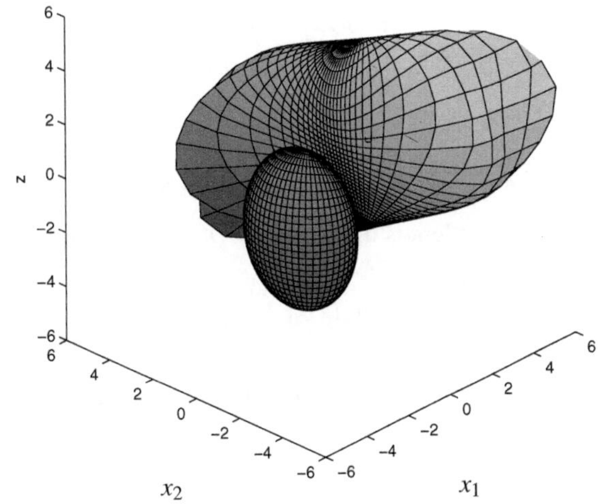

(b)

Fig. 4.8 Example 4.16. Boundaries of the estimates of the DA (solid) found with variable LFs of degree 2 (a) and 4 (b) by maximizing an approximation of the volume for (4.31).

Table 4.4 Example 4.16. Estimates found with variable LFs by maximizing an approximation of the volume.

System	Equilibrium point	LF	Cost trace$(V)/c$	No. of iterations (4.72)–(4.73)
(4.3)	0_2	$v_2(x)$	1.035	16
(4.3)	0_2	$v_4(x)$	0.441	19
(4.3)	0_2	$v_6(x)$	0.208	8
(4.4)	y_{eq1}	$v_2(x)$	1.037	14
(4.4)	y_{eq1}	$v_4(x)$	0.558	29
(4.4)	y_{eq1}	$v_6(x)$	0.366	11
(4.4)	y_{eq2}	$v_2(x)$	45.502	8
(4.4)	y_{eq2}	$v_4(x)$	3.752	31
(4.4)	y_{eq2}	$v_6(x)$	3.073	8
(4.31)	y_{eq1}	$v_2(x)$	0.614	32
(4.31)	y_{eq1}	$v_4(x)$	0.181	16
(4.31)	y_{eq2}	$v_2(x)$	0.962	11
(4.31)	y_{eq2}	$v_4(x)$	0.272	20

Specifically, let $w(x)$ be a positive definite polynomial. This strategy amounts to solving

$$\sup_{c,d,v} d \quad \text{s.t.} \begin{cases} \mathscr{W}(d) \subseteq \mathscr{V}(c) \\ \mathscr{V}(c) \subseteq \mathscr{X} \\ v(x) > 0 \;\; \forall x \in \mathbb{R}_0^n \\ v \in \mathscr{P}_n \end{cases} \tag{4.74}$$

where

$$\mathscr{W}(d) = \{x \in \mathbb{R}^n : \; w(x) \leq d\}. \tag{4.75}$$

One way to handle this problem is through the SOS index. To this end, let us define

$$p_w = -q_w(x)' \begin{pmatrix} v(x) - c \\ d - w(x) \end{pmatrix} \tag{4.76}$$

where $q_w \in \mathscr{P}_n^2$. A lower bound to the solution of (4.74) is given by

$$\sup_{c,d,q,q_w,v} \; d \quad \text{s.t.} \begin{cases} \lambda_{qua}(p) > 0 \\ \lambda_{pol}(q_1) > 0 \\ \lambda_{qua}(q_2) > 0 \\ \lambda_{pol}(p_w) > 0 \\ \lambda_{pol}(q_{wi}) > 0 \;\; \forall i = 1,2 \\ \lambda_{qua}(v) > 0. \end{cases} \tag{4.77}$$

Let us define the SMR matrices

$$
\begin{aligned}
P_w(c,d,Q_w,V)+L_w(\alpha_w) &= CSMR_{qua}(p_w)\\
Q_w &= \operatorname{diag}(Q_{w1},Q_{w2})\\
Q_{wi} &= SMR_{pol}(q_{wi})\ \forall i=1,2.
\end{aligned} \tag{4.78}
$$

Problem (4.77) can be rewritten as

$$
\begin{aligned}
&\sup_{c,d,Q,Q_w,V,\alpha,\alpha_w} \quad d\\
&\text{s.t.} \begin{cases}
P(c,Q,V)+L(\alpha) > 0\\
Q > 0\\
P_w(c,d,Q_w,V)+L_w(\alpha_w) > 0\\
Q_w > 0\\
V > 0\\
\operatorname{trace}(Q_1) = 1\\
\operatorname{trace}(Q_{w1}) = 1.
\end{cases}
\end{aligned} \tag{4.79}
$$

We have that the first and the second inequalities of (4.79) are BMIs due to the products of Q with V and c, and to the products of Q_w with V, c and d.

One way to address (4.79) is as follows:

1. solve (4.79) for a fixed V;
2. update V through a gradient-like method.

Let us observe that solving (4.79) for a fixed V amounts, first, to determining the largest c that satisfies the constraints (4.79), and, second, to determining the largest d that satisfies these constraints. The former step can be done through the GEVP (4.30), and a similar GEVP can be built in order to solve the latter step. Also, the parameter V can be replaced by the parameter $\mathbf{v}$ in (4.64) in order to reduce the number of variables used to updated $v(x)$.

Another way to address problem (4.79) consists of alternatively freezing some of the variables in the BMIs. In particular, let us consider the case of constant Q_{w1}. For fixed V with $V > 0$ one solves (4.72). The current estimate of the DA is $\mathscr{V}(c)$ where c is updated with its optimal value in (4.72). Then, we also update Q with its optimal value in (4.72), and for such a fixed value of Q we solve

$$
\begin{aligned}
&\sup_{d,Q_{w2},V,\alpha,\alpha_w} \quad d\\
&\text{s.t.} \begin{cases}
P(c,Q,V)+L(\alpha) > 0\\
P_w(c,d,Q_w,V)+L_w(\alpha_w) > 0\\
Q_{w2} > 0\\
V > 0
\end{cases}
\end{aligned} \tag{4.80}
$$

which can be done through a GEVP similarly to (4.30). The current estimate of the DA is $\mathscr{V}(c)$ where $v(x)$ is updated with the optimal value of V in (4.80) according to (4.63). Finally, the procedure is repeated using in (4.72) the found V. The cost d is monotonically non-decreasing in the procedure, which is terminated when the cost increase is lesser than a chosen threshold.

Example 4.17. Let us consider problem (4.74) with the choice $w(x) = \|x\|^2$ in order to obtain optimal estimates of the DA for the systems studied in Example 4.16. Hence, we address problem (4.79) by solving iteratively (4.72) and (4.80) in the case of polynomial LFs of degree 2, 4 and 6. The found LFs, that we denote with $v_2(x)$, $v_4(x)$ and $v_6(x)$, are obtained by using the following initializations:

1. in the case of $v_2(x)$, the LF used in Examples 4.4–4.6;
2. in the case of $v_4(x)$ and $v_6(x)$, the LFs $v_2(x)$ and $v_4(x)$, respectively.

Table 4.5 and Figures 4.9–4.10 show the obtained results. In Figure 4.9a, the dotted line is the boundary of the DA found with the trajectory reverse method [48]. Observe that this is almost coincident with the boundary of the estimate $\mathscr{V}(1)$ provided by the found $v_6(x)$, which is

$$\begin{aligned} v_6(x) = {} & 0.397x_1^2 + 0.381x_1x_2 + 0.213x_2^2 + 0.000x_1^3 + 0.001x_1^2x_2 - 0.000x_1x_2^2 \\ & + 0.001x_2^3 - 0.067x_1^4 - 0.089x_1^3x_2 + 0.020x_1^2x_2^2 + 0.008x_1x_2^3 - 0.001x_2^4 \\ & - 0.000x_1^5 - 0.000x_1^4x_2 - 0.000x_1^3x_2^2 - 0.001x_1^2x_2^3 - 0.000x_1x_2^4 - 0.000x_2^5 \\ & + 0.007x_1^6 - 0.001x_1^5x_2 - 0.004x_1^4x_2^2 + 0.001x_1^3x_2^3 + 0.002x_1^2x_2^4 \\ & + 0.001x_1x_2^5 + 0.000x_2^6. \end{aligned}$$

Table 4.5 Example 4.17. Estimates found with variable LFs by enlarging fixed shape sets.

System	Equilibrium point	LF	Cost d	No. of iterations (4.72) and (4.80)
(4.3)	0_2	$v_2(x)$	1.397	12
(4.3)	0_2	$v_4(x)$	2.128	15
(4.3)	0_2	$v_6(x)$	2.341	14
(4.4)	y_{eq1}	$v_2(x)$	0.975	6
(4.4)	y_{eq1}	$v_4(x)$	0.975	3
(4.4)	y_{eq1}	$v_6(x)$	0.975	2
(4.4)	y_{eq2}	$v_2(x)$	0.034	7
(4.4)	y_{eq2}	$v_4(x)$	0.37	32
(4.4)	y_{eq2}	$v_6(x)$	0.382	8
(4.31)	y_{eq1}	$v_2(x)$	2.137	17
(4.31)	y_{eq1}	$v_4(x)$	2.26	9
(4.31)	y_{eq2}	$v_2(x)$	1.936	7
(4.31)	y_{eq2}	$v_4(x)$	1.948	5

□

4.4.3 Establishing Global Asymptotical Stability

In Sections 4.4.1–4.4.2 we have seen that establishing estimates of the DA by using variable LFs involve nonconvex optimization problems. This happens due to the product between the multiplier $q(x)$ and the LF $v(x)$ in $p(x)$ in (4.7). However, if

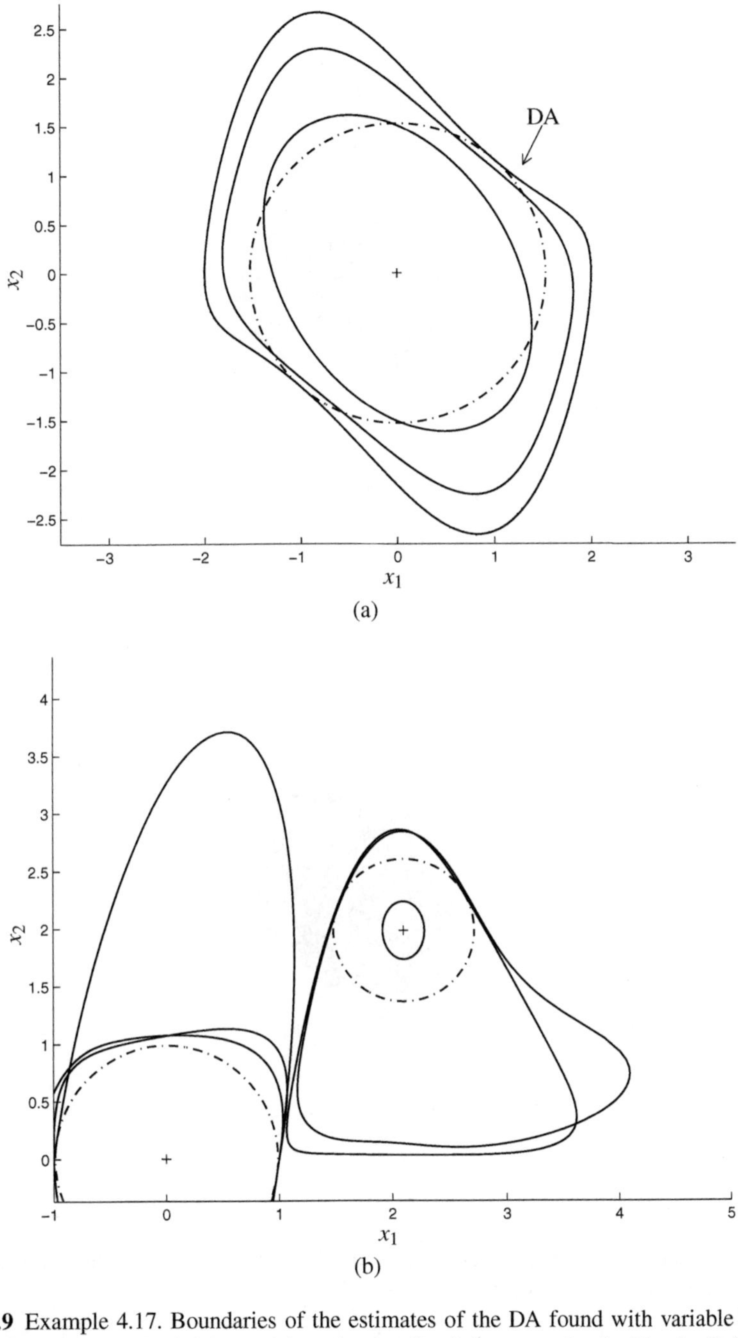

Fig. 4.9 Example 4.17. Boundaries of the estimates of the DA found with variable LFs of degree 2 (smallest), 4 and 6 (largest) by enlarging fixed shape sets for the Van der Pol circuit (4.3) (a) and the predator-prey system (4.4) (b). The "+" denote the equilibrium points. The dashdot lines show the boundary of the sets $\mathscr{W}(d)$ found with $v_6(x)$. In (a), the dotted line is the boundary of the DA.

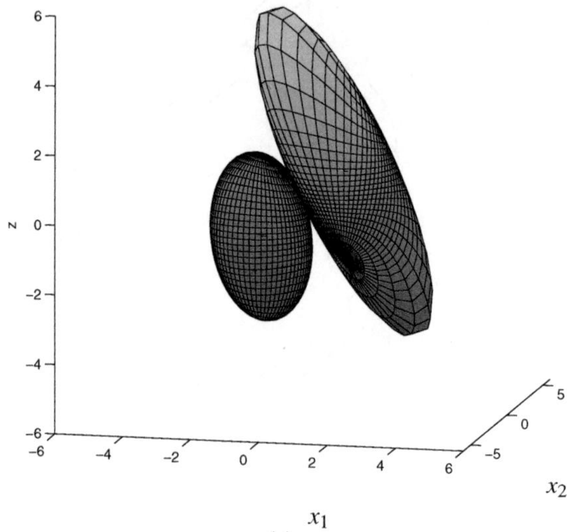

(a)

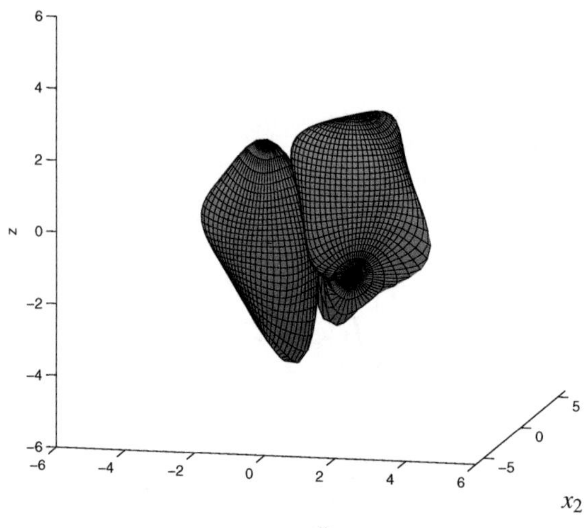

(b)

Fig. 4.10 Example 4.17. Boundaries of the estimates of the DA (solid) found with variable LFs of degree 2 (a) and 4 (b) by enlarging fixed shape sets for (4.31).

the candidate estimate of the DA is the whole space (i.e., $\mathbb{R}^n$), the strategies previously presented boil down to convex optimization problems in the form of an LMI feasibility test.

Indeed, since $\mathscr{D} = \mathbb{R}^n$ is obtained for $c = \infty$, the strategy amounts to finding $v \in \mathscr{P}_n$ (if any) such that $v(x)$ and $-\dot{v}(x)$ are positive definite, or in other words, a polynomial LF proving global asymptotical stability of the origin according to Theorem 3.2. Analogously to the previous sections, we can investigate this condition through SOS polynomials as explained in the following result.

Corollary 4.5. *Let $v \in \mathscr{P}_n$ be a locally quadratic polynomial to determine, and define*

$$\begin{aligned} V &= SMR_{qua}(v) \\ P_g(V) + L(\alpha) &= CSMR_{qua}(p_g) \end{aligned} \tag{4.81}$$

where

$$p_g(x) = -\dot{v}(x). \tag{4.82}$$

Suppose that there exist V and α such that the following LMIs hold:

$$\begin{cases} V > 0 \\ P_g(V) + L(\alpha) > 0 \\ \text{trace}(V) = 1. \end{cases} \tag{4.83}$$

Then, $v(x)$ provided by such a V is a global LF for the origin of (4.1), *and $\mathscr{D} = \mathbb{R}^n$.*

Proof. Suppose that (4.83) holds. This implies that $v(x)$ and $-\dot{v}(x)$ are positive definite with $v(x)$ radially unbounded, and from Theorem 3.2 we conclude the proof. □

Let us observe that, similarly to the LMI conditions derived in this chapter, a necessary condition for (4.83) to hold is that the quadratic part $v_2(x)$ of $v(x)$ is a LF for the system linearized at the origin, i.e. (4.14) holds with V_2 defined by (4.15)–(4.16). Moreover, (4.83) also requires that the highest degree forms of $v(x)$ and $-\dot{v}(x)$ are positive definite. These requirements can be removed as explained in Section 7.4.

Example 4.18. Let us consider the system

$$\begin{cases} \dot{x}_1 = -x_1 + x_3 - 4x_2^2 x_3 - x_1^5 \\ \dot{x}_2 = -x_2 + 2x_1 x_3^4 - x_2^5 \\ \dot{x}_3 = x_2 - 2x_3 + x_1^5 + 3x_1^2 x_2^3 - x_3^5. \end{cases} \tag{4.84}$$

For this system we have

$$A = \begin{pmatrix} -1 & 0 & 1 \\ 0 & -1 & 0 \\ 0 & 1 & -2 \end{pmatrix}$$

and hence the origin is locally asymptotically stable since A is Hurwitz. Let us investigate global asymptotical stability of the origin through Corollary 4.5. By using a variable LF of degree 2 we find that (4.83) cannot be satisfied for any V and α.

Instead, by using a variable LF of degree 4, we find that (4.83) can be fulfilled, in particular the found global LF is described by

$$V = \begin{pmatrix} 0.127 & 0.004 & 0.035 & 0.000 & 0.000 & 0.000 & -0.000 & -0.000 & 0.000 \\ \star & 0.203 & 0.019 & 0.000 & -0.000 & 0.000 & -0.000 & 0.000 & -0.000 \\ \star & \star & 0.144 & -0.000 & 0.000 & -0.000 & -0.000 & 0.000 & 0.000 \\ \star & \star & \star & 0.074 & 0.027 & 0.033 & -0.018 & 0.011 & -0.020 \\ \star & \star & \star & \star & 0.137 & -0.004 & -0.025 & -0.009 & -0.005 \\ \star & \star & \star & \star & \star & 0.116 & -0.014 & -0.009 & -0.010 \\ \star & \star & \star & \star & \star & \star & 0.073 & -0.025 & -0.030 \\ \star & \star & \star & \star & \star & \star & \star & 0.065 & -0.025 \\ \star & \star & \star & \star & \star & \star & \star & \star & 0.062 \end{pmatrix}.$$

□

4.5 Controller Design

In this section we consider the problem of designing controllers for enlarging the DA. Specifically, we consider the presence of an input and an output in (4.1) as follows:

$$\begin{cases} \dot{x}(t) = f(x(t)) + g(x(t))u(t) \\ y(t) = h(x(t)) \\ x(0) = x_{init}. \end{cases} \tag{4.85}$$

Indeed, $u(t) \in \mathbb{R}^{n_u}$ is the input, $y(t) \in \mathbb{R}^{n_y}$ is the output, $g \in \mathscr{P}_n^{n \times n_u}$ and $h \in \mathscr{P}_n^{n_y}$.

Without loss of generality, let us consider that the equilibrium point of interest for $u = 0_{n_u}$ is the origin. We focus on a control strategy of the form

$$u = k(y) \tag{4.86}$$

where $k \in \mathscr{P}_{n_y}^{n_u}$. Hence, $k(y)$ has to satisfy

$$f(0_n) + g(0_n)k(h(0_n)) = 0_n. \tag{4.87}$$

Let us parametrize $k(y)$ as

$$k(y) = Kb_{pol}(y, \partial k) \tag{4.88}$$

where ∂k is the degree of the controller, and $K \in \mathbb{R}^{n_u \times d_{pol}(n_y, \partial k)}$ is a matrix to be determined. Condition (4.87) turns into

$$f(0_n) + g(0_n)Kb_{pol}(h(0_n), \partial k) = 0_n. \tag{4.89}$$

We consider the presence of constraints on the input to be designed, for simplicity expressed as

$$K \in \mathscr{K} \tag{4.90}$$

where

$$\mathscr{K} = \left\{ K : K_{i,j} \in (k_{i,j}^-, k_{i,j}^+) \right\} \tag{4.91}$$

for some $k_{i,j}^-, k_{i,j}^+ \in \mathbb{R}$.

Let us consider first the case of a given LF $v(x)$. The temporal derivative of $v(x)$ along the trajectories of (4.85) is given by

$$\dot{v}(x) = \nabla v(x)' \left(f(x) + g(x)k(h(x)) \right). \tag{4.92}$$

Corollary 4.6. *Let $v \in \mathscr{P}_n$ be positive definite, and let $c \in \mathbb{R}$ be positive. Define the matrices in* (4.10)*, and let $P(K,Q)$ be the SMR matrix of $p(x)$ in* (4.10) *with $\dot{v}(x)$ given by* (4.92) *where the dependence on K has been made explicit. Suppose that there exist K, Q and α such that the following inequalities hold:*

$$\begin{cases} P(K,Q) + L(\alpha) > 0 \\ Q > 0 \\ K \in \mathscr{K} \\ \text{trace}(Q_1) = 1. \end{cases} \tag{4.93}$$

Then, $v(x)$ is a LF for the origin of (4.85)*, and $\mathscr{V}(c) \subseteq \mathscr{D}$.*

Proof. Analogous to the proof of Corollary 4.2. □

Corollary 4.6 provides a condition for computing a controller $k(y)$ which ensures that $\mathscr{V}(c)$ is an inner estimate of the DA. This condition is in general nonconvex since the first inequality is a BMI due to the product of K and Q_1. One way to cope of this problem is to fix Q_1, for which the optimization problem boils down to an LMI feasibility test.

Similarly to the uncontrolled system (4.1) for which we have defined the LEDA, it is useful to define for the controlled system (4.85) the controlled LEDA, i.e. the maximum LEDA achievable for some admissible controller. Specifically, let us define the quantity

$$\begin{aligned} \gamma_{con} = & \sup_{c,K} c \\ & \text{s.t.} \begin{cases} \mathscr{V}(c) \subseteq \mathscr{X} \\ K \in \mathscr{K}. \end{cases} \end{aligned} \tag{4.94}$$

Then, $\mathscr{V}(\gamma_{con})$ is said the maximum LEDA for (4.85) provided by the LF $v(x)$.

From Corollary 4.6 one can define a natural lower bound of γ_{con} as

$$\begin{aligned} \hat{\gamma}_{con} = & \sup_{c,K,Q,\alpha} c \\ & \text{s.t.} \begin{cases} P(c,K,Q) + L(\alpha) > 0 \\ Q > 0 \\ \text{trace}(Q_1) = 1 \\ K \in \mathscr{K} \end{cases} \end{aligned} \tag{4.95}$$

where $P(c,K,Q)$ be the SMR matrix of $p(x)$ in (4.10) with $\dot{v}(x)$ given by (4.92) where the dependence on K and c has been made explicit. This lower bound can be computed via a one-parameter sweep on the scalar c using the condition provided in Corollary 4.6 for any fixed c. Another way to obtain the lower bound $\hat{\gamma}_{con}$ is via a GEVP similarly to the case of uncontrolled system considered in Theorem 4.3.

Example 4.19. Let us consider the system

$$\begin{cases} \dot{x}_1 = -x_1 + x_2 - x_1^2 - x_2^3 + x_1 u \\ \dot{x}_2 = -2x_2 - x_1^2 + u \\ y = x_1. \end{cases} \tag{4.96}$$

The problem consists of designing a polynomial controller as in (4.86) with $\mathcal{K}$ defined by

$$\mathcal{K} = \{K = (k_1, k_2, \ldots)' : \ k_i \in (-1,1)\} \tag{4.97}$$

with $v(x) = x_1^2 + x_2^2$.

First of all, let us observe that (4.89) to holds if and only if $k(0) = 0$. By choosing $b_{pol}(y, \partial k) = (1, y, \ldots, y^{\partial k})'$, this is equivalent to

$$k_1 = 0.$$

With $m_k = 0$ we obtain the lower bound of γ_{con} and the corresponding controller

$$\hat{\gamma}_{con} = 0.513, \ \ k(y) = 0$$

(observe that this corresponds to an uncontrolled system since $k(y) = 0$ due to the above constraint on k_1). With $\partial k = 1$ we find

$$\hat{\gamma}_{con} = 3.574, \ \ k(y) = y$$

while with $\partial k = 2$ we obtain

$$\hat{\gamma}_{con} = 6.158, \ \ k(y) = 0.942y - 0.120y^2.$$

By using Theorem 4.4 we also conclude that these bounds are tight for the found controllers. In particular, for the case $\partial k = 2$, we find

$$\mathcal{M}_1 = \{(1.425, -2.032)'\}.$$

Figure 4.11 shows the estimates $\mathcal{V}(\hat{\gamma}_{con})$ corresponding to the three found controllers, the curve $\dot{v}(x) = 0$ corresponding to the controller obtained for $\partial k = 2$, and the point in $\mathcal{M}_1$ ("□" marker). □

The case of variable LF can be addressed similarly to Section 4.4 via a sequence of convex optimization problems obtained by iteratively fixing some of the variables in the condition of Corollary 4.6. Also in this case one can proceed either by

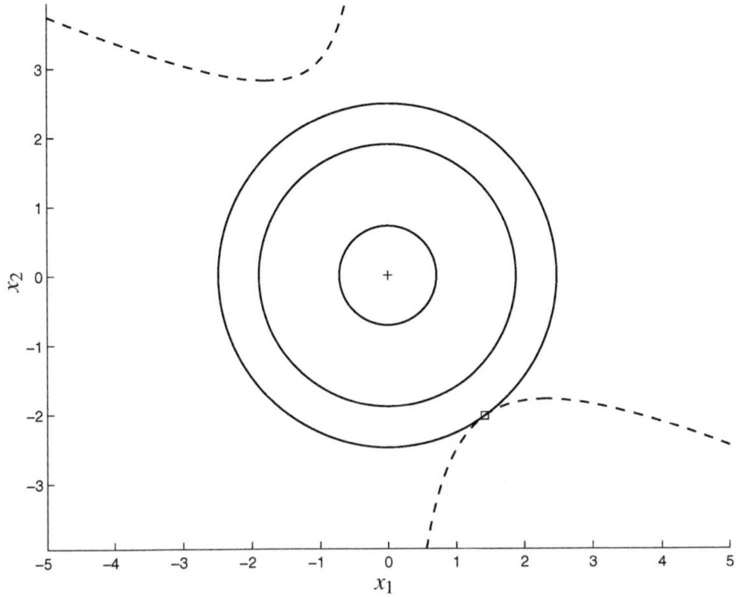

Fig. 4.11 Example 4.19. Boundary of $\mathscr{V}(\hat{\gamma}_{con})$ corresponding to the three found controllers (solid), and curve $\dot{v}(x) = 0$ corresponding to the controller obtained for $\partial k = 2$ (dashed). The "+" denotes the equilibrium point.

maximizing the volume of the estimate or by enlarging the estimate with fixed shape sets.

Example 4.20. Let us consider the system in Example 4.19 and the problem of designing a polynomial controller of degree 2 with variable LFs. Table 4.6 shows the results obtained by maximizing an approximation of the volume analogously to Example 4.16, while Table 4.7 shows the results obtained by enlarging the estimate with fixed shape sets defined by $w(x) = \|x\|^2$ analogously to Example 4.17.

Table 4.6 Example 4.20. Controller design with variable LFs by maximizing an approximation of the volume.

LF	Cost trace$(V)/c$	Controller $k(y)$	No. of iterations (4.72)–(4.73)
$v_2(x)$	0.24	$y+0.036y^2$	9
$v_4(x)$	0.024	$0.684y-0.160y^2$	20
$v_6(x)$	0.006	$y-0.236y^2$	9

□

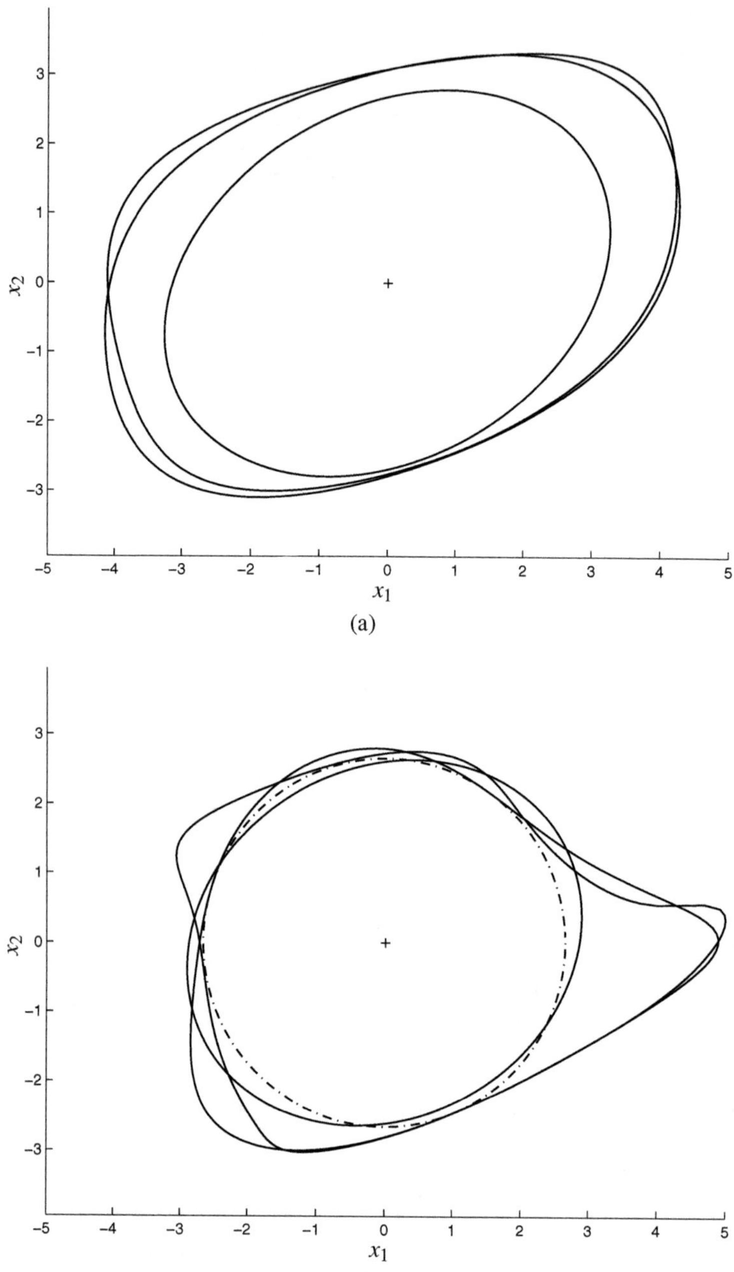

Fig. 4.12 Example 4.20. Controller design with variable LFs: maximizing an approximation of the volume (a) and enlarging the estimate with fixed shape sets (b). The "+" denotes the equilibrium point. The dashdot line shows the boundary of the set $\mathscr{W}(d)$ found with $v_6(x)$.

Table 4.7 Example 4.20. Controller design with variable LFs by enlarging the estimate with fixed shape sets.

LF	Cost d	Controller $k(y)$	No. of iterations (4.72) and (4.80)
$v_2(x)$	6.36	$y+0.008y^2$	3
$v_4(x)$	7.046	$y-0.385y^2$	6
$v_6(x)$	7.050	$y-0.373y^2$	4

4.6 Notes and References

The estimation of the DA in polynomial systems has been studied since long time, see e.g. [48, 71] and references therein.

The use of SOS programming described in Section 4.2 for computing the LEDA provided by polynomial LFs and for verifying their exactness was introduced in [16]. Alternative LMI techniques were proposed e.g. in [83, 112, 50], while non-LMI techniques were considered e.g. in [113]. One of the first contributions addressing the computation of the LEDA is [96].

The estimation of the DA via quadratic LFs described in Section 4.3 was proposed in [111, 36, 31].

The computation of optimal estimates of the DA described in Section 4.4 was considered in [29] for the case of quadratic LFs, and in [62] for the case of polynomial LFs. Other LMI techniques were proposed by exploiting simulations [115] and composite LFs [110]. Non-LMI techniques were proposed e.g. in [45].

The design of controllers for enlarging the DA described in Section 4.5 was proposed in [16]. Related techniques were proposed e.g. in [61] for optimal control of polynomial systems using SOS programming.

Chapter 5
RDA in Uncertain Polynomial Systems

This chapter addresses the problem of estimating the RDA of common equilibrium points of uncertain polynomial systems by using polynomial LFs and SOS programming. In particular, the chapter considers dynamical systems whose dynamics is described by a polynomial function of the state with coefficients depending polynomially on a time-invariant uncertain vector constrained over a semialgebraic set. It is shown that a condition for establishing whether a sublevel set of a common LF is an estimate of the RDA can be obtained in terms of an LMI feasibility test by exploiting SOS parameter-dependent polynomials and their representation through the SMR. Then, a GEVP is proposed for computing a candidate of the largest of such estimates, and a sufficient and necessary condition for establishing the tightness of the candidate is provided. This methodology is hence extended to the case of polynomial LFs with coefficients depending polynomially on the uncertainty. For this case, the computation of parameter-dependent estimates is discussed. Moreover, a strategy is proposed for estimating the intersection of such estimates in order to provide a common estimate that is independent on the uncertainty. Lastly, the chapter addresses the computation of optimal estimates of the RDA by using variable LFs, the problem of establishing robust global asymptotical stability of the common equilibrium point, and the design of polynomial static output controllers for enlarging the RDA.

5.1 Uncertain Polynomial Systems

With the term uncertain polynomial system we refer to a dynamical system where the temporal derivative of the state is a polynomial function of the state and of the uncertainty. In this chapter we address the problem of estimating and controlling the RDA of a common equilibrium point of uncertain polynomial systems. As explained in Section 3.7, we can assume without loss of generality that the common equilibrium point under consideration is the origin. Hence, we consider the uncertain polynomial system described by

G. Chesi: Domain of Attraction, LNCIS 415, pp. 151–196, 2011.
springerlink.com

$$\begin{cases} \dot{x}(t) = f(x(t), \theta) \\ x(0) = x_{init} \\ \theta \in \Theta \\ f(0_n, \theta) = 0_n \ \forall \theta \in \Theta \end{cases} \tag{5.1}$$

where $x \in \mathbb{R}^n$ is the state, $t \in \mathbb{R}$ is the time, $\dot{x}(t)$ is the temporal derivative of the state, $x_{init} \in \mathbb{R}^n$ is the initial condition, $\theta \in \mathbb{R}^{n_\theta}$ is the uncertainty, $\Theta \subset \mathbb{R}^r$ is the set of admissible uncertainties, and $f \in \mathscr{P}^n_{n,n_\theta}$. We consider the case where Θ is a semialgebraic set expressed as

$$\Theta = \{\theta \in \mathbb{R}^{n_\theta} : \ a(\theta) \geq 0, \ \ b(\theta) = 0\} \tag{5.2}$$

for some $a \in \mathscr{P}^{n_a}_{n_\theta}$ and $b \in \mathscr{P}^{n_b}_{n_\theta}$. In the sequel we will drop the dependence on the time for ease of notation.

System (5.1) has at least one common equilibrium point (the origin). Depending on $f(x, \theta)$, (5.1) may have other equilibrium points, either common or dependent on the uncertainty. The RDA of the origin is limited by these equilibria: in fact, by denoting with $\mathscr{X}_0(\theta)$ the set of equilibrium points of (5.1) except the origin, one has that

$$\mathscr{R} \cap \mathscr{X}_0 = \emptyset \tag{5.3}$$

where

$$\mathscr{X}_0 = \bigcup_{\theta \in \Theta} \mathscr{X}_0(\theta). \tag{5.4}$$

Example 5.1. Let us consider the mass-spring system in Figure 5.1. By denoting with y the position of the mass, the system can be described by

$$m\ddot{y} + c\dot{y} + h(y) = 0$$

where m is the mass, c is the friction coefficient, and $h(y)$ is the restoring force of the spring. In the case of softening spring, $h(y)$ is given by

$$h(y) = h_1 y(1 - h_2 y^2)$$

under the condition that $h_2 y^2 < h_1$, where $h_1, h_2 \in \mathbb{R}$ are positive [65]. We consider the case where h_1 is variable and the other parameters are fixed, in particular

$$m = 1, \ \ c = 0.5, \ \ h_2 = 0.25, \ \ h_1 \in [1, 2].$$

By defining $x_1 = y$, $x_2 = \dot{y}$ and $\theta = h_1$, we can express this system as

$$\begin{cases} \dot{x}_1 = x_2 \\ \dot{x}_2 = -\theta x_1 - 0.5 x_2 + 0.25 \theta x_1^3 \\ \theta \in \Theta = [1, 2]. \end{cases} \tag{5.5}$$

This is an uncertain polynomial system, moreover the origin is a common equilibrium point. The set Θ can be expressed as in (5.2) with

$$a(\theta) = -2 + 3\theta - \theta^2.$$

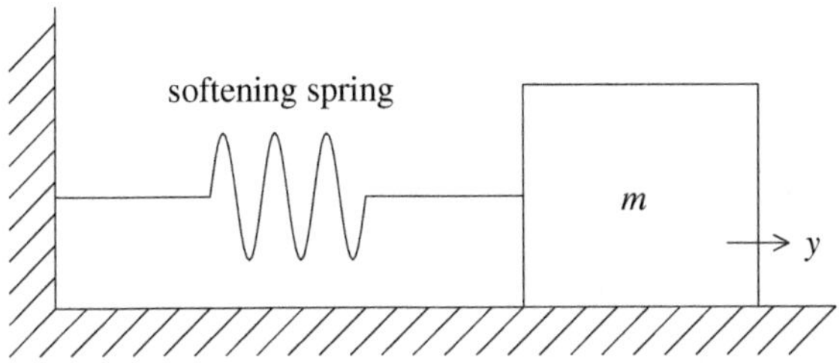

Fig. 5.1 Example 5.1 (mass-spring system). A mechanical system with a mass and a softening spring. □

Example 5.2. Let us consider the model of an electric arc

$$\begin{cases} \dot{y}_1 = -\frac{R}{L}y_1 - \frac{1}{L}y_2 + \frac{V}{L} \\ \dot{y}_2 = -\frac{1}{C}y_1 - \frac{1}{C}y_2y_3 \\ \dot{y}_3 = y_3\left(\frac{1}{Q_0}y_2^2y_3 - \frac{1}{\tau}\right) \end{cases}$$

where y_1 is the circuit current, y_2 is the arc voltage, and y_3 is the arc conductance [48]. The quantities R, L and C are the resistance, inductance and capacitance of an RLC circuit connected to the arc, V is the voltage of the arc, and Q_0 and τ are parameters of the arc. The equilibrium point $(0,V,0)'$ corresponds to the condition of extinguished arc. Let us shift this equilibrium point into the origin by defining $x_1 = y_1, x_2 = y_2 - V$ and $x_3 = y_3$. We obtain:

$$\begin{cases} \dot{x}_1 = -\frac{R}{L}x_1 - \frac{1}{L}x_2 \\ \dot{x}_2 = -\frac{1}{C}x_1 - \frac{1}{C}(x_2 + V)x_3 \\ \dot{x}_3 = x_3\left(\frac{1}{Q_0}(x_2 + V)^2x_3 - \frac{1}{\tau}\right). \end{cases}$$

This is an uncertain polynomial system, moreover the origin is a common equilibrium point. We consider the case where C is variable and the other parameters are fixed, in particular

$$R = 21 \cdot 10^{-3}, \ L = 68.2 \cdot 10^{-6}, \ V = 424, \ Q_0 = 50 \cdot 10^{-3}, \ \tau = 5 \cdot 10^{-6}$$
$$C \in [1 \cdot 10^{-6}, 5 \cdot 10^{-6}].$$

By defining $\theta = C_0/C$ with $C_0 = 5\cdot 10^{-6}$, the system can be rewritten as

$$\begin{cases} \dot{x}_1 = -\frac{R}{L}x_1 - \frac{1}{L}x_2 \\ \dot{x}_2 = -\theta(x_1 - (x_2+V))x_3 \\ \dot{x}_3 = x_3\left(\frac{1}{Q_0}(x_2+V)^2 x_3 - \frac{1}{\tau}\right) \\ \theta \in \Theta = [1,5]. \end{cases} \quad (5.6)$$

The set Θ can be expressed as in (5.2) with

$$a(\theta) = -5 + 6\theta - \theta^2.$$

□

5.2 Estimates via Common LFs

In this section we describe a method for computing estimates of the RDA based on common LFs and SOS programming.

5.2.1 *Establishing Estimates*

The following result provides a condition for establishing whether a sublevel set of a given function is an inner estimate of the RDA by testing the positivity of some functions.

Theorem 5.1. *Let $v : \mathbb{R}^n \to \mathbb{R}$ be positive definite, continuously differentiable and radially unbounded, and let $c \in \mathbb{R}$ be positive. Suppose that there exist $q : \mathbb{R}^n \times \mathbb{R}^{n_\theta} \to \mathbb{R}^2$, $r : \mathbb{R}^n \times \mathbb{R}^{n_\theta} \to \mathbb{R}^{n_a}$ and $s : \mathbb{R}^n \times \mathbb{R}^{n_\theta} \to \mathbb{R}^{n_b}$ such that*

$$\left.\begin{array}{l} p(x,\theta) > 0 \\ q(x,\theta) > 0 \\ r(x,\theta) \geq 0 \end{array}\right\} \ \forall x \in \mathbb{R}_0^n \ \forall \theta \in \mathbb{R}^{n_\theta} \quad (5.7)$$

where

$$p(x,\theta) = -q(x,\theta)'\begin{pmatrix} \dot{v}(x,\theta) \\ c - v(x) \end{pmatrix} - r(x,\theta)'a(\theta) - s(x,\theta)'b(\theta) \quad (5.8)$$

and

$$\dot{v}(x,\theta) = \nabla v(x)' f(x,\theta). \quad (5.9)$$

Then, $v(x)$ is a common LF for the origin of (5.1)*, and $\mathcal{V}(c) \subseteq \mathcal{R}$.*

Proof. Suppose that (5.7) holds, and let $\theta \in \Theta$. Then, from the first inequality it follows that

$$\begin{aligned} 0 &< -q(x,\theta)'\begin{pmatrix} \dot{v}(x,\theta) \\ c - v(x) \end{pmatrix} - r(x,\theta)'a(\theta) - s(x,\theta)'b(\theta) \\ &\leq -q(x,\theta)'\begin{pmatrix} \dot{v}(x,\theta) \\ c - v(x) \end{pmatrix} \end{aligned}$$

since $r(x,\theta) \geq 0$ from the third inequality, and since $a(\theta) \geq 0$ and $b(\theta) = 0$ from definition of Θ. Next, let $x \in \mathscr{V}(c) \setminus \{0_n\}$. Then, it follows that

$$\begin{aligned} 0 &< -q_1(x,\theta)\dot{v}(x,\theta) - q_2(x,\theta)(c - v(x)) \\ &\leq -q_1(x,\theta)\dot{v}(x,\theta) \end{aligned}$$

since $q_2(x) > 0$ from the second inequality in (5.7), and since $v(x) \leq c$. Moreover, since $q_1(x,\theta) > 0$ from the second inequality in (5.7), this implies that $\dot{v}(x,\theta) < 0$. Hence, it follows that (3.60) holds. Since $c > 0$, this implies that (3.46) holds, i.e. $v(x)$ is a common Lyapunov for the origin. From Theorem 3.13 we conclude that $\mathscr{V}(c) \subseteq \mathscr{R}$. □

Theorem 5.1 provides a condition for establishing whether $\mathscr{V}(c)$ is included in $\mathscr{R}$ through the introduction of the auxiliary functions $q(x,\theta)$, $r(x,\theta)$ and $s(x,\theta)$. This condition does not require a priori knowledge of the fact whether $v(x)$ is a common LF for the origin of (5.1): indeed, it is easy to see that (5.7) cannot be satisfied for any positive c if $v(x)$ is not a common LF for the origin of (5.1).

In the sequel we will consider the case where the functions $q(x,\theta)$, $r(x,\theta)$, $s(x,\theta)$ and $v(x)$ are polynomial. In such a case, the condition of Theorem 5.1 can be checked through the SOS index. To this end, let us observe that

$$\text{(5.7) holds} \Rightarrow p(x,\theta),\ q_2(x,\theta) \text{ and } r_i(x,\theta),\ i = 1,\ldots,n_a, \text{ are locally quadratic polynomials in } x \text{ for all } \theta \in \mathbb{R}^{n_\theta}.$$

This means that, a necessary condition for the SOS index of these polynomials to be positive, is that such a SOS index is defined with respect to the power vector $b_{lin}(x,\cdot) \otimes b_{pol}(\theta,\cdot)$. The following result summarizes this discussion.

Corollary 5.1. *Let $v \in \mathscr{P}_n$ be positive definite, and let $c \in \mathbb{R}$ be positive. Suppose that there exist $q \in \mathscr{P}^2_{n,n_\theta}$, $r \in \mathscr{P}^{n_a}_{n,n_\theta}$ and $s \in \mathscr{P}^{n_b}_{n,n_\theta}$ such that*

$$\begin{cases} \lambda_{quapol}(p) > 0 \\ \lambda_{polpol}(q_1) > 0 \\ \lambda_{quapol}(q_2) > 0 \\ \lambda_{quapol}(r_i) \geq 0 \ \forall i = 0,\ldots,n_a \end{cases} \tag{5.10}$$

where $p(x,\theta)$ is as in (5.8). Then, $v(x)$ is a common LF for the origin of (5.1), and $\mathscr{V}(c) \subseteq \mathscr{R}$.

Proof. Suppose that (5.10) holds. This implies that

$$
\begin{array}{r}
p(x,\theta) > 0 \;\; \forall x \in \mathbb{R}_0^n \; \forall \theta \in \mathbb{R}^{n_\theta} \\
q_1(x,\theta) > 0 \;\; \forall x \in \mathbb{R}^n \; \forall \theta \in \mathbb{R}^{n_\theta} \\
q_2(x,\theta) > 0 \;\; \forall x \in \mathbb{R}_0^n \; \forall \theta \in \mathbb{R}^{n_\theta} \\
r_i(x,\theta) \geq 0 \;\; \forall x \in \mathbb{R}_0^n \; \forall \theta \in \mathbb{R}^{n_\theta} \; \forall i = 0,\ldots,n_a.
\end{array}
$$

Consequently, (5.7) holds, and from Theorem 5.1 we conclude the proof. □

Corollary 5.1 provides a condition for establishing whether $\mathscr{V}(c)$ is an inner estimate of $\mathscr{R}$ through the SOS index. This condition can be checked through an LMI feasibility test. Indeed, with $q \in \mathscr{P}^2_{n,n_\theta}$, $r \in \mathscr{P}^{n_a}_{n,n_\theta}$, $s \in \mathscr{P}^{n_b}_{n,n_\theta}$ and $p(x,\theta)$ as in (5.8) let us define the SMR matrices

$$
\begin{aligned}
P(Q,R,\mathbf{s}) + L(\alpha) &= CSMR_{quapol}(p) \\
Q &= \operatorname{diag}(Q_1,Q_2) \\
Q_1 &= SMR_{polpol}(q_1) \\
Q_2 &= SMR_{quapol}(q_2) \\
R &= \operatorname{diag}(R_1,\ldots,R_{n_a}) \\
R_i &= SMR_{quapol}(r_i) \;\; \forall i = 1,\ldots,n_a \\
\mathbf{s} &= (\mathbf{s}_1',\ldots,\mathbf{s}_{n_b}')' \\
\mathbf{s}_i &= COE_{quapol}(s_i) \;\; \forall i = 1,\ldots,n_b.
\end{aligned}
\tag{5.11}
$$

Corollary 5.2. *Let $v \in \mathscr{P}_n$ be positive definite, and let $c \in \mathbb{R}$ be positive. Define the quantities in* (5.11). *Suppose that there exist Q, R, $\mathbf{s}$ and α such that the following LMIs hold:*

$$
\begin{cases}
P(Q,R,\mathbf{s}) + L(\alpha) > 0 \\
Q > 0 \\
R \geq 0 \\
\operatorname{trace}(Q_1) = 1.
\end{cases}
\tag{5.12}
$$

Then, $v(x)$ is a common LF for the origin of (5.1)*, and $\mathscr{V}(c) \subseteq \mathscr{R}$.*

Proof. Suppose that (5.12) holds. This implies that (5.10) holds, and from Corollary (5.1) we conclude the proof. □

Corollary 5.2 shows how the condition (5.10) can be converted into an LMI feasibility test. Let us observe that the constraint $\operatorname{trace}(Q_1) = 1$ normalizes the variables involved in the test: in fact,

$$
\begin{array}{c}
(5.12) \text{ holds for some } Q,\, R,\, \mathbf{s} \text{ and } \alpha \\
\Updownarrow \\
(5.12) \text{ holds for } \rho Q,\, \rho R,\, \rho\mathbf{s} \text{ and } \rho\alpha \text{ for all } \rho > 0.
\end{array}
$$

Let us consider the selection of the degrees of the polynomials in $q(x,\theta)$, $r(x,\theta)$ and $s(x,\theta)$. A possibility is to choose them in order to maximize the degrees of freedom in (5.12) for fixed degrees of $p(x,\theta)$. This is equivalent to require that the degrees of $q_1(x,\theta)\dot{v}(x,\theta)$, $q_2(x,\theta)(c-v(x))$, $r(x,\theta)'a(\theta)$ and $s(x,\theta)'b(\theta)$, rounded to the smallest following even integers, are equal, i.e.

$$2\left\lceil\frac{\partial^x q_1\dot{v}}{2}\right\rceil = 2\left\lceil\frac{\partial^x q_2(c-v)}{2}\right\rceil = 2\left\lceil\frac{\partial^x r_i a_i}{2}\right\rceil = \partial^x s_j b_j$$
$$2\left\lceil\frac{\partial^\theta q_1\dot{v}}{2}\right\rceil = 2\left\lceil\frac{\partial^\theta q_2(c-v)}{2}\right\rceil = 2\left\lceil\frac{\partial^\theta r_i a_i}{2}\right\rceil = \partial^\theta s_j b_j$$

for all $i=1,\ldots,n_a$ and for all $j=1,\ldots,n_b$. This can be achieved as follows:

- choose even degrees for $q_1(x,\theta)$ denoted by $2k_x = \partial^x q_1$ and $2k_\theta = \partial^\theta q_1$;
- set the degrees of $q_2(x,\theta)$ as $\partial^x q_2 = 2\left\lceil\frac{\partial^x f - 1}{2}\right\rceil + 2k_x$ and $\partial^\theta q_2 = 2\left\lceil\frac{\max\{\partial^\theta f + 2k_\theta, \partial a, \partial b\}}{2}\right\rceil$;
- set the degrees of $r_i(x,\theta)$ as $\partial^x r_i = \partial v + \partial^x q_2$ and $\partial^\theta r_i = \partial^\theta q_2 - 2\left\lceil\frac{\partial a_i}{2}\right\rceil$;
- set the degrees of $s_i(x,\theta)$ as $\partial^x s_i = \partial^x r_i$ and $\partial^\theta s_i = \partial^\theta q_2 - \partial b_i$;
- hence, the degrees of $p(x,\theta)$ are given by $\partial^x p = \partial v + \partial^x q_2$ and $\partial^\theta p = \partial^\theta q_2$.

(5.13)

The number of scalar variables in the LMIs (5.12) is given by

$$\begin{aligned}\omega_3 = {} & l_{quapol}(n,n_\theta,\partial^x p,\partial^\theta p) + t_{polpol}(n,n_\theta,2k_x,2k_y) \\ & + t_{quapol}(n,n_\theta,\partial^x q_2,\partial^\theta q_2) + \Sigma_{i=1}^{n_a} t_{quapol}(n,n_\theta,\partial^x r_i,\partial^\theta r_i) \\ & + \Sigma_{i=1}^{n_b}(d_{lin}(n,\partial^x s_i) - n)\, d_{pol}(n_\theta,\partial^\theta s_i) - 1.\end{aligned} \tag{5.14}$$

Tables 5.1–5.2 show this number in some cases according to the choice (5.13). In the sequel we will assume that the degrees of the polynomials in $q(x,\theta)$, $r(x,\theta)$ and $s(x,\theta)$ are chosen according to (5.13) unless specified otherwise.

Table 5.1 Number of scalar variables in the LMIs (5.12) for $n=2$, $n_a=n_\theta$, $n_b=0$, $\partial^\theta f \in \{1,2\}$, $\partial a_i \in \{1,2\}$ and $k_x = k_\theta = 0$.

$(n_\theta=1)$ $\partial^x f \setminus \partial v$	2	4	6
2 or 3	44	151	395
4 or 5	196	440	896

$(n_\theta=2)$ $\partial^x f \setminus \partial v$	2	4	6
2 or 3	99	339	882
4 or 5	438	981	1992

Table 5.2 Number of scalar variables in the LMIs (5.12) for $n = 3$, $n_a = n_\theta$, $n_b = 0$, $\partial^\theta f \in \{1,2\}$, $\partial a_i \in \{1,2\}$ and $k_x = k_\theta = 0$.

$(n_\theta = 1)$			
$\partial^x f \setminus \partial v$	2	4	6
2 or 3	144	712	2479
4 or 5	862	2629	6970

$(n_\theta = 2)$			
$\partial^x f \setminus \partial v$	2	4	6
2 or 3	327	1598	5522
4 or 5	1931	5855	15461

5.2.2 *Choice of the LF*

Let us observe that a necessary condition for (5.12) to hold is that

$$\begin{cases} V_2 > 0 \\ V_2 A(\theta) + A(\theta)' V_2 < 0 \;\; \forall \theta \in \Theta \\ A(\theta) = \mathrm{lp}(f(\cdot,\theta)) \end{cases} \tag{5.15}$$

where $\mathrm{lp}(f(\cdot,\theta))$ is the state matrix of the linearized system in (3.53), and $V_2 = V_2'$ defines the quadratic part of $v(x)$ according to (4.15)–(4.16). Condition (5.15) holds if and only if the linearized system is robustly asymptotically stable (i.e., $A(\theta)$ is Hurwitz for all $\theta \in \Theta$) and $v_2(x)$ is a common LF for it. Hence, the latter is a requirement for (5.12) to hold. This requirement can be removed with a strategy similar to that presented in Section 7.4 for the case of uncertainty-free polynomial systems.

Based on this discussion, a simple way to choose a common LF $v(x)$ for estimating the RDA is as follows:

- find V_2 (if any) that satisfies (5.15);
- define $v_2(x) = x'V_2x$;
- define $v(x)$ as in (4.16) where $v_3(x), v_4(x), \ldots$ are arbitrarily chosen under the constraint that $v(x)$ is positive definite. (5.16)

Let us observe that one can establish whether $v(x)$ is positive definite through the generalized SOS index according to Corollary 2.2. Also, let us observe that one can search for a matrix V_2 satisfying (5.15) with the technique presented in Section 2.2.4 as explained in the following result.

Theorem 5.2. *Define* $A(\theta)$ *as in* (5.15). *Suppose that there exist* $Q, R, \mathbf{S}, V_2$ *and* α *satisfying the constraints*

$$\begin{cases} P(Q,R,\mathbf{S},V_2) + L(\alpha) > 0 \\ Q > 0 \\ R \geq 0 \\ V_2 > 0 \\ \mathrm{trace}(Q) = 1 \\ \mathrm{trace}(V_2) = 1 \end{cases} \tag{5.17}$$

where

$$\begin{aligned} P(Q,R,\mathbf{S},V_2)+L(\alpha) &= CSMR_{mat}(G) \\ Q &= SMR_{pol}(q) \\ R &= \operatorname{diag}(R_1,\ldots,R_{n_a}) \\ R_i &= SMR_{mat}(H_i)\ \forall i=1,\ldots,n_a \\ \mathbf{S} &= (\mathbf{S}_1,\ldots,\mathbf{S}_{n_b}) \\ \mathbf{S}_i &= COE_{mat}(J_i)\ \forall i=1,\ldots,n_b \end{aligned} \tag{5.18}$$

and $q \in \mathscr{P}_{n_\theta}$, $H_i = H_i', J_i = J_i' \in \mathscr{P}_{n_\theta}^{n\times n}$, *and*

$$G(\theta) = -q(\theta)\left(V_2A(\theta)+A(\theta)'V_2\right) - \sum_{i=1}^{n_a} H_i(\theta)a_i(\theta) - \sum_{i=1}^{n_b} J_i(\theta)b_i(\theta). \tag{5.19}$$

Then, the linearized system is robustly asymptotically stable. Moreover, (5.15) *holds, i.e.* $v_2(x) = x'V_2x$ *is a common LF for the linearized system.*

Proof. Suppose that (5.17) holds. This implies that

$$\left.\begin{array}{r} G(\theta) > 0 \\ q(\theta) > 0 \\ H_i(\theta) \geq 0\ \forall i=1,\ldots,n_a \\ V_2 > 0 \end{array}\right\} \forall \theta \in \mathbb{R}^{n_\theta}.$$

Let $\theta \in \Theta$. Since $a(\theta) \geq 0$ and $b(\theta) = 0$ it follows that

$$\begin{aligned} 0 &< -q(\theta)\left(V_2A(\theta)+A(\theta)'V_2\right) - \textstyle\sum_{i=1}^{n_a} H_i(\theta)a_i(\theta) - \sum_{i=1}^{n_b} J_i(\theta)b_i(\theta) \\ &\leq -q(\theta)\left(V_2A(\theta)+A(\theta)'V_2\right). \end{aligned}$$

Since $q(\theta) > 0$, this implies that $V_2A(\theta)+A(\theta)'V_2 < 0$. We hence conclude the proof by observing that $V_2 > 0$. □

Let us observe that the first constraint in (5.17) is a BMI due to the product of Q with V_2. A simple way to transform (5.17) into a set of LMIs consists of fixing Q under the constraints $Q > 0$ and $\operatorname{trace}(Q) = 1$. In the case $\partial q = 0$, this simply means that $q(\theta) = 1$ and $Q = 1$.

Example 5.3. For the mass-spring system (5.5), let us consider the problem of establishing whether $\mathscr{V}(c)$ is a subset of the RDA of the origin with the choice

$$\begin{aligned} v(x) &= 3x_1^2 + x_1x_2 + 2x_2^2 \\ c &= 1. \end{aligned}$$

For this system we have

$$A(\theta) = \begin{pmatrix} 0 & \theta_1+\theta_2 \\ -2\theta_1-\theta_2 & -0.5\theta_1-0.5\theta_2 \end{pmatrix}$$

and hence $A(\theta)$ is Hurwitz for all $\theta \in \Theta$. Moreover, $v(x)$ is a common LF for the origin. In fact, (5.15) holds with

$$V_2 = \begin{pmatrix} 3 & 0.5 \\ \star & 2 \end{pmatrix}.$$

Let us use Corollary 5.2 choosing $k_x = k_\theta = 0$ in (5.13). Since $\partial f^x = 3$ and $\partial f^\theta = 1$, we have

$$\partial^x q_2 = 2,\ \partial^\theta q_2 = 2,\ \partial^x p = 4,\ \partial^\theta p = 2.$$

The LMIs in (5.12) have 44 scalar variables. We find that these LMIs hold, hence concluding that $\mathscr{V}(c) \subseteq \mathscr{R}$. □

5.2.3 *Parameter-Dependent LEDA and LERDA*

The condition provided in Corollary 5.2 is built for a chosen scalar c, which determines the extension of the estimate $\mathscr{V}(c)$. Clearly, it would be helpful if one could search for the largest c for which the LMIs in (5.12) are feasible. In fact, this would allow one to obtain the largest estimate of the RDA that can be obtained through the common LF $v(x)$.

In order to address this problem, let us start by introducing the following definition.

Definition 5.1 (Parameter-Dependent LEDA). Let $v(x)$ be a common LF for the origin of (5.1), and define

$$\begin{aligned} \gamma(\theta) = \ & \sup_c c \\ & \text{s.t. } \mathscr{V}(c) \subseteq \mathscr{X}(\theta) \end{aligned} \tag{5.20}$$

where $\mathscr{X}(\theta)$ is given by (3.61). Then, $\mathscr{V}(\gamma(\theta))$ is said the *parameter-dependent LEDA* of the origin provided by the common LF $v(x)$. □

In other words, the parameter-dependent LEDA is the LEDA provided by a chosen common LF $v(x)$ and parametrized by the uncertainty θ. The largest estimate of the RDA that can be obtained through the common LF $v(x)$ is defined as follows.

Definition 5.2 (LERDA). Let $v(x)$ be a common LF for the origin of (5.1), and define

$$\begin{aligned} \gamma^\# = \ & \sup_c c \\ & \text{s.t. } \mathscr{V}(c) \subseteq \mathscr{X}(\theta)\ \forall \theta \in \Theta \end{aligned} \tag{5.21}$$

where $\mathscr{X}(\theta)$ is given by (3.61). Then, $\mathscr{V}(\gamma^\#)$ is said the *LERDA* of the origin provided by the common LF $v(x)$. □

Clearly, the LERDA coincides with the intersection of all the LEDAs provided by the common LF $v(x)$ for any admissible value of the uncertainty, i.e.

$$\mathscr{V}(\gamma^\#) = \bigcap_{\theta \in \Theta} \mathscr{V}(\gamma(\theta)) \tag{5.22}$$

where $\gamma(\theta)$ is as in (5.20). Hence, one has that

$$\gamma^{\#} = \inf_{\theta \in \Theta} \gamma(\theta). \tag{5.23}$$

From Corollary 5.2 one can define a natural lower bound of $\gamma^{\#}$ as

$$\begin{aligned} \hat{\gamma}^{\#} = & \sup_{c,Q,R,\mathbf{s},\alpha} \; c \\ & \text{s.t.} \left\{ \begin{array}{r} P(c,Q,R,\mathbf{s}) + L(\alpha) > 0 \\ Q > 0 \\ R \geq 0 \\ \text{trace}(Q_1) = 1 \end{array} \right. \end{aligned} \tag{5.24}$$

where $P(c,Q,R,\mathbf{s})$ is the SMR matrix of $p(x,\theta)$ in (5.11) where the dependence on c has been made explicit.

A way to compute $\hat{\gamma}^{\#}$ is via a one-parameter sweep on the scalar c in Corollary 5.2, which can be conducted for instance via a bisection algorithm.

Another way to obtain $\hat{\gamma}^{\#}$ is via a GEVP. Indeed, let us define the polynomial

$$p_1(x,\theta) = -q(x,\theta)' \begin{pmatrix} \dot{v}(x,\theta) \\ -v(x) \end{pmatrix} - r(x,\theta)'a(\theta) - s(x,\theta)'b(\theta) \tag{5.25}$$

and its SMR matrix as

$$P_1(Q,R,\mathbf{s}) + L(\alpha) = CSMR_{quapol}(p_1). \tag{5.26}$$

Let us introduce the following preliminary result.

Theorem 5.3. *Let $v(x)$ and V be related as in* (4.23)*, and $q_2(x,\theta)$ and Q_2 be related as in* (5.11)*. For $\mu \in \mathbb{R}$ define*

$$p_2(x,\theta) = q_2(x,\theta)(1 + \mu v(x)) \tag{5.27}$$

and

$$P_2 = T_2' (\text{diag}(1, \mu V) \otimes Q_2) T_2 \tag{5.28}$$

where T_2 is the matrix satisfying

$$\begin{aligned} & \begin{pmatrix} 1 \\ b_{lin}(x, \partial v/2) \end{pmatrix} \otimes b_{lin}(x, \partial^x q_2/2) \otimes b_{pol}(\theta, \partial^\theta q_2/2) \\ & = T_2 b_{pol}(\theta, \partial^\theta p_2/2) \otimes b_{lin}(x, \partial^x p_2/2). \end{aligned} \tag{5.29}$$

Then, P_2 satisfies

$$P_2 = SMR_{quapol}(p_2). \tag{5.30}$$

Moreover,

$$\mu V > 0 \text{ and } Q_2 > 0 \;\Rightarrow\; P_2 > 0. \tag{5.31}$$

Proof. For $x \in \mathbb{R}_0^n$ and $\theta \in \mathbb{R}^{n_\theta}$ define $y = b_{pol}(\theta, \partial^\theta p_2/2) \otimes b_{lin}(x, \partial^x p_2/2)$, and let us pre- and post-multiply P_2 by y' and y, respectively. We get:

$$\begin{aligned}
& y'P_2 y \\
&= y'T_2{}'(\mathrm{diag}(1,\mu V) \otimes Q_2)T_2 y \\
&= \left(\begin{pmatrix} 1 \\ b_{lin}(x, \partial v/2) \end{pmatrix} \otimes b_{lin}(x, \partial^x q_2/2) \otimes b_{pol}(\theta, \partial^\theta q_2/2) \right)' (\mathrm{diag}(1,\mu V) \otimes Q_2) \\
&\quad \cdot \begin{pmatrix} 1 \\ b_{lin}(x, \partial v/2) \end{pmatrix} \otimes b_{lin}(x, \partial^x q_2/2) \otimes b_{pol}(\theta, \partial^\theta q_2/2) \\
&= (1 + \mu v(x)) q_2(x, \theta) \\
&= p_2(x, \theta)
\end{aligned}$$

i.e. (5.30) holds. Then, let us observe that $\mu V > 0$ and $Q_2 > 0$ implies that

$$(\mathrm{diag}(1,\mu V) \otimes Q_2) > 0.$$

Moreover, T_2 is a full column matrix, and hence it follows that (5.31) holds. □

Theorem 5.3 provides an explicit expression for an SMR matrix of $p_2(x,\theta)$. As explained in the theorem, this matrix is guaranteed to be positive definite provided that μV and Q_2 are positive definite. The following result shows how $\hat{\gamma}^\#$ can be obtained via a quasi-convex optimization problem.

Theorem 5.4. *Let $v \in \mathscr{P}_n$ be positive definite, and let $\mu \in \mathbb{R}$ be positive. Define the quantities in* (4.23) *and* (5.26)*. Assume that $V > 0$, select $\partial^x q_2$ and $\partial^\theta q_2$ according to* (5.13)*, and let $P_2(Q_2)$ be the SMR matrix of $p_2(x,\theta)$ in* (5.28) *where the dependence on Q_2 has been made explicit. Then, $\hat{\gamma}^\#$ in* (5.24) *is given by*

$$\hat{\gamma}^\# = -\frac{z^*}{1 + \mu z^*} \tag{5.32}$$

where z^* is the solution of the GEVP

$$\begin{aligned}
z^* = & \inf_{Q,R,\mathrm{s},z,\alpha} \; z \\
& \text{s.t.} \begin{cases} zP_2(Q_2) + P_1(Q,R,\mathrm{s}) + L(\alpha) > 0 \\ Q > 0 \\ R \geq 0 \\ 1 + \mu z > 0 \\ \mathrm{trace}(Q_1) = 1. \end{cases}
\end{aligned} \tag{5.33}$$

Proof. First of all, observe that $v(x)$ is positive definite since $V > 0$. Now, suppose that the constraints in (5.33) hold. For $x \in \mathbb{R}_0^n$ and $\theta \in \mathbb{R}^{n_\theta}$ define $y = b_{pol}(\theta, \partial^\theta p_2/2) \otimes b_{lin}(x, \partial^x p_2/2)$, and let us pre- and post-multiply the first inequality in (5.33) by y' and y, respectively. We get:

$$
\begin{aligned}
0 &< y'\left(zP_2(Q_2)+P_1(Q,R,\mathbf{s})+L(\alpha)\right)y \\
&= zp_2(x,\theta)+p_1(x,\theta) \\
&= zq_2(x,\theta)(1+\mu v(x)) - q(x,\theta)'\begin{pmatrix}\dot{v}(x,\theta)\\ -v(x)\end{pmatrix} - r(x,\theta)'a(\theta) - s(x,\theta)'b(\theta) \\
&= -q_1(x,\theta)\dot{v}(x,\theta) - q_2(x,\theta)(-z-z\mu v(x)-v(x)) - r(x,\theta)'a(\theta) - s(x,\theta)'b(\theta) \\
&= -q_1(x,\theta)\dot{v}(x,\theta) - (1+\mu z)q_2(x,\theta)\left(\frac{-z}{1+\mu z} - v(x)\right) - r(x,\theta)'a(\theta) - s(x,\theta)'b(\theta).
\end{aligned}
$$

Hence, the first inequality in (5.24) coincides with the first inequality in (5.33) whenever $q_2(x,\theta)$ and c are replaced by

$$
\begin{aligned}
q_2(x,\theta) &\rightarrow q_2(x,\theta)(1+\mu z) \\
c &\rightarrow \frac{-z}{1+\mu z}.
\end{aligned}
$$

Since $1+\mu z$ is positive, it follows that the constraints in (5.24) are equivalent to those in (5.33), and hence (5.32) holds. □

Theorem 5.4 states that the LERDA provided by a common LF $v(x)$ can be found by solving the GEVP (5.33) provided that $v(x)$ admits a positive definite SMR matrix V.

In the case where Θ is a polytope and $\partial^\theta f = 1$, the LERDA can be further characterized as explained in the following result.

Theorem 5.5. *Let $v \in \mathscr{P}_n$ be positive definite. Suppose that Θ is a bounded convex polytope and $\partial^\theta f = 1$. Then, $\gamma^\#$ is given by*

$$
\gamma^\# = \min_{\theta\in\mathrm{ver}(\Theta)} \gamma(\theta). \tag{5.34}
$$

Proof. Since $\partial^\theta f = 1$, it follows that $\dot{v}(x,\theta)$ is affine linear in θ. Consequently, if Θ is a bounded convex polytope, it follows that, for all $x\in\mathbb{R}^n$, $\dot{v}(x,\theta)$ achieves its maximum over Θ at the vertices of Θ, i.e.

$$
\max_{\theta\in\Theta} \dot{v}(x,\theta) = \max_{\theta\in\mathrm{ver}(\Theta)} \dot{v}(x,\theta) \;\; \forall x\in\mathbb{R}^n.
$$

Since

$$
\begin{aligned}
\gamma^\# &= \sup_c \; c \\
&\quad \text{s.t. } \mathscr{V}(c) \subseteq \{x\in\mathbb{R}^n:\ \dot{v}(x,\theta) < 0\ \forall\theta\in\Theta\}\cup\{0_n\}
\end{aligned}
$$

one has that

$$
\begin{aligned}
\gamma^\# &= \min_{\theta\in\mathrm{ver}(\Theta)} \sup_c \; c \\
&\quad \text{s.t. } \mathscr{V}(c) \subseteq \{x\in\mathbb{R}^n:\ \dot{v}(x,\theta) < 0\}\cup\{0_n\}
\end{aligned}
$$

which is (5.34). □

From Theorem 5.5 a lower bound of $\gamma^\#$ can be directly computed with the techniques described in Chapter 4 as

$$\min_{\theta \in \text{ver}(\Theta)} \hat{\gamma}(\theta) \quad (5.35)$$

where $\hat{\gamma}(\theta)$ is the lower bound in (4.20) parametrized by θ.

Example 5.4. For the mass-spring system (5.5), let us consider the problem of determining $\hat{\gamma}^\#$ in (5.24) with $v(x) = 3x_1^2 + x_1x_2 + 2x_2^2$. Let us use Theorem 4.3 with $\partial^x q_1 = \partial^\theta q_1 = 0$. The GEVP (4.30) has 45 scalar variables. We simply choose $\mu = 1$, and we obtain $\hat{\gamma}^\# = 2.127$. Figure 5.2a shows the estimate $\mathscr{V}(\hat{\gamma}^\#)$ and the curve $\dot{v}(x,\theta) = 0$ for some values of θ. As we can see, the estimate $\mathscr{V}(\hat{\gamma}^\#)$ is tangent to one of these curves, hence implying that $\hat{\gamma}^\#$ is tight, i.e. $\hat{\gamma}^\# = \gamma^\#$. Figure 5.2b shows $\hat{\gamma}^\#$ and $\gamma(\theta)$ calculated for some values of θ.

Let us observe that, since in this case $\Theta = [1,2]$ is a bounded convex polytope and $\partial^\theta f = 1$, we can also use Theorem 5.5. Indeed, as shown by Figure 5.2b, $\hat{\gamma}^\# = \gamma^\#$ is achieved at one of the two vertices of Θ (specifically, for $\theta = 1$). □

Example 5.5. Let us consider the system

$$\begin{cases} \dot{x}_1 = 2x_1 + x_2 - (2 + 4\theta - 5\theta^2)x_1^2 - x_2^3 \\ \dot{x}_2 = -(3-\theta)x_2 - x_1^3 + x_2^2 \\ \theta \in \Theta = [0,1]. \end{cases} \quad (5.36)$$

The origin of this uncertain polynomial systems is a common equilibrium point. The set Θ can be expressed as in (5.2) with

$$a(\theta) = \theta - \theta^2.$$

Let us consider the problem of estimating the RDA of the origin, in particular determining $\hat{\gamma}^\#$ in (5.24) with the choice

$$v(x) = x_1^2 + x_2^2.$$

The GEVP (5.33) has 45 scalar variables. We find $\hat{\gamma}^\# = 0.502$. Figure 5.3a shows the estimate $\mathscr{V}(\hat{\gamma}^\#)$ and the curve $\dot{v}(x,\theta) = 0$ for some values of θ. As we can see, the estimate $\mathscr{V}(\hat{\gamma}^\#)$ is tangent to one of these curves, hence implying that $\hat{\gamma}^\#$ is tight, i.e. $\hat{\gamma}^\# = \gamma^\#$. Figure 5.3b shows $\hat{\gamma}^\#$ and $\gamma(\theta)$ calculated for some values of θ. □

Example 5.6. Let us consider the system

$$\begin{cases} \dot{x}_1 = -2x_1 - x_2 - (1-\theta_2)x_1^2 + \theta_1 x_1 x_2 \\ \dot{x}_2 = -2x_1 - 3x_2 - \theta_2 x_1 x_2 + (2+\theta_1)x_2^2 \\ \theta \in \Theta = \{\theta \in \mathbb{R}^2 : \|\theta\| \leq 1\}. \end{cases} \quad (5.37)$$

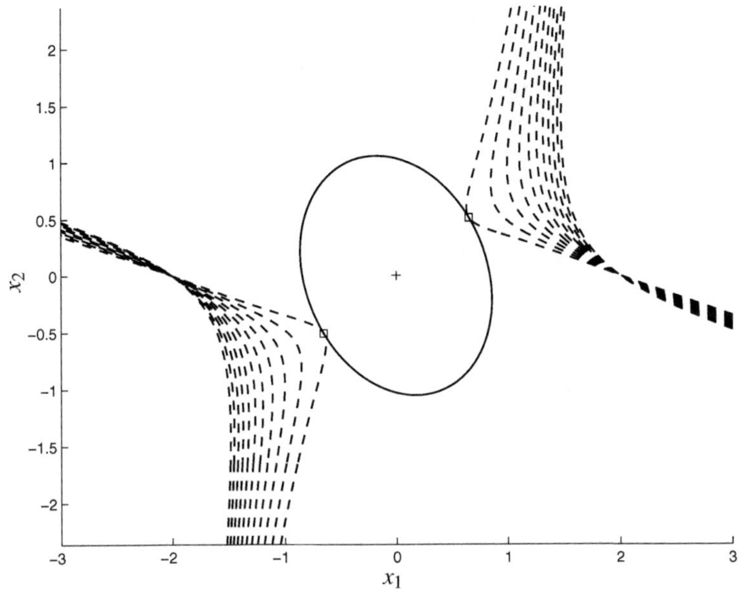

(a)

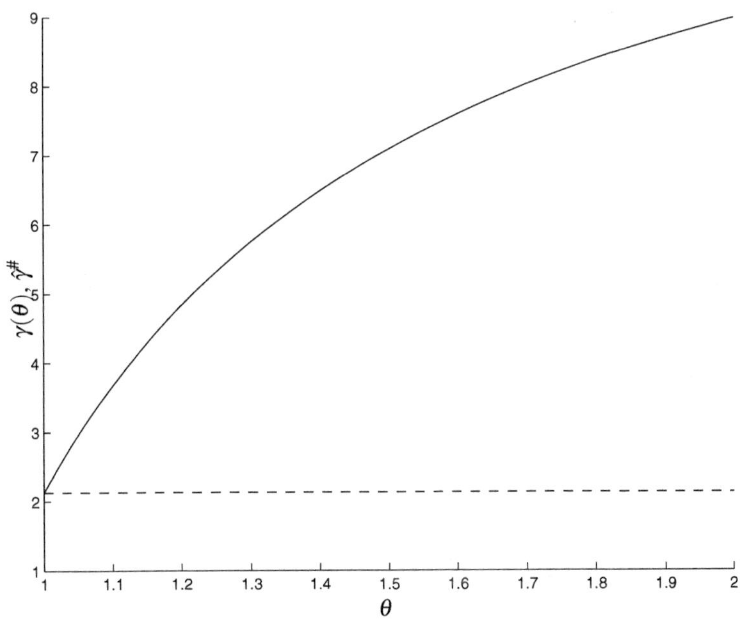

(b)

Fig. 5.2 Example 5.4 (mass-spring system). (a) Boundary of $\mathscr{V}(\hat{\gamma}^{\#})$ (solid) and curve $\dot{v}(x,\theta) = 0$ for some values of θ (dashed). The "+" denotes the common equilibrium point. (b) Lower bound $\gamma(\theta)$ calculated for some values of θ (solid) and lower bound $\hat{\gamma}^{\#}$ (dashed).

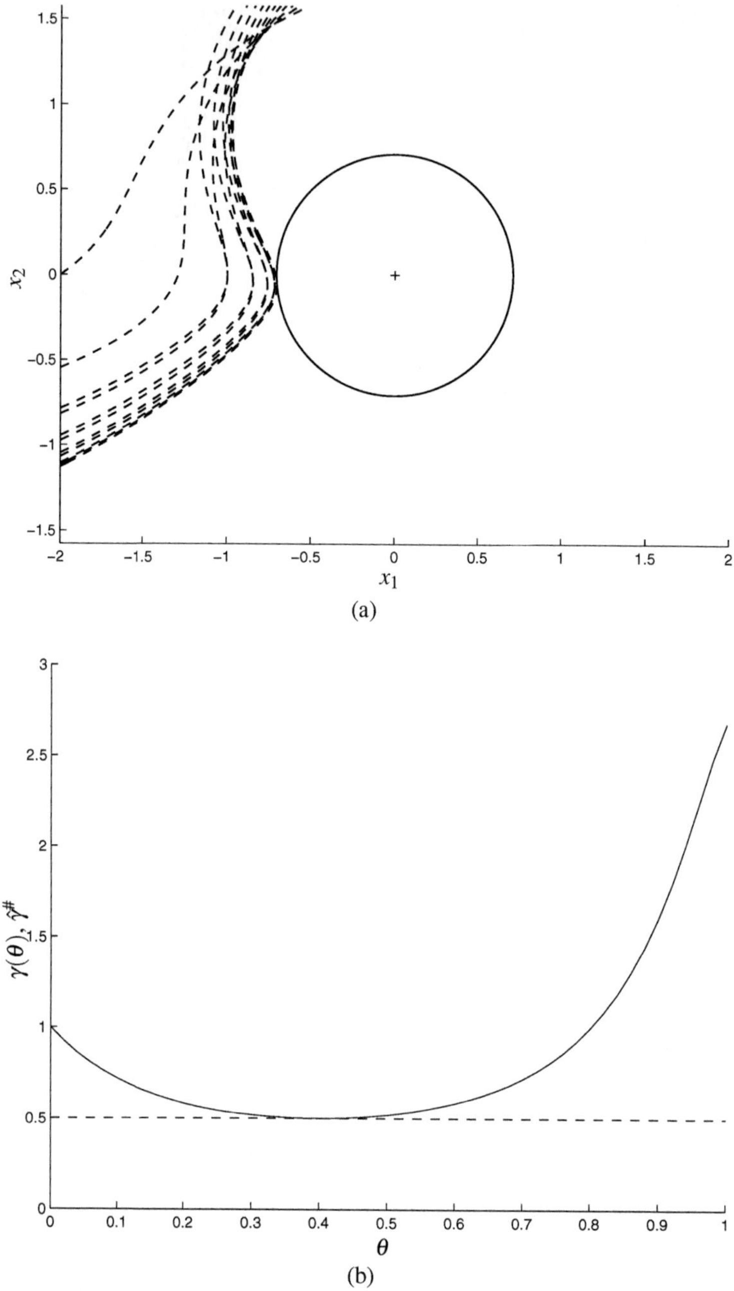

Fig. 5.3 Example 5.5. (a) Boundary of $\mathscr{V}(\hat{\gamma}^{\#})$ (solid) and curve $\dot{v}(x,\theta) = 0$ for some values of θ (dashed). The "+" denotes the common equilibrium point. (b) Lower bound $\gamma(\theta)$ calculated for some values of θ (solid) and lower bound $\hat{\gamma}^{\#}$ (dashed).

The origin of this uncertain polynomial systems is a common equilibrium point. The set Θ can be expressed as in (5.2) with

$$a(\theta) = 1 - \|\theta\|^2.$$

Let us consider the problem of estimating the RDA of the origin, in particular determining $\hat{\gamma}^\#$ in (5.24) with the choice

$$v(x) = 3x_1^2 + x_1x_2 + x_2^2.$$

The GEVP (5.33) has 85 scalar variables. We find $\hat{\gamma}^\# = 0.303$. Figure 5.4a shows the estimate $\mathscr{V}(\hat{\gamma}^\#)$ and the curve $\dot{v}(x,\theta) = 0$ for some values of θ. As we can see, the estimate $\mathscr{V}(\hat{\gamma}^\#)$ is tangent to one of these curves, hence implying that $\hat{\gamma}^\#$ is tight, i.e. $\hat{\gamma}^\# = \gamma^\#$. Figure 5.4b shows $\hat{\gamma}^\#$ and $\gamma(\theta)$ calculated for some values of θ. □

5.2.4 Estimate Tightness

Once that the lower bound $\hat{\gamma}^\#$ has been found, a natural question concerns its tightness. The following result provides a necessary and sufficient condition for answering to this question.

Theorem 5.6. *Let $\gamma^\#$ and $\hat{\gamma}^\#$ be as in* (5.21) *and* (5.24)*, respectively. Suppose that $0 < \gamma^\# < \infty$, and define*

$$\mathscr{M} = \left\{(x,\theta) \in \mathbb{R}^n \times \mathbb{R}^{n_\theta} : \ b_{pol}(\theta, \partial^\theta p/2) \otimes b_{lin}(x, \partial p/2) \in \ker(M)\right\} \tag{5.38}$$

where

$$M = P(\hat{\gamma}^\#, Q^*, R^*, \mathbf{s}^*) + L(\alpha^*) \tag{5.39}$$

and $P(\hat{\gamma}^\#, Q^*, R^*, \mathbf{s}^*) + L(\alpha^*)$ is the found optimal value of $P(c,Q,R,\mathbf{s}) + L(\alpha)$ in (5.24) *(or, equivalently, the found optimal value of $zP_2(Q_2) + P_1(Q,R,\mathbf{s}) + L(\alpha)$ in* (5.33)*). Define also the set*

$$\mathscr{M}_1 = \left\{(x,\theta) \in \mathscr{M} : \ \theta \in \Theta \ \text{ and } \ v(x) = \hat{\gamma}^\# \ \text{ and } \ \dot{v}(x,\theta) = 0\right\}. \tag{5.40}$$

Then,

$$\hat{\gamma}^\# = \gamma^\# \iff \mathscr{M}_1 \neq \emptyset. \tag{5.41}$$

Proof. "⇒" Suppose that $\hat{\gamma}^\# = \gamma^\#$, and let (x^*, θ^*) be the tangent point between the surface $\dot{v}(x,\theta) = 0$ and the sublevel set $\mathscr{V}(\gamma^\#)$ with $\theta^* \in \Theta$, i.e.

$$\begin{aligned} \dot{v}(x^*, \theta^*) &= 0 \\ v(x^*) &= \gamma^\# \\ \theta^* &\in \Theta. \end{aligned}$$

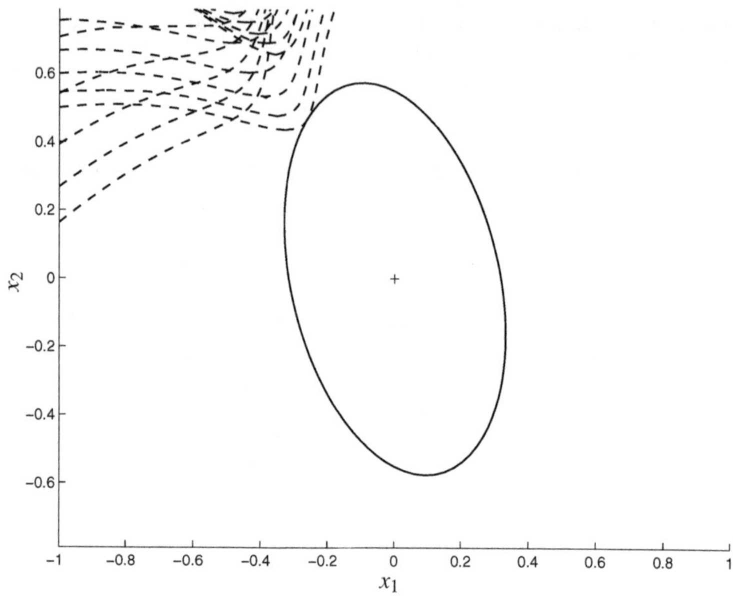

(a)

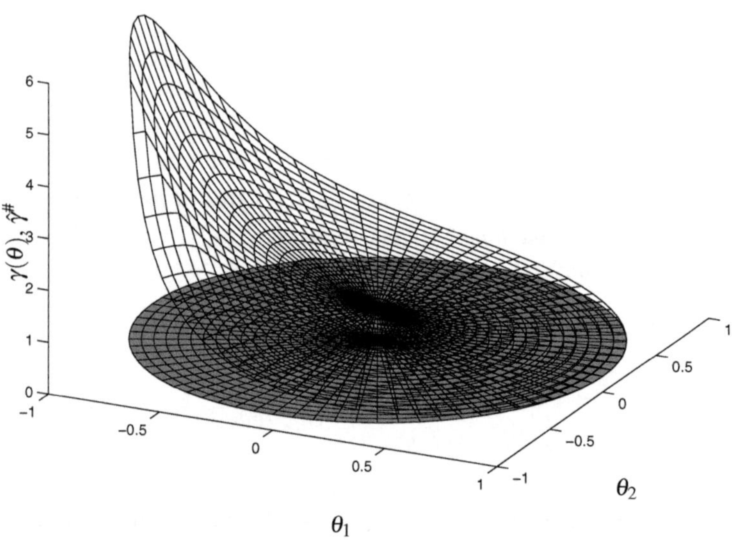

(b)

Fig. 5.4 Example 5.6. (a) Boundary of $\mathscr{V}(\hat{\gamma}^{\#})$ (solid) and curve $\dot{v}(x,\theta) = 0$ for some values of θ (dashed). The "+" denotes the common equilibrium point. (b) Lower bound $\gamma(\theta)$ calculated for some values of θ (transparent surface) and lower bound $\hat{\gamma}^{\#}$ (grey surface).

Define $y = b_{pol}(\theta^*, \partial^\theta p/2) \otimes b_{lin}(x^*, \partial p/2)$. Pre- and post-multiplying M by y' and y, respectively, we get

$$\begin{aligned} 0 &\le y'My \\ &= -q(x^*,\theta^*)' \begin{pmatrix} \dot{v}(x^*,\theta^*) \\ \gamma^\# - v(x^*) \end{pmatrix} - r(x^*,\theta^*)'a(\theta^*) - s(x^*,\theta^*)'b(\theta^*) \\ &= -r(x^*,\theta^*)'a(\theta^*) \end{aligned}$$

since M is positive semidefinite for definition of $\hat{\gamma}^\#$, $\dot{v}(x^*,\theta^*) = 0$, $\gamma^\# - v(x^*) = 0$ and $b(\theta^*) = 0$. Now, since $r(x^*,\theta^*) \ge 0$ and $a(\theta^*) \ge 0$, it follows that

$$r(x^*,\theta^*) = 0, \quad a(\theta^*) = 0, \quad y'My = 0.$$

This implies that

$$y \in \ker(M)$$

i.e. $(x^*,\theta^*) \in \mathscr{M}$. Hence, $(x^*,\theta^*) \in \mathscr{M}_1$, i.e. $\mathscr{M}_1 \neq \emptyset$.

"$\Leftarrow$" Suppose that $\mathscr{M}_1 \neq \emptyset$ and let $(x,\theta) \in \mathscr{M}_1$. It follows that $\theta \in \Theta$, $v(x) = \hat{\gamma}^\#$ and $\dot{v}(x,\theta) = 0$. Since $\hat{\gamma}^\#$ is a lower bound of $\gamma^\#$, this implies that x is a tangent point between the surface $\dot{v}(x,\theta) = 0$ and the sublevel set $\mathscr{V}(\hat{\gamma}^\#)$. Hence, $\hat{\gamma}^\# = \gamma^\#$. □

Theorem 5.6 provides a condition for establishing whether the lower bound provided by (5.24) is tight. In particular, this happens if the set $\mathscr{M}_1$ in (5.40) is nonempty. This set can be found via trivial substitution from the set $\mathscr{M}$ in (5.38) (at least whenever $\mathscr{M}$ is finite), which can be computed as explained in Section 1.5.2 after (1.202).

Example 5.7. For the mass-spring system (5.5), let us consider the problem of establishing tightness of $\hat{\gamma}^\#$ found in Example 5.4 with $v(x) = 3x_1^2 + x_1x_2 + 2x_2^2$. We find that

$$\mathscr{M}_1 = \{((0.652, 0.510)', 1), ((-0.652, -0.510)', 1)\}$$

and from Theorem 5.41 we conclude that $\hat{\gamma}^\# = \gamma^\#$. Figure 5.2a show the x-component of the points in $\mathscr{M}_1$ ("□" markers). □

5.3 Estimates via Parameter-Dependent LFs

This section considers the problem of estimating the RDA via polynomial LFs that depend on the uncertainty.

5.3.1 Establishing Parameter-Dependent Estimates

The following result provides a condition for establishing whether a parameter-dependent sublevel set is an inner estimate of the RDA.

Theorem 5.7. *Let* $v : \mathbb{R}^n \times \mathbb{R}^{n_\theta} \to \mathbb{R}$*, with* $v(x,\theta)$ *positive definite, continuously differentiable and radially unbounded in* x *for all* $\theta \in \Theta$*, and let* $c : \mathbb{R}^{n_\theta} \to \mathbb{R}$*, with* $c(\theta) > 0$ *for all* $\theta \in \Theta$*. Suppose that there exist* $q : \mathbb{R}^n \times \mathbb{R}^{n_\theta} \to \mathbb{R}^2$*,* $r : \mathbb{R}^n \times \mathbb{R}^{n_\theta} \to \mathbb{R}^{n_a}$ *and* $s : \mathbb{R}^n \times \mathbb{R}^{n_\theta} \to \mathbb{R}^{n_b}$ *such that*

$$\left.\begin{array}{r} p(x,\theta) > 0 \\ q(x,\theta) > 0 \\ r(x,\theta) \geq 0 \end{array}\right\} \quad \forall x \in \mathbb{R}^n_0 \; \forall \theta \in \mathbb{R}^{n_\theta} \tag{5.42}$$

where

$$p(x,\theta) = -q(x,\theta)' \begin{pmatrix} \dot{v}(x,\theta) \\ c(\theta) - v(x,\theta) \end{pmatrix} - r(x,\theta)'a(\theta) - s(x,\theta)'b(\theta) \tag{5.43}$$

and

$$\dot{v}(x,\theta) = \nabla v(x,\theta)' f(x,\theta). \tag{5.44}$$

Then, $v(x,\theta)$ *is a parameter-dependent LF for the origin of* (5.1)*, and*

$$\mathscr{V}(c(\theta),\theta) \subseteq \mathscr{R} \;\; \forall \theta \in \Theta. \tag{5.45}$$

Proof. Suppose that (5.42) holds, and let $\theta \in \Theta$. Then, from the first inequality it follows that

$$\begin{aligned} 0 &< -q(x,\theta)' \begin{pmatrix} \dot{v}(x,\theta) \\ c(\theta) - v(x,\theta) \end{pmatrix} - r(x,\theta)'a(\theta) - s(x,\theta)'b(\theta) \\ &\leq -q(x,\theta)' \begin{pmatrix} \dot{v}(x,\theta) \\ c(\theta) - v(x,\theta) \end{pmatrix} \end{aligned}$$

since $r(x,\theta) \geq 0$ from the third inequality in (5.42), and since $a(\theta) \geq 0$ and $b(\theta) = 0$ from definition of Θ. Next, let $x \in \mathscr{V}(c(\theta),\theta) \setminus \{0_n\}$. Then, it follows that

$$\begin{aligned} 0 &< -q_1(x,\theta)\dot{v}(x,\theta) - q_2(x,\theta)(c(\theta) - v(x,\theta)) \\ &\leq -q_1(x,\theta)\dot{v}(x,\theta) \end{aligned}$$

since $q_2(x,\theta) > 0$ from the second inequality in (5.42), and since $v(x,\theta) \leq c(\theta)$. Moreover, since $q_1(x,\theta) > 0$ from the second inequality in (5.42), this implies that $\dot{v}(x,\theta) < 0$. Hence, it follows that (3.64) holds. Since $c(\theta) > 0$, this implies that (3.49) holds, i.e. $v(x,\theta)$ is a parameter-dependent Lyapunov for the origin. From Theorem 3.14 we conclude that (5.45) holds. □

Theorem 5.7 provides a condition for establishing whether $\mathscr{V}(c(\theta),\theta)$ is included in $\mathscr{R}$ for all admissible values of θ. This condition does not require a priori knowledge of the fact whether $v(x,\theta)$ is a parameter-dependent LF for the origin of (5.1): indeed, it is easy to see that (5.7) cannot be satisfied if $v(x,\theta)$ is not a parameter-dependent LF for the origin of (5.1).

In the sequel we will consider the case where the functions $c(\theta)$, $q(x,\theta)$, $r(x,\theta)$, $s(x,\theta)$ and $v(x,\theta)$ are polynomial. In such a case, the condition of Theorem 5.7 can be checked through the SOS index similarly to the condition of Theorem 5.1 as explained hereafter.

Corollary 5.3. *Let $v \in \mathscr{P}_{n,n_\theta}$ be such that $v(x,\theta)$ is positive definite in x for all $\theta \in \Theta$, and let $c \in \mathscr{P}_{n_\theta}$ be such that $c(\theta) > 0$ for all $\theta \in \Theta$. Let $p(x,\theta)$ as in* (5.43), *and suppose that there exist $q \in \mathscr{P}^2_{n,n_\theta}$, $r \in \mathscr{P}^{n_a}_{n,n_\theta}$ and $s \in \mathscr{P}^{n_b}_{n,n_\theta}$ such that*

$$\begin{cases} \lambda_{quapol}(p) > 0 \\ \lambda_{polpol}(q_1) > 0 \\ \lambda_{quapol}(q_2) > 0 \\ \lambda_{quapol}(r_i) \geq 0 \;\; \forall i = 0,\ldots,n_a. \end{cases} \tag{5.46}$$

Then, $v(x,\theta)$ is a parameter-dependent LF for the origin of (5.1), *and* (5.45) *holds.*

Proof. Suppose that (5.46) holds. Analogously to the proof of Corollary 5.1, this implies that (5.42) holds, and from Theorem 5.7 we conclude the proof. □

Corollary 5.3 provides a condition for establishing whether $\mathscr{V}(c(\theta),\theta)$ is an inner estimate of $\mathscr{R}$ for all admissible values of θ through the SOS index. The following result explains how this condition can be checked through an LMI feasibility test.

Corollary 5.4. *Let $v \in \mathscr{P}_{n,n_\theta}$ be such that $v(x,\theta)$ is positive definite in x for all $\theta \in \Theta$, and let $c \in \mathscr{P}_{n_\theta}$ be such that $c(\theta) > 0$ for all $\theta \in \Theta$. Define the quantities in* (5.11) *with $p(x,\theta)$ as in* (5.43). *Suppose that there exist Q, R, $\mathbf{s}$ and α such that the following LMIs hold:*

$$\begin{cases} P(Q,R,\mathbf{s}) + L(\alpha) > 0 \\ Q > 0 \\ R \geq 0 \\ \operatorname{trace}(Q_1) = 1. \end{cases} \tag{5.47}$$

Then, $v(x,\theta)$ is a parameter-dependent LF for the origin of (5.1), *and* (5.45) *holds.*

Proof. Suppose that (5.47) holds. This implies that (5.46) holds, and from Corollary (5.3) we conclude the proof. □

Let us consider the selection of the degrees of $q(x,\theta)$, $r(x,\theta)$ and $s(x,\theta)$. A possibility is to choose them in order to maximize the degrees of freedom in (5.47) for fixed degrees of $p(x,\theta)$. This is equivalent to require that the degrees of $q_1(x,\theta)\dot{v}(x,\theta)$, $q_2(x,\theta)(c(\theta)-v(x,\theta))$, $r(x,\theta)'a(\theta)$ and $s(x,\theta)'b(\theta)$, rounded to the smallest following even integers, are equal. This can be achieved as follows:

$$\begin{array}{|l|}\hline
\bullet \text{ choose even degrees for } q_1(x,\theta) \text{ denoted by} \\
\quad 2k_x = \partial^x q_1 \text{ and } 2k_\theta = \partial^\theta q_1; \\
\bullet \text{ set the degrees of } q_2(x,\theta) \text{ as} \\
\quad \partial^x q_2 = 2\left\lceil \dfrac{\partial^x f - 1}{2} \right\rceil + 2k_x \\
\quad \partial^\theta q_2 = 2\left\lceil \dfrac{\max\{\partial^\theta f + \partial^\theta v + 2k_\theta, \partial a, \partial b, \partial c\} - \max\{\partial^\theta v, \partial c\}}{2} \right\rceil; \\
\bullet \text{ define} \\
\quad m_\theta = 2\left\lceil \dfrac{\partial^\theta q_2 + \max\{\partial^\theta v, \partial c\}}{2} \right\rceil; \\
\bullet \text{ set the degrees of } r_i(x,\theta) \text{ as} \\
\quad \partial^x r_i = \partial^x v + \partial^x q_2 \text{ and } \partial^\theta r_i = 2m_\theta - 2\left\lceil \dfrac{\partial a_i}{2} \right\rceil; \\
\bullet \text{ set the degrees of } s_i(x,\theta) \text{ as} \\
\quad \partial^x s_i = \partial^x r_i \text{ and } \partial^\theta s_i = 2m_\theta - \partial b_i; \\
\bullet \text{ hence, the degrees of } p(x,\theta) \text{ are given by} \\
\quad \partial^x p = \partial^x v + \partial^x q_2 \text{ and } \partial^\theta p = 2m_\theta. \\ \hline
\end{array} \tag{5.48}$$

The number of scalar variables in the LMIs (5.47) is as in (5.14). Tables 5.3–5.4 show this number in some cases according to the choice (5.48). Let us observe that (5.48) coincides with (5.13) whenever $\partial^\theta v = \partial c = 0$. In the sequel we will assume that the degrees of the polynomials in $q(x,\theta)$, $r(x,\theta)$ and $s(x,\theta)$ are chosen according to (5.48) unless specified otherwise.

Table 5.3 Number of scalar variables in the LMIs (5.47) for $n=2$, $n_a = n_\theta$, $n_b = 0$, $\partial^\theta v = 1$, $\partial^\theta f = 1$, $\partial a_i \in \{1,2\}$ and $k_x = k_\theta = 0$.

$(n_\theta = 1)$ $\partial^x f \setminus \partial^x v$	2	4	6
2 or 3	37	144	388
4 or 5	156	400	856

$(n_\theta = 2)$ $\partial^x f \setminus \partial^x v$	2	4	6
2 or 3	81	321	864
4 or 5	333	876	1887

Table 5.4 Number of scalar variables in the LMIs (5.47) for $n=3$, $n_a = n_\theta$, $n_b = 0$, $\partial^\theta v = 1$, $\partial^\theta f = 1$, $\partial a_i \in \{1,2\}$ and $k_x = k_\theta = 0$.

$(n_\theta = 1)$ $\partial^x f \setminus \partial^x v$	2	4	6
2 or 3	129	697	2464
4 or 5	736	2503	6844

$(n_\theta = 2)$ $\partial^x f \setminus \partial^x v$	2	4	6
2 or 3	288	1559	5483
4 or 5	1598	5522	15128

5.3.2 Choice of the LF

Let us observe that a necessary condition for (5.47) to hold in the case of parameter-dependent LFs is that

$$\left.\begin{array}{r} V_2(\theta) > 0 \\ V_2(\theta)A(\theta) + A(\theta)'V_2(\theta) < 0 \\ A(\theta) = \mathrm{lp}(f(\cdot,\theta)) \end{array}\right\} \forall \theta \in \Theta \tag{5.49}$$

where $V_2(\theta) = V_2(\theta)'$ defines the quadratic part of $v(x,\theta)$, i.e.

$$v_2(x,\theta) = x'V_2(\theta)x \tag{5.50}$$

and $v(x,\theta)$ is expressed as

$$v(x,\theta) = \sum_{i=2}^{\partial^x v} v_i(x,\theta) \tag{5.51}$$

where $v_i(x,\theta)$ is a homogeneous polynomial of degree i in x for all fixed $\theta \in \mathbb{R}^{n_\theta}$. Condition (5.15) holds if and only if the linearized system is robustly asymptotically stable (i.e., $A(\theta)$ is Hurwitz for all $\theta \in \Theta$) and $v_2(x,\theta)$ is a parameter-dependent LF for it. Hence, the latter is a requirement for (5.47) to hold. This requirement can be removed with a strategy similar to that presented in Section 7.4 for the case of uncertainty-free polynomial systems.

Based on this discussion, a simple way to choose a parameter-dependent LF $v(x,\theta)$ for estimating the RDA is as follows:

• find $V_2(\theta)$ (if any) that satisfies (5.49); • define $v_2(x,\theta) = x'V_2(\theta)x$; • define $v(x,\theta)$ as in (5.51) where $v_3(x,\theta), v_4(x,\theta), \ldots$ are arbitrarily chosen under the constraint that $v(x,\theta)$ is positive definite in x for all $\theta \in \Theta$.	(5.52)

Let us observe that one can establish whether $v(x,\theta)$ is positive definite in x for all $\theta \in \Theta$ with an extension of the technique derived in Section 2.2.4 for establishing positivity of matrix polynomials. The details of this extension will be given in (5.86)–(5.87).

Also, let us observe that one can search for a matrix $V_2(\theta)$ satisfying (5.49) with the technique presented in Section 2.2.4 as explained in the following result.

Theorem 5.8. *Define* $A(\theta)$ *as in* (5.15). *Suppose that there exist* $Q, R, \mathbf{S}, W_2$ *and* α *satisfying the constraints*

$$\begin{cases} P(Q,R,\mathbf{S},W_2)+L(\alpha)>0 \\ Q>0 \\ R\geq 0 \\ W_2>0 \\ \text{trace}(Q)=1 \\ \text{trace}(W_2)=1 \end{cases} \tag{5.53}$$

where

$$\begin{aligned} P(Q,R,\mathbf{S},W_2)+L(\alpha) &= CSMR_{mat}(G) \\ Q &= SMR_{pol}(q) \\ R &= \text{diag}(R_1,\ldots,R_{n_a}) \\ R_i &= SMR_{mat}(H_i)\ \forall i=1,\ldots,n_a \\ \mathbf{S} &= (\mathbf{S}_1,\ldots,\mathbf{S}_{n_b}) \\ \mathbf{S}_i &= COE_{mat}(J_i)\ \forall i=1,\ldots,n_b \\ W_2 &= SMR_{mat}(V_2) \end{aligned} \tag{5.54}$$

and $q\in\mathscr{P}_{n_\theta}$, $H_i=H_i',J_i=J_i'\in\mathscr{P}_{n_\theta}^{n\times n}$, *and*

$$G(\theta)=-q(\theta)\left(V_2(\theta)A(\theta)+A(\theta)'V_2(\theta)\right)-\sum_{i=1}^{n_a}H_i(\theta)a_i(\theta)-\sum_{i=1}^{n_b}J_i(\theta)b_i(\theta). \tag{5.55}$$

Then, the linearized system is robustly asymptotically stable. Moreover, (5.49) *holds, i.e.* $v_2(x,\theta)=x'V_2(\theta)x$ *is a parameter-dependent LF for the linearized system.*

Proof. Suppose that (5.53) holds. This implies that

$$\left.\begin{array}{l} G(\theta)>0 \\ q(\theta)>0 \\ H_i(\theta)\geq 0\ \forall i=1,\ldots,n_a \\ V_2(\theta)>0 \end{array}\right\}\ \forall\theta\in\mathbb{R}^{n_\theta}.$$

Let $\theta\in\Theta$. Since $a(\theta)\geq 0$ and $b(\theta)=0$ it follows that

$$\begin{aligned} 0 &< -q(\theta)\left(V_2(\theta)A(\theta)+A(\theta)'V_2(\theta)\right)-\textstyle\sum_{i=1}^{n_a}H_i(\theta)a_i(\theta)-\sum_{i=1}^{n_b}J_i(\theta)b_i(\theta) \\ &\leq -q(\theta)\left(V_2(\theta)A(\theta)+A(\theta)'V_2(\theta)\right). \end{aligned}$$

Since $q(\theta)>0$, this implies that $V_2(\theta)A(\theta)+A(\theta)'V_2(\theta)<0$. We hence conclude the proof by observing that $V_2(\theta)>0$. □

Similarly to (5.17), the first constraint in (5.53) is a BMI due to the product of Q with W_2. A simple way to transform (5.53) into a set of LMIs consists of fixing Q under the constraints $Q>0$ and $\text{trace}(Q)=1$.

Example 5.8. For the electric arc system (5.6), let us consider the problem of establishing whether $\mathscr{V}(c(\theta),\theta)$ is a subset of the RDA of the origin for all $\theta\in\Theta$, for

$$\begin{aligned} v(x) &= 10^{-6}(0.013+2\theta)x_1^2+10^{-9}(-0.2+\theta)x_1x_2 \\ &\quad +10^{-9}(147.6-0.2\theta)x_2^2+33500x_3^2 \\ c(\theta) &= 0.3. \end{aligned} \tag{5.56}$$

Let us use Corollary 5.4 choosing $k_x = k_\theta = 0$ in (5.48). Since $\partial f^x = 4$ and $\partial f^\theta = 1$, we have

$$\partial^x q_2 = 4, \;\; \partial^\theta q_2 = 0, \;\; \partial^x p = 6, \;\; \partial^\theta p = 2.$$

The LMIs in (5.12) have 742 scalar variables, and we find that these LMIs hold for some Q, R and α. Hence, $\mathscr{V}(c(\theta),\theta) \subseteq \mathscr{R}$ for all $\theta \in \Theta$. □

5.3.3 Parameter-Dependent LEDA

The condition provided in Corollary 5.4 is built for a chosen polynomial $c(\theta)$, which determines the extension of the parameter-dependent sublevel set $\mathscr{V}(c(\theta),\theta)$. Clearly, it would be helpful if one could enlarge $\mathscr{V}(c(\theta),\theta)$ by searching for larger $c(\theta)$ for which the LMIs in (5.47) are feasible. To this end, let us introduce the following definition.

Definition 5.3 (Parameter-Dependent LEDA (continued)). Let $v(x,\theta)$ be a parameter-dependent LF for the origin of (5.1), and define

$$\begin{aligned} \gamma(\theta) = \; & \sup_c \; c \\ & \text{s.t.} \;\; \mathscr{V}(c,\theta) \subseteq \mathscr{Z}(\theta) \end{aligned} \tag{5.57}$$

where $\mathscr{Z}(\theta)$ is given by (3.61). Then, $\mathscr{V}(\gamma(\theta))$ is said the *parameter-dependent LEDA* of the origin provided by the parameter-dependent LF $v(x,\theta)$. □

Let us observe that estimating $\gamma(\theta)$ amounts to searching for a function $c(\theta)$ (for instance, a polynomial with chosen degree) that approximates as close as possible $\gamma(\theta)$. In particular, by requiring that $c(\theta)$ is a lower bound of $\gamma(\theta)$, one should solve

$$\begin{aligned} & \inf_c \int_\Theta \gamma(\theta) - c(\theta) d\theta \\ & \text{s.t.} \; \begin{cases} c(\theta) \leq \gamma(\theta) \;\; \forall \theta \in \Theta \\ \partial c = constant. \end{cases} \end{aligned} \tag{5.58}$$

Unfortunately, this problem is nontrivial since $\gamma(\theta)$ is unknown.

A more tractable problem consists of determining the lower bound $c(\theta)$ of $\gamma(\theta)$ that maximizes a given cost independent on $\gamma(\theta)$. Specifically, we define this lower bound as

$$\gamma_w(\theta) = c^*(\theta) \tag{5.59}$$

where $c^*(\theta)$ is the optimal value of $c(\theta)$ in the optimization problem

$$\begin{aligned} \omega_w = & \sup_c \; w' COE_{pol}(c) \\ & \text{s.t.} \left\{ \begin{array}{c} c(\theta) \leq \gamma(\theta) \;\; \forall \theta \in \Theta \\ \partial c = constant \end{array} \right. \end{aligned} \tag{5.60}$$

and w is a chosen vector that defines the cost.

Let us observe that, depending on w, one can enlarge $c(\theta)$ according to different criteria. For instance, by selecting w such that

$$w' COE_{pol}(c) = c(\theta_0) \tag{5.61}$$

one maximizes $c(\theta_0)$. Also, by selecting w such that

$$w' COE_{pol}(c) = \int_\Theta c(\theta) d\theta \tag{5.62}$$

one maximizes the average of $c(\theta)$ over Θ.

The function $\gamma_w(\theta)$ can be estimated by exploiting SOS polynomials. Specifically, let us define

$$p_c(\theta) = q_c(\theta) c(\theta) - r_c(\theta)' a(\theta) - s_c(\theta)' b(\theta) \tag{5.63}$$

for some $q_c \in \mathscr{P}_{n_\theta}$, $r_c \in \mathscr{P}_{n_\theta}^{n_a}$ and $s_c \in \mathscr{P}_{n_\theta}^{n_b}$.

Corollary 5.5. *Let $v \in \mathscr{P}_{n,n_\theta}$ be such that $v(x,\theta)$ is positive definite in x for all $\theta \in \Theta$. Let us define the optimization problem*

$$\begin{aligned} \hat{\omega}_w = & \sup_{c,q,q_c,r,r_c,s,s_c} \; w' COE_{pol}(c) \\ & s.t. \left\{ \begin{array}{c} \lambda_{quapol}(p) > 0 \\ \lambda_{polpol}(q_1) > 0 \\ \lambda_{quapol}(q_2) > 0 \\ \lambda_{quapol}(r_i) \geq 0 \;\; \forall i = 0, \ldots, n_a \\ \lambda_{pol}(p_c) > 0 \\ \lambda_{pol}(q_c) > 0 \\ \lambda_{pol}(r_{ci}) \geq 0 \;\; \forall i = 0, \ldots, n_a \\ \partial c = constant. \end{array} \right. \end{aligned} \tag{5.64}$$

Then, $\hat{\omega}_w \leq \omega_w$, and

$$\hat{\gamma}_w(\theta) \leq \gamma_w(\theta) \leq \gamma(\theta) \;\; \forall \theta \in \Theta \tag{5.65}$$

where

$$\hat{\gamma}_w(\theta) = c^*(\theta) \tag{5.66}$$

and $c^*(\theta)$ is the optimal value of $c(\theta)$ in (5.64).

Proof. Suppose that the inequalities in (5.64) hold. Then, the first three inequalities coincide with those in (5.47), and hence (5.45) holds. This implies that (5.65) holds and $\hat{\omega}_w \leq \omega_w$. Lastly, let us observe that the last three inequalities in (5.64) imply that $c(\theta) > 0$ for all $\theta \in \Theta$. □

The optimization problem (5.64) is in general nonconvex since the first inequality is a BMI due to the product of $q_2(\theta)$ and $c(\theta)$, and the second inequality is a BMI due to the product of $q_c(\theta)$ and $c(\theta)$. One way to cope of this problem is to fix $q_2(\theta)$ and $q_c(\theta)$, as the optimization problem boils down to an SDP.

Let us observe that, if one chooses $\partial c = 0$, then $\hat{\gamma}_w(\theta)$ can be computed with a GEVP similar to (5.4) introduced for the case of common LFs.

Example 5.9. For the mass-spring system (5.5), let us consider the problem of determining $\hat{\gamma}_w(\theta)$ with

$$v(x) = 2\theta x_1^2 + (1.5 - 0.5\theta)x_1 x_2 + 2x_2^2.$$

Let us use Corollary 5.5 choosing $k_x = k_\theta = 2$ in (5.48). We solve (5.64) for the cases $\partial c = 0, 1$ choosing w according to (5.62), which for the present case turns out to be

$$w = \begin{cases} 1 & \text{with } \partial c = 0 \\ (1, 1.5)' & \text{with } \partial c = 1. \end{cases}$$

We obtain

$$\hat{\gamma}_w(\theta) = \begin{cases} 5.590 & \text{with } \partial c = 0 \\ 5.487 + 0.104\theta & \text{with } \partial c = 1. \end{cases}$$

The SDP (5.64) has 470 scalar variables for $\partial c = 1$. Figure 5.5a shows the estimate $\mathcal{V}(\hat{\gamma}_w(\theta))$ for $\partial c = 1$ and the curve $\dot{v}(x, \theta) = 0$ for some values of θ. Figure 5.5b shows $\hat{\gamma}_w(\theta)$ for $\partial c = 1$ and $\gamma(\theta)$ calculated for some values of θ. □

Example 5.10. For the electric arc system (5.6), let us consider the problem of determining $\hat{\gamma}_w(\theta)$ with $v(x, \theta)$ as in (5.56). Let us choose $\partial c = 0$. In this case, $\hat{\gamma}_w(\theta)$ can be computed with a GEVP similar to (5.4), and we obtain

$$\hat{\gamma}_w(\theta) = 0.577.$$

This GEVP has 743 scalar variables. Figure 5.6a shows the estimate $\mathcal{V}(\hat{\gamma}_w(\theta))$ and the surface $\dot{v}(x, \theta) = 0$ for some values of θ. Figure 5.6b shows $\hat{\gamma}_w(\theta)$ and $\gamma(\theta)$ calculated for some values of θ. □

Example 5.11. For the system (5.36), let us consider the problem of determining $\hat{\gamma}_w(\theta)$ with

$$v(x) = x_1^2 + 2\theta x_1 x_2 + (1 + 4\theta)x_2^2.$$

Let us use Corollary 5.5. We solve (5.64) for the cases $\partial c = 0, 1, 2$ choosing w according to (5.62), which for the present case turns out to be

$$w = \begin{cases} 1 & \text{with } \partial c = 0 \\ (1, 0.5)' & \text{with } \partial c = 1 \\ (1, 0.5, 0.333)' & \text{with } \partial c = 2. \end{cases}$$

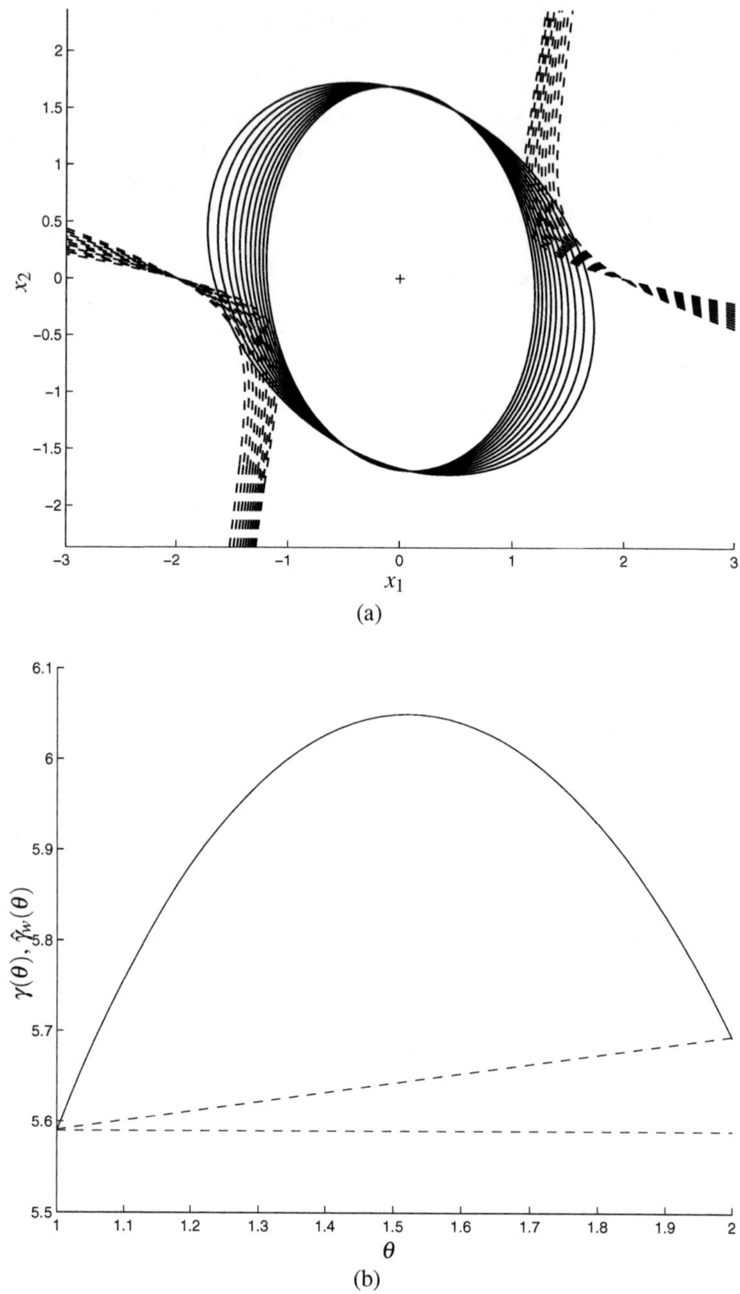

Fig. 5.5 Example 5.9 (mass-spring system). (a) Boundary of $\mathscr{V}(\hat{\gamma}_w(\theta), \theta)$ for $\partial c = 1$ (solid) and curve $\dot{v}(x, \theta) = 0$ for some values of θ (dashed). The "+" denotes the common equilibrium point. (b) Lower bound $\gamma(\theta)$ calculated for some values of θ (solid) and lower bounds $\hat{\gamma}_w(\theta)$ for $\partial c = 0, 1$ (dashed).

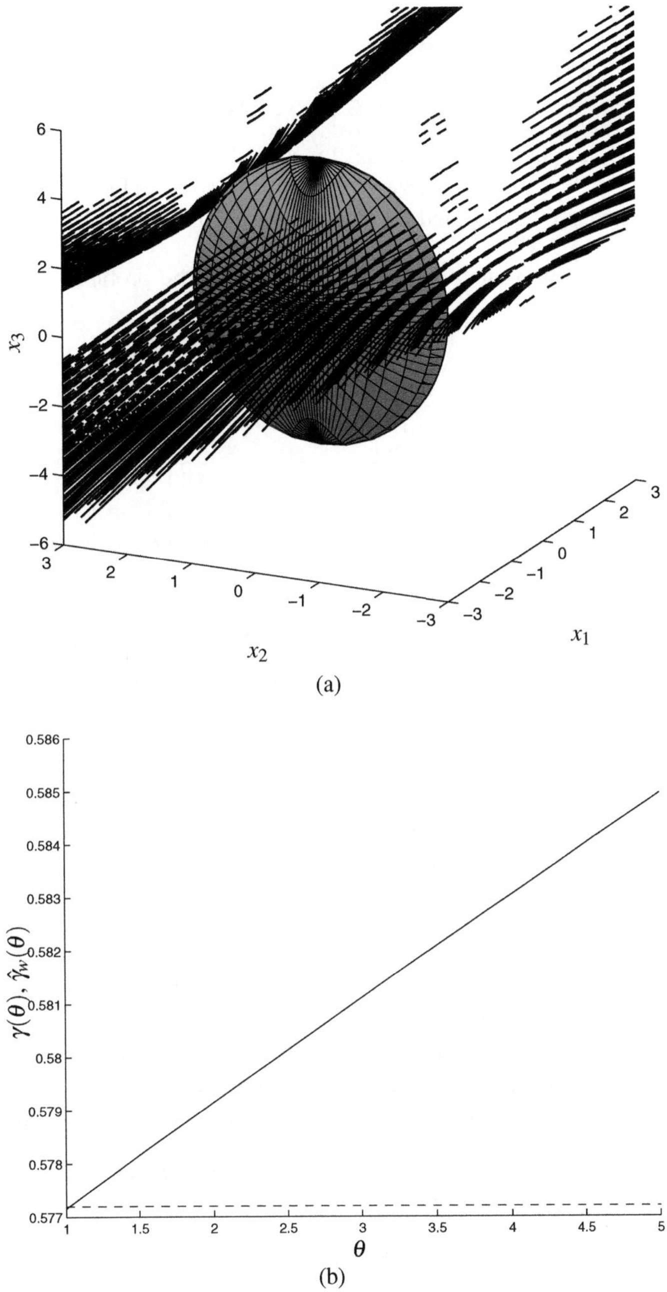

Fig. 5.6 Example 5.10 (electric arc system). (a) Boundary of $\mathscr{V}(\hat{\gamma}_w(\theta), \theta)$ (solid) and surface $\dot{v}(x, \theta) = 0$ for some values of θ (dashed). (b) Lower bound $\gamma(\theta)$ calculated for some values of θ (solid) and lower bound $\hat{\gamma}_w(\theta)$ (dashed).

We obtain

$$\hat{\gamma}_w(\theta) = \begin{cases} 0.552 & \text{with } \partial c = 0 \\ 0.394 + 0.345\theta & \text{with } \partial c = 1 \\ 0.918 - 1.825\theta + 2.224\theta & \text{with } \partial c = 2. \end{cases}$$

The SDP (5.64) has 957 scalar variables for $\partial c = 2$. Figure 5.7a shows the estimate $\mathscr{V}(\hat{\gamma}_w(\theta))$ for $\partial c = 2$ and the curve $\dot{v}(x,\theta) = 0$ for some values of θ. Figure 5.7b shows $\hat{\gamma}_w(\theta)$ and $\gamma(\theta)$ calculated for some values of θ. □

Example 5.12. For the system (5.37), let us consider the problem of determining $\hat{\gamma}_w(\theta)$ with

$$v(x) = (3 - \theta_1)x_1^2 + x_1x_2 + (2 + \theta_2)x_2^2.$$

Let us use Corollary 5.5. We solve (5.64) for the cases $\partial c = 0,1,2$ choosing w according to (5.62), which for the present case turns out to be

$$w = \begin{cases} 3.142 & \text{with } \partial c = 0 \\ (3.142, 0, 0)' & \text{with } \partial c = 1 \\ (3.142, 0, 0, 0.785, 0, 0.785)' & \text{with } \partial c = 2. \end{cases}$$

We obtain

$$\hat{\gamma}_w(\theta) = \begin{cases} 0.462 & \text{with } \partial c = 0 \\ 1.182 - 1.130\theta_1 + 0.347\theta_2 & \text{with } \partial c = 1 \\ 1.752 - 2.927\theta_1 + 0.407\theta_2 + 1.622\theta_1^2 & \\ \quad - 0.705\theta_1\theta_2 - 0.852\theta_2^2 & \text{with } \partial c = 2. \end{cases}$$

The SDP (5.64) has 1593 scalar variables for $\partial c = 2$. Figure 5.8a shows the estimate $\mathscr{V}(\hat{\gamma}_w(\theta))$ for $\partial c = 2$ and the curve $\dot{v}(x,\theta) = 0$ for some values of θ. Figure 5.8b shows $\hat{\gamma}_w(\theta)$ for $\partial c = 2$ and $\gamma(\theta)$ calculated for some values of θ. □

5.3.4 LERDA

In the previous subsections we have described a technique to compute the parameter-dependent LEDA. Let us observe that such an estimate, $\mathscr{V}(c(\theta),\theta)$, depends on θ, and hence one may need to consider the common region of $\mathscr{V}(c(\theta),\theta)$ as θ varies over Θ if θ is not measurable. Such a common region is defined as follows.

Definition 5.4 (LERDA (continued)). Let $v(x,\theta)$ be a parameter-dependent LF for the origin of (5.1), and define

$$\mathscr{I} = \bigcap_{\theta \in \Theta} \mathscr{V}(\gamma(\theta),\theta) \tag{5.67}$$

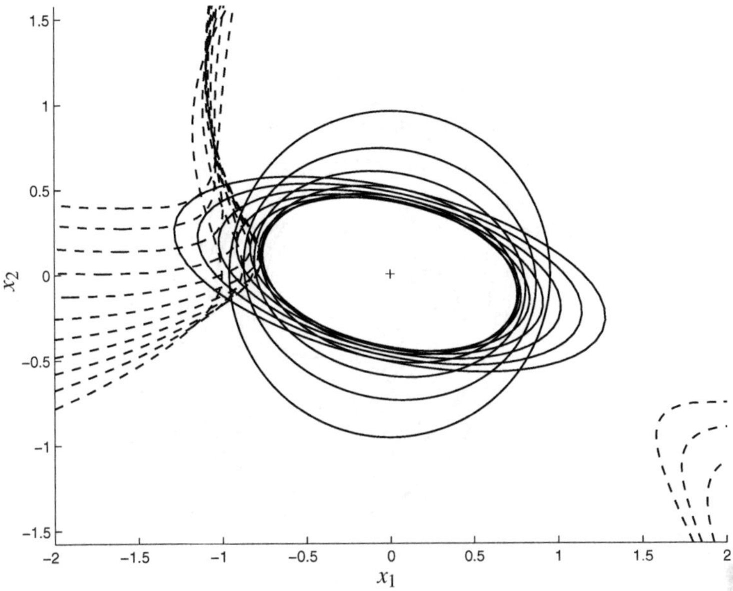

(a)

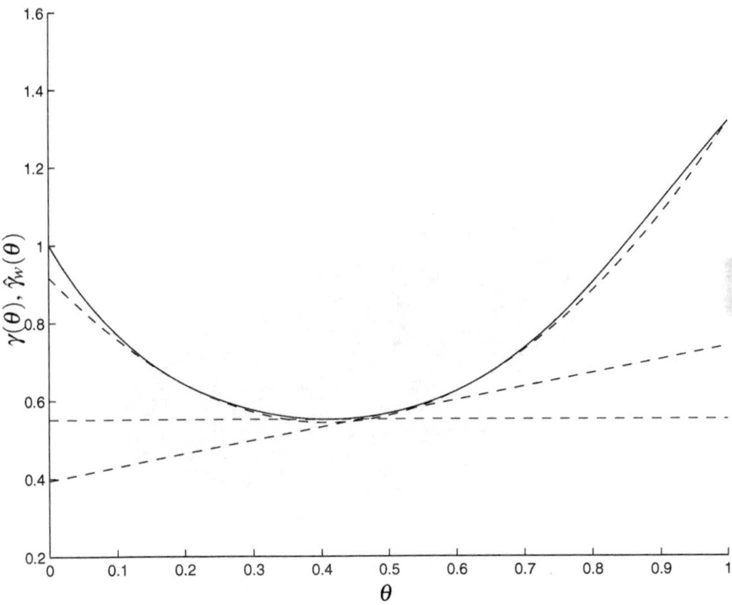

(b)

Fig. 5.7 Example 5.11. (a) Boundary of $\mathscr{V}(\hat{\gamma}_w(\theta), \theta)$ for $\partial c = 2$ (solid) and curve $\dot{v}(x, \theta) = 0$ for some values of θ (dashed). The "+" denotes the common equilibrium point. (b) Lower bound $\gamma(\theta)$ calculated for some values of θ (solid) and lower bounds $\hat{\gamma}_w(\theta)$ for $\partial c = 0, 1, 2$ (dashed).

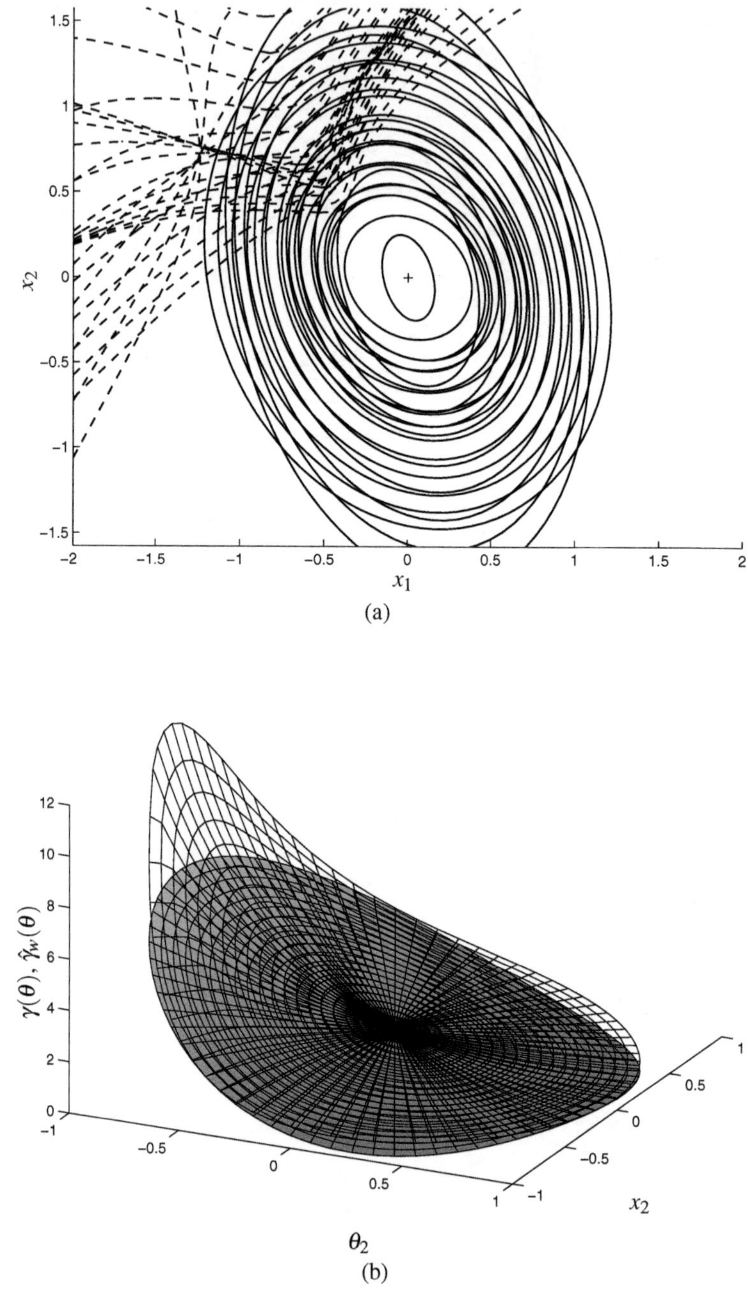

Fig. 5.8 Example 5.12. (a) Boundary of $\mathscr{V}(\hat{\gamma}_w(\theta), \theta)$ for $\partial c = 2$ (solid) and curve $\dot{v}(x, \theta) = 0$ for some values of θ (dashed). The "+" denotes the common equilibrium point. (b) Lower bound $\gamma(\theta)$ calculated for some values of θ (transparent surface) and lower bound $\hat{\gamma}_w(\theta)$ for $\partial c = 2$ (grey surface).

where $\gamma(\theta)$ given by (5.20). Then, $\mathscr{I}$ is said the *LERDA* of the origin provided by the parameter-dependent LF $v(x,\theta)$. □

Here we consider the problem of estimating $\mathscr{I}$. Specifically, we look for estimates of $\mathscr{I}$ of the form

$$\mathscr{W}(d) = \{x \in \mathbb{R}^n : \; w(x) \leq d\} \tag{5.68}$$

where $w \in \mathscr{P}_n$ is positive definite, which can be either fixed or variable.

First, let us consider the case where $w(x)$ is fixed. The problem amounts to determining the largest d for which $\mathscr{W}(d)$ is included in $\mathscr{I}$, i.e.

$$\begin{aligned} \delta = & \sup_{d} \; d \\ & \text{s.t. } \; \mathscr{W}(d) \subseteq \mathscr{V}(c(\theta),\theta) \;\; \forall \theta \in \Theta. \end{aligned} \tag{5.69}$$

To this end, let us define the polynomial

$$p_w(x,\theta) = -q_w(x,\theta)' \begin{pmatrix} v(x,\theta) - c(\theta) \\ d - w(x) \end{pmatrix} - r_w(x,\theta)' a(\theta) - s_w(x,\theta)' b(\theta) \tag{5.70}$$

where $q_w \in \mathscr{P}^2_{n,n_\theta}$, $r_w \in \mathscr{P}^{n_a}_{n,n_\theta}$ and $s_w \in \mathscr{P}^{n_b}_{n,n_\theta}$. Similarly to the previous results in this chapter, a lower bound of δ can be found through the SOS index as

$$\begin{aligned} \hat{\delta} = & \sup_{q_w, r_w, s_w, d} \; d \\ & \text{s.t. } \begin{cases} \lambda_{polpol}(p_w) > 0 \\ \lambda_{polpol}(q_{wi}) > 0 \;\; \forall i = 1,2 \\ \lambda_{polpol}(r_{wi}) \geq 0 \;\; \forall i = 1,\ldots,n_a. \end{cases} \end{aligned} \tag{5.71}$$

Computing $\hat{\delta}$ amounts to solving an optimization problem with matrix inequality constraints. Specifically, let us define

$$\begin{aligned} P_w(d, Q_w, R_w, \mathbf{s}_w) + L_w(\alpha_w) &= CSMR_{polpol}(p_w) \\ Q_w &= \operatorname{diag}(Q_{w1}, Q_{w2}) \\ Q_{wi} &= SMR_{polpol}(q_{wi}) \;\; \forall i = 1,2 \\ R_w &= \operatorname{diag}(R_{w1}, \ldots, R_{wn_a}) \\ R_{wi} &= SMR_{polpol}(r_{wi}) \;\; \forall i = 1,\ldots,n_a \\ \mathbf{s}_w &= (\mathbf{s}'_{w1}, \ldots, \mathbf{s}'_{wn_b})' \\ \mathbf{s}_{wi} &= COE_{polpol}(s_{wi}) \;\; \forall i = 1,\ldots,n_b. \end{aligned} \tag{5.72}$$

We have that

$$\begin{aligned} \hat{\delta} = & \sup_{d, Q_w, R_w, \mathbf{s}_w, \alpha_w} \; d \\ & \text{s.t. } \begin{cases} P_w(d, Q_w, R_w, \mathbf{s}_w) + L_w(\alpha_w) > 0 \\ Q_w > 0 \\ R_w \geq 0 \\ \operatorname{trace}(Q_{w1}) = 1. \end{cases} \end{aligned} \tag{5.73}$$

A way to compute $\hat{\delta}$ is to fix Q_{w2} and solve (5.73) that in such a case boils down to an SDP. However this solution may require to increase the degree of Q_{w1}, and consequently the degree of $p_w(x,\theta)$, which may significantly increase the computational burden.

Another way to compute $\hat{\delta}$ where Q_{w2} does not need to be fixed is via a one-parameter sweep on the scalar d in Corollary 5.2, which can be conducted for instance via a bisection algorithm.

A further way to obtain $\hat{\delta}$ where Q_{w2} does not need to be fixed is via a GEVP. Indeed, let us define the polynomial

$$p_{w1}(x,\theta) = -q_w(x,\theta)' \begin{pmatrix} v(x,\theta) - c(\theta) \\ -w(x) \end{pmatrix} - r_w(x,\theta)'a(\theta) - s_w(x,\theta)'b(\theta) \quad (5.74)$$

and let us introduce its complete SMR matrix as

$$P_{w1}(Q_w, R_w, \mathbf{s}_w) + L_w(\alpha_w) = CSMR_{polpol}(p_{w1}). \quad (5.75)$$

Define

$$p_{w2}(x,\theta) = q_{w2}(x,\theta)(1+\mu w(x)) \quad (5.76)$$

and the SMR matrix

$$W = SMR_{qua}(w). \quad (5.77)$$

Similarly to Theorem 5.3, let us define

$$P_{w2}(Q_{w2}) = T_3' \left(\begin{pmatrix} 1 & 0 \\ 0 & \mu W \end{pmatrix} \otimes Q_{w2} \right) T_3 \quad (5.78)$$

where T_3 is defined by

$$\begin{aligned} &\begin{pmatrix} 1 \\ b_{lin}(x, \partial w/2) \end{pmatrix} \otimes b_{pol}(x, \partial^x q_{w2}/2) \otimes b_{pol}(\theta, \partial^\theta q_{w2}/2) \\ &= T_3 b_{pol}(x, \partial^x p_{w2}/2) \otimes b_{pol}(\theta, \partial^\theta p_{w2}/2). \end{aligned} \quad (5.79)$$

Assume that $w(x)$ can be expressed as in (5.77) with $W > 0$ and that $\partial^x q_{w2} = \partial^x p_{w2}$ and $\partial^\theta q_{w2} = \partial^\theta p_{w2}$. Then, similarly to Theorem (5.4), $\hat{\delta}$ is given by

$$\hat{\delta} = -\frac{z^*}{1+\mu z^*} \quad (5.80)$$

where z^* is the solution of the GEVP

$$z^* = \inf_{Q_w, R_w, \mathbf{s}_w, z, \alpha_w} z$$
$$\text{s.t.} \begin{cases} zP_{w2}(Q_{w2}) + P_{w1}(Q_w, R_w, \mathbf{s}_w) + L_w(\alpha_w) > 0 \\ Q_w > 0 \\ R_w \geq 0 \\ 1+\mu a > 0 \\ \text{trace}(Q_{w1}) = 1. \end{cases} \quad (5.81)$$

Next, let us consider the case of variable $w(x)$. The problem amounts to determining the polynomial $w(x)$ that provides the least conservative estimate of $\mathscr{I}$. The conservatism can be measured in various ways analogously to what has been done in Section 4.4, for instance one can search for the estimate with the largest volume:

$$\begin{aligned} \sup_{w} \ & \mathrm{vol}(\mathscr{W}(1)) \\ \text{s.t.} \ & \mathscr{W}(1) \subseteq \mathscr{V}(c(\theta),\theta) \ \forall \theta \in \Theta. \end{aligned} \tag{5.82}$$

One way to address this problem consists of parametrizing the candidate function $w(x)$ and, for any fixed $w(x)$, solve the GEVP (5.81), evaluate the volume of $\mathscr{W}(\hat{\delta})$, and update $w(x)$ through a gradient-like method.

Another way consists of solving a sequence of LMI problems obtained by alternatively freezing some of the variables in the BMIs produced by letting $w(x)$ vary where the objective consists of minimizing the trace of the SMR matrix of $w(x)$. Specifically, by fixing Q_{w2} with $Q_{w2} > 0$, the problem amounts to solving the SDP

$$\begin{aligned} \inf_{Q_{w1},R_w,\mathbf{s}_w,W,\alpha_w} \ & \mathrm{trace}(W) \\ \text{s.t.} \ & \begin{cases} P_w(Q_w,R_w,\mathbf{s}_w,W)+L_w(\alpha_w) > 0 \\ Q_{w1} > 0 \\ R_w \geq 0 \\ W > 0 \end{cases} \end{aligned} \tag{5.83}$$

where $P_w(Q_w,R_w,\mathbf{s}_w,W)+L_w(\alpha_w)$ is the SMR matrix of $p_w(x,\theta)$ with $d=1$ and $w(x)$ related to W by (5.77). Observe that since Q_{w2} is fixed, the constraint $1 = \mathrm{trace}(Q_{w1})$ must not be included.

Example 5.13. For the mass-spring system (5.5), let us consider the problem of estimating the LERDA $\mathscr{I}$ for the parameter-dependent LEDA found in Example 5.9 with $\partial c = 1$. First, we use the fixed polynomial $w(x) = x_1^2 + x_2^2$ for generating the estimate $\mathscr{W}(d)$. By using the GEVP (5.81) with $\partial^x q_{w1} = \partial^\theta q_{w1} = 0$ (and the degrees of the other polynomials consequently selected similarly to (5.48)) we find

$$\hat{\delta} = 1.413.$$

Second, we estimate $\mathscr{I}$ with a variable polynomial $w(x)$ of degree 2, in particular solving (5.83) with $\partial^x q_{w1} = \partial^\theta q_{w1} = 2$ and $Q_{w2} = I$. We find the estimate $\mathscr{W}(1)$ with

$$w(x) = 0.703x_1^2 + 0.091x_1x_2 + 0.363x_2^2.$$

Figure 5.9 shows the found estimates and the boundary of $\mathscr{V}(\hat{\gamma}_w(\theta),\theta)$ for $\partial c = 1$. As we can see, the found estimates of the LERDA are tangent to this boundary, hence implying that the estimates are tight. Let us also observe that the boundary of $\mathscr{W}(1)$ found with variable $w(x)$ coincides with the LERDA $\mathscr{I}$ since the dashed ellipse overlaps with $\mathscr{V}(\hat{\gamma}_w(\theta),\theta)$. □

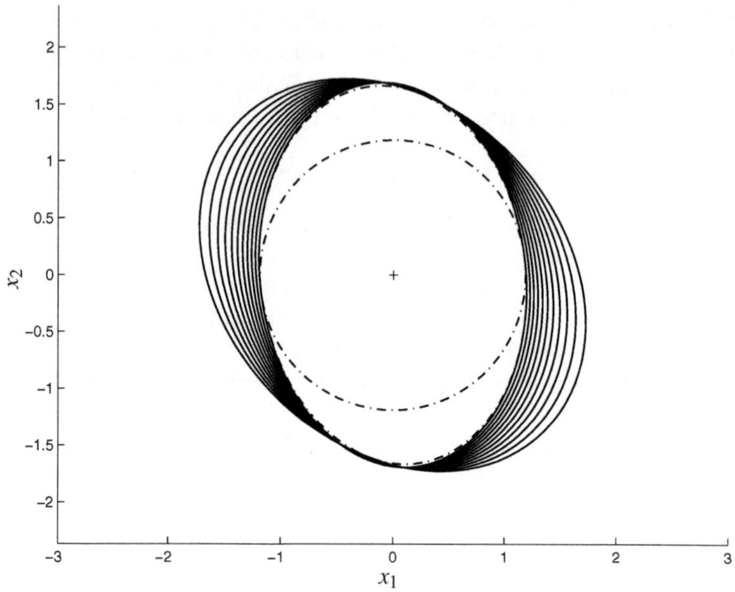

(a)

Fig. 5.9 Example 5.13 (mass-spring system). Boundary of $\mathscr{V}(\hat{\gamma}_w(\theta),\theta)$ for $\partial c = 1$ for some values of θ (solid), boundary of $\mathscr{W}(\hat{\delta})$ found with fixed $w(x)$ (dashed circle) and boundary of $\mathscr{W}(1)$ found with variable $w(x)$ (dashed ellipse, coincident with some solid lines). The "+" denotes the common equilibrium point.

Example 5.14. For the electric arc system (5.6), let us consider the problem of estimating $\mathscr{I}$ for the parameter-dependent LEDA found in Example 5.10. First, we use the fixed polynomial $w(x) = 10^3(x_1^2 + x_2^2 + x_3^2)$ for generating the estimate $\mathscr{W}(d)$. By using the GEVP (5.81) we find

$$\hat{\delta} = 0.392.$$

Second, we estimate $\mathscr{I}$ with a variable polynomial $w(x)$ of degree 2 similarly to Example 5.13. We find the estimate $\mathscr{W}(1)$ with

$$w(x) = 10^3(1.781x_1^2 + 0.007x_1x_2 + 4.537x_2^2 + 1.004x_3^2).$$

Figure 5.10 shows the found estimate in the latter case. □

Example 5.15. For the system (5.36), let us consider the problem of estimating $\mathscr{I}$ for the parameter-dependent LEDA found in Example 5.11 with $\partial c = 2$. We proceed similarly to Example 5.13, and we find

$$\hat{\delta} = 0.191$$

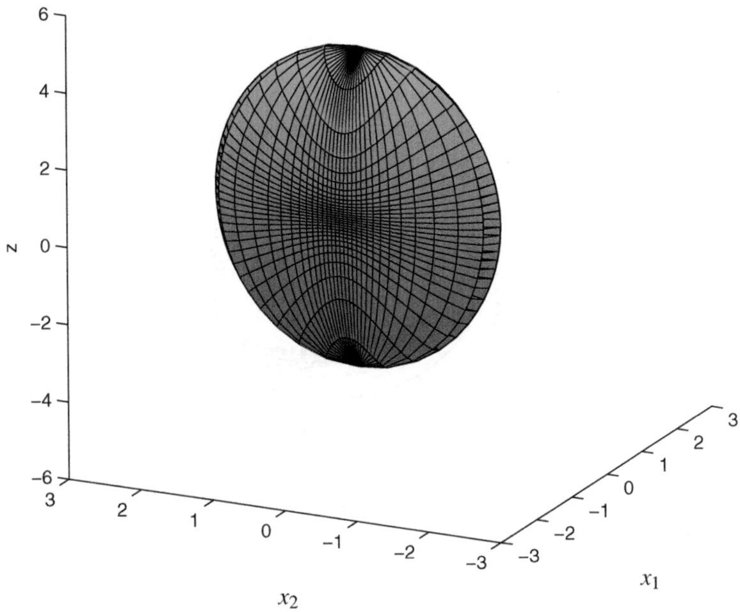

(a)

Fig. 5.10 Example 5.14 (electric arc system). Boundary of $\mathscr{W}(1)$ found with variable $w(x)$.

for the case of fixed $w(x)$, and

$$w(x) = 1.882x_1^2 + 1.799x_1x_2 + 5.480x_2^2$$

for the case of variable $w(x)$. Figure 5.11 shows the found estimates. Also in this case the found estimates of the LERDA are tangent to the parameter-dependent LEDA, hence implying that the estimates are tight. Let us also observe that the boundary of $\mathscr{W}(1)$ found with variable $w(x)$ coincides with the LERDA $\mathscr{I}$ since the dashed ellipse overlaps with $\mathscr{V}(\hat{\gamma}_w(\theta), \theta)$. □

Example 5.16. For the system (5.37), let us consider the problem of estimating $\mathscr{I}$ for the parameter-dependent LEDA found in Example 5.12 with $\partial c = 0$. We proceed similarly to Example 5.13, and we find

$$\hat{\delta} = 0.112$$

for the case of fixed $w(x)$, and

$$w(x) = 8.658x_1^2 + 2.165x_1x_2 + 6.494x_2^2$$

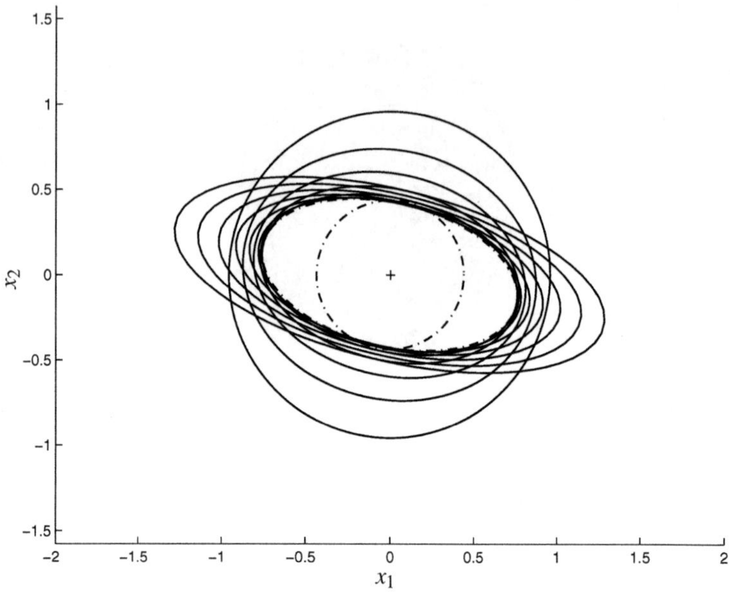

(a)

Fig. 5.11 Example 5.15. Boundary of $\mathscr{V}(\hat{\gamma}_w(\theta),\theta)$ for $\partial c = 2$ for some values of θ (solid), boundary of $\mathscr{W}(\hat{\delta})$ found with fixed $w(x)$ (dashed circle) and boundary of $\mathscr{W}(1)$ found with variable $w(x)$ (dashed ellipse, coincident with some solid lines). The "+" denotes the common equilibrium point.

for the case of variable $w(x)$. Figure 5.12 shows the found estimates. Also in this case the found estimates of the LERDA are tangent to the parameter-dependent LEDA, hence implying that the estimates are tight.

Let us observe that $\mathscr{I} = 0_2$ with $\partial c = 2$. In fact, the found $\hat{\gamma}_w(\theta)$ with $\partial c = 2$ vanishes for some $\theta \in \Theta$, and hence the LERDA boils down to the common equilibrium point. □

5.4 Optimal Estimates

In the previous sections we have considered the estimation of the RDA through fixed LFs, either common or parameter-dependent. Here we consider the case of variable LFs, in order to obtain estimates that are optimal according to some specified criteria. As in the case of uncertainty-free polynomial systems treated in Chapter 4, we consider the following criteria: first, maximizing the volume of the estimate, and second, enlarging the estimate with fixed shape sets.

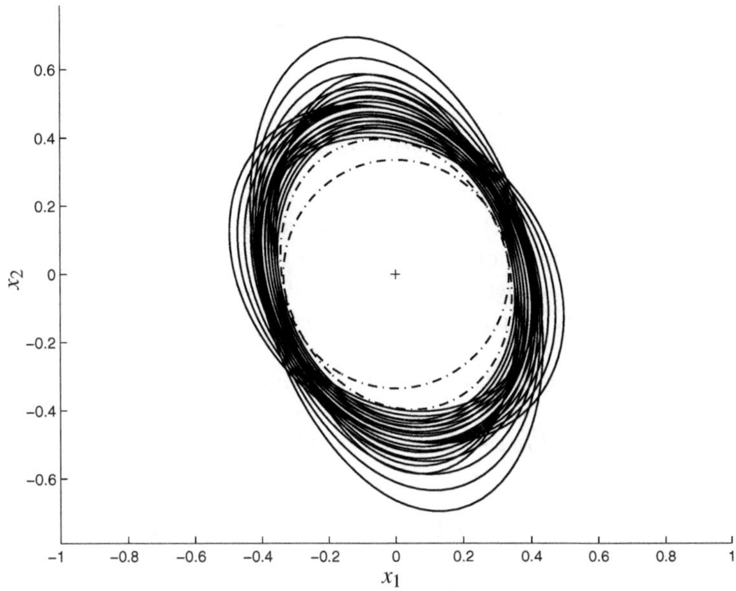

(a)

Fig. 5.12 Example 5.16. Boundary of $\mathscr{V}(\hat{\gamma}_w(\theta), \theta)$ for $\partial c = 0$ for some values of θ (solid), boundary of $\mathscr{W}(\hat{\delta})$ found with fixed $w(x)$ (dashed circle) and boundary of $\mathscr{W}(1)$ found with variable $w(x)$ (dashed ellipse). The "+" denotes the common equilibrium point.

5.4.1 *Maximizing the Volume of the Estimate*

Let us consider the first criterion. In the case of common LFs, this amounts to maximizing the volume of the LERDA in Definition 5.2, i.e.

$$\begin{array}{ll} \sup\limits_{v} & \mathrm{vol}(\mathscr{V}(\gamma^{\#})) \\ \text{s.t.} & \begin{cases} v(x) > 0 \ \ \forall x \in \mathbb{R}_0^n \\ v \in \mathscr{P}_n \end{cases} \end{array} \tag{5.84}$$

where $\gamma^{\#}$ defines the LERDA provided by $v(x)$ according to (5.21). In the case of parameter-dependent LFs, a natural extension of this criterion is to maximize the volume of the LERDA in Definition 5.4, i.e.

$$\begin{array}{ll} \sup\limits_{v} & \mathrm{vol}(\mathscr{I}) \\ \text{s.t.} & \begin{cases} v(x,\theta) > 0 \ \ \forall x \in \mathbb{R}_0^n \ \forall \theta \in \Theta \\ v \in \mathscr{P}_{n,n_\theta} \end{cases} \end{array} \tag{5.85}$$

where $\mathscr{I}$ given in (5.67) is the common region of $\mathscr{V}(c(\theta), \theta)$ for $\theta \in \Theta$. One way to address these problems is:

1. parametrizing the LF, either common or parameter-dependent;
2. for any fixed LF, estimating either $\gamma^\#$ or $\mathscr{I}$ as described in the previous sections of this chapter;
3. evaluating the volume of the found estimate, i.e. either $\mathrm{vol}(\mathscr{V}(\gamma^\#))$ or $\mathrm{vol}(\mathscr{I})$;
4. updating the LF through a gradient-like method.

Another way consists of alternatively freezing some of the variables in the BMIs obtained by letting either $v(x)$ or $v(x,\theta)$ vary, hence solving a sequence of LMI problems. The case of common LF is similar to what has been derived in Section 4.4.1. For the case of parameter-dependent LF, the positive definiteness of $v(x,\theta)$ for $\theta \in \Theta$ is imposed by introducing the polynomial

$$p_v(x,\theta) = q_v(x,\theta)v(x,\theta) - r_v(x,\theta)'a(\theta) - s_v(x,\theta)'b(\theta) \tag{5.86}$$

where $q_v \in \mathscr{P}_{n,n_\theta}$, $r_v \in \mathscr{P}_{n,n_\theta}^{n_a}$ and $s_v \in \mathscr{P}_{n,n_\theta}^{n_b}$. Indeed, $v(x,\theta)$ for $\theta \in \Theta$ is positive definite for $\theta \in \Theta$ if

$$\begin{cases} \lambda_{quapol}(p_v) > 0 \\ \lambda_{polpol}(q_v) > 0 \\ \lambda_{quapol}(r_{vi}) \geq 0 \ \forall i = 1,\ldots,n_a. \end{cases} \tag{5.87}$$

5.4.2 Enlarging the Estimate with Fixed Shape Sets

Let us consider now the second criterion, which consists of enlarging the estimate with fixed shape sets. Specifically, let $w \in \mathscr{P}_n$ be positive definite, and let $\mathscr{W}(d)$ be the fixed shape set used to enlarge the estimate of the RDA where $d \in \mathbb{R}$. In the case of common LFs this strategy amounts to solving

$$\begin{aligned} &\sup_{c,d,v} d \\ &\text{s.t.} \begin{cases} \mathscr{V}(c) \subseteq \mathscr{X}(\theta) \ \forall \theta \in \Theta \\ \mathscr{W}(d) \subseteq \mathscr{V}(c) \\ v(x) > 0 \ \forall x \in \mathbb{R}_0^n \\ v \in \mathscr{P}_n \end{cases} \end{aligned} \tag{5.88}$$

while in the case of parameter-dependent LFs this strategy can be formulated as

$$\begin{aligned} &\sup_{c,d,v} d \\ &\text{s.t.} \begin{cases} \mathscr{V}(c(\theta),\theta) \subseteq \mathscr{X}(\theta) \ \forall \theta \in \Theta \\ \mathscr{W}(d) \subseteq \mathscr{V}(c,\theta) \ \forall \theta \in \Theta \\ v(x,\theta) > 0 \ \forall x \in \mathbb{R}_0^n \ \forall \theta \in \Theta \\ v \in \mathscr{P}_{n,n_\theta}. \end{cases} \end{aligned} \tag{5.89}$$

Clearly, in (5.88)–(5.89) $\mathscr{W}(d)$ represents the LERDA provided by variable LF $v(x)$ or $v(x,\theta)$ and by the chosen $w(x)$.

One way to address these problems is:

1. finding the largest d for a fixed LF using the techniques presented in this chapter;
2. update the LF through a gradient-like method.

Another way consists of alternatively freezing some of the variables in the BMIs obtained by letting either $v(x)$ or $v(x,\theta)$ vary, hence solving a sequence of LMI problems.

Example 5.17. Let us consider problem (5.89) with the choice $w(x) = \|x\|^2$ in order to obtain optimal estimates of the RDA for the mass-spring system (5.5). We address this problem by alternatively freezing some of the variables in the obtained BMIs in the case of parameter-dependent polynomial LFs $v(x,\theta)$ with $\partial^x v = 2,4$ and $\partial^\theta v = 1$. The found LFs, that we denote with $v_2(x,\theta)$ and $v_4(x,\theta)$, are obtained by using the following initializations:

1. in the case of $v_2(x,\theta)$, the LF used in Example 5.9;
2. in the case of $v_4(x,\theta)$, the LF $v_2(x,\theta)$.

Table 5.5 and Figure 5.13 show the obtained results.

Table 5.5 Example 5.17 (mass-spring system). Estimates found with variable parameter-dependent LFs by enlarging fixed shape sets.

LF	Cost d	No. of iterations for solving (5.89)
$v_2(x)$	2.191	11
$v_4(x)$	2.708	4

□

5.4.3 Establishing Robust Global Asymptotical Stability

The strategies derived in Sections 4.4.1–4.4.2 for estimating the RDA by using variable LFs involve nonconvex optimization problems. This happens due to the product between the multiplier $q(x,\theta)$ and either the common LF $v(x)$ or the parameter-dependent LF $v(x,\theta)$ in $p(x,\theta)$. However, if the candidate estimate of the RDA is the whole space (i.e., $\mathbb{R}^n$), the strategies previously presented boil down to convex optimization problems in the form of an LMI feasibility test analogously to the case of uncertainty-free polynomial systems treated in Section 4.4.3.

Specifically, in such a case the strategy amounts to finding either a global common LF or a global parameter-dependent LF proving robust global asymptotical stability of the origin according to Theorems 3.8 or 3.10. Such LFs can be searched for through SOS polynomials similarly to what has been done in Sections 4.4.1–4.4.2. The following result provides the detail for the general case of parameter-dependent LF.

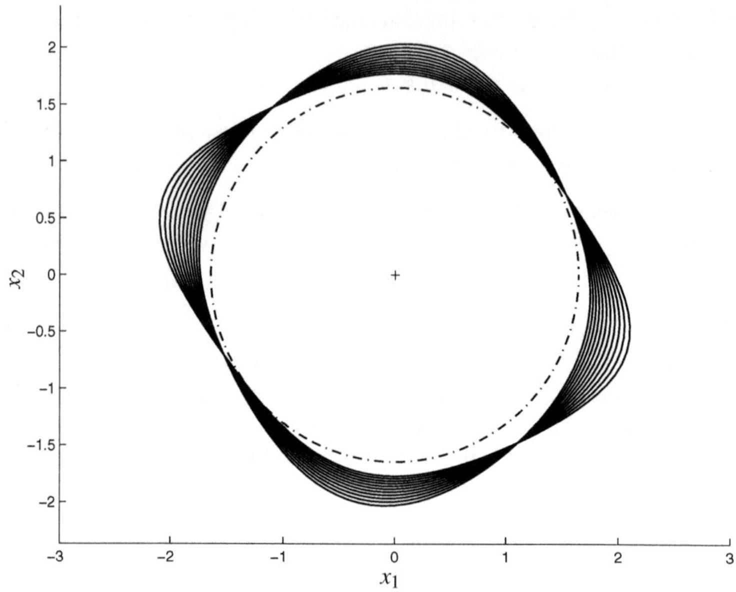

Fig. 5.13 Example 5.17 (mass-spring system). Boundary of $\mathscr{V}(c(\theta),\theta)$ corresponding to the found $v_4(x,\theta)$ for some values of θ (solid). The "+" denotes the common equilibrium point. The dashdot line shows the boundary of the LERDA $\mathscr{W}(d)$.

Corollary 5.6. *Suppose there exists* $r_g, r_v \in \mathscr{P}^{n_a}_{n,n_\theta}$, $s_g, s_v \in \mathscr{P}^{n_b}_{n,n_\theta}$ *and* $v \in \mathscr{P}_{n,n_\theta}$ *such that*

$$\begin{cases} \lambda_{quapol}(p_g) > 0 \\ \lambda_{quapol}(r_{gi}) \geq 0 \;\; \forall i = 0,\ldots,n_a \\ \lambda_{quapol}(p_v) > 0 \\ \lambda_{quapol}(r_{vi}) \geq 0 \;\; \forall i = 1,\ldots,n_a \end{cases} \tag{5.90}$$

where

$$p_g(x,\theta) = -\dot{v}(x,\theta) - r_g(x,\theta)'a(\theta) - s_g(x,\theta)'b(\theta) \tag{5.91}$$

and $p_v(x,\theta)$ *is as* (5.86) *for* $q_v(x,\theta) = 1$. *Then,* $v(x)$ *is a global parameter-dependent LF for the origin of* (5.1)*, and* $\mathscr{R} = \mathbb{R}^n$.

Proof. Suppose that (5.90) holds. This implies that $v(x,\theta) > 0$ and $-\dot{v}(x,\theta) > 0$ for all $x \in \mathbb{R}^n_0$ with $v(x)$ radially unbounded for all $\theta \in \mathbb{R}^{n_\theta}$, and from Theorem 3.10 we conclude the proof. □

Let us observe that, similarly to the LMI conditions derived in this chapter, a necessary condition for (5.90) to hold is that the quadratic part $v_2(x,\theta)$ of $v(x,\theta)$ is a parameter-dependent LF for the system linearized at the origin, i.e. (5.49) holds with $V_2(\theta)$ defined by (5.50)–(5.51). Moreover, (5.90) also requires that the highest degree forms of $v(x,\theta)$ and $-\dot{v}(x,\theta)$ are positive definite for all $\theta \in \Theta$. These requirements can be removed with a strategy similar to that presented in Section 7.4 for the case of uncertainty-free polynomial systems.

Example 5.18. Let us consider the system

$$\begin{cases} \dot{x}_1 = x_2 - x_2^2 - x_1^3 \\ \dot{x}_2 = -(1+4\theta)x_1 - x_2 - \theta x_1^2 x_2 - x_2^3 \\ \theta \in \Theta = [0,1]. \end{cases} \tag{5.92}$$

For this system we have

$$A(\theta) = \begin{pmatrix} 0 & 1 \\ -1-4\theta & -1 \end{pmatrix}$$

and hence the origin is robustly locally asymptotically stable since $A(\theta)$ is Hurwitz for all $\theta \in \Theta$. Let us investigate robust global asymptotical stability of the origin through Corollary 5.6. By using a variable common LF $v(x,\theta)$ with $\partial^x v = 2$ and $\partial^\theta v = 0$ we find that (5.90) cannot be satisfied. Instead, by using a variable parameter-dependent LF $v(x,\theta)$ with $\partial^x v = 2$ and $\partial^\theta v = 1$, we find that (5.90) can be fulfilled, in particular the found global parameter-dependent LF is

$$v(x,\theta) = (1+1.813\theta)x_1^2 + (0.537+0.059\theta)x_1x_2 + (0.707-0.143\theta)x_2^2. \quad \square$$

5.5 Controller Design

In this section we consider the problem of designing controllers for enlarging the RDA. Specifically, we consider the presence of an input and an output in (5.1) as follows:

$$\begin{cases} \dot{x}(t) = f(x(t),\theta) + g(x(t),\theta)u(t) \\ y(t) = h(x(t),\theta) \\ x(0) = x_{init} \\ \theta \in \Theta. \end{cases} \tag{5.93}$$

Indeed, $u(t) \in \mathbb{R}^{n_u}$ is the input, $y(t) \in \mathbb{R}^{n_y}$ is the output, $g \in \mathscr{P}_{n,n_\theta}^{n \times n_u}$ and $h \in \mathscr{P}_{n,n_\theta}^{n_y}$.

Without loss of generality, let us consider that the equilibrium point of interest for $u = 0_{n_u}$ is the origin. We focus on a control strategy of the form

$$\begin{aligned} u &= k(y) \\ k(y) &= Kb_{pol}(y, \partial k) \end{aligned} \tag{5.94}$$

where ∂k is the degree of the controller, and $K \in \mathbb{R}^{n_u \times d_{pol}(n_y, \partial k)}$ is a matrix to be determined. This matrix is constrained according to

$$\begin{aligned} K &\in \mathscr{K} \\ \mathscr{K} &= \left\{ K : K_{i,j} \in (k_{i,j}^-, k_{i,j}^+) \right\} \end{aligned} \tag{5.95}$$

for some $k_{i,j}^-, k_{i,j}^+ \in \mathbb{R}$. Moreover, K has to satisfy

$$f(0_n,\theta) + g(0_n,\theta)K b_{pol}(h(0_n,\theta),\partial k) = 0_n \ \ \forall \theta \in \Theta. \tag{5.96}$$

Let us consider the case of a given common LF $v(x)$. The temporal derivative of $v(x)$ along the trajectories of (5.93) is given by

$$\dot{v}(x,\theta) = \nabla v(x)' \left(f(x,\theta) + g(x,\theta)k(h(x,\theta))\right). \tag{5.97}$$

A condition for computing a controller $k(y)$ which ensures that $\mathscr{V}(c)$ is an inner estimate of the RDA can be readily obtained by letting K vary inside $\mathscr{K}$ in Corollary 5.2. The condition so obtained contains a BMI due to the product of K with Q_1. One way to cope of this problem is to fix Q_1, for which the optimization problem boils down to an LMI feasibility test.

Similarly to the case of uncontrolled system (5.1) for which we have defined the LERDA, it is useful to define for the controlled system (5.93) the controlled LERDA, i.e. the maximum LERDA achievable for some admissible controller. This is given by $\mathscr{V}(\gamma_{con}^{\#})$ where

$$\begin{array}{rl} \gamma_{con}^{\#} = & \sup\limits_{c} \ c \\ & \text{s.t.} \ \begin{cases} \mathscr{V}(c) \subseteq \mathscr{X}(\theta) \ \ \forall \theta \in \Theta \\ \quad K \in \mathscr{K}. \end{cases} \end{array} \tag{5.98}$$

A lower bound $\hat{\gamma}_{con}^{\#}$ of $\gamma_{con}^{\#}$ can be found through Theorem 5.4 by letting K vary inside $\mathscr{K}$ in the GEVP (5.33) and by fixing Q_1 in order to avoid the generation of a BMI due to the product between Q_1 and K.

Example 5.19. Let us consider the system

$$\begin{cases} \dot{x}_1 = x_2 \\ \dot{x}_2 = -(2+\theta)x_1 - x_2 + x_2^2 + \theta x_1^2 x_2 + u \\ y = x_1 - 3x_2 \\ \theta \in \Theta = [-1,1]. \end{cases} \tag{5.99}$$

The problem consists of designing a polynomial controller as in (4.86) with $\mathscr{K}$ defined by

$$\mathscr{K} = \left\{K = (k_1,k_2,\ldots)' : \ k_i \in (-1,1)\right\} \tag{5.100}$$

with $v(x) = 2x_1^2 + x_1x_2 + x_2^2$.

First of all, let us observe that (5.96) to holds if and only if $k(0) = 0$. By choosing $b_{pol}(y,\partial k) = (1,y,\ldots,y^{\partial k})'$, this is equivalent to

$$k_1 = 0.$$

With $\partial k = 0$ we obtain the lower bound of $\gamma_{con}^{\#}$ and the corresponding controller

$$\hat{\gamma}_{con}^{\#} = 0.025, \ \ k(y) = 0$$

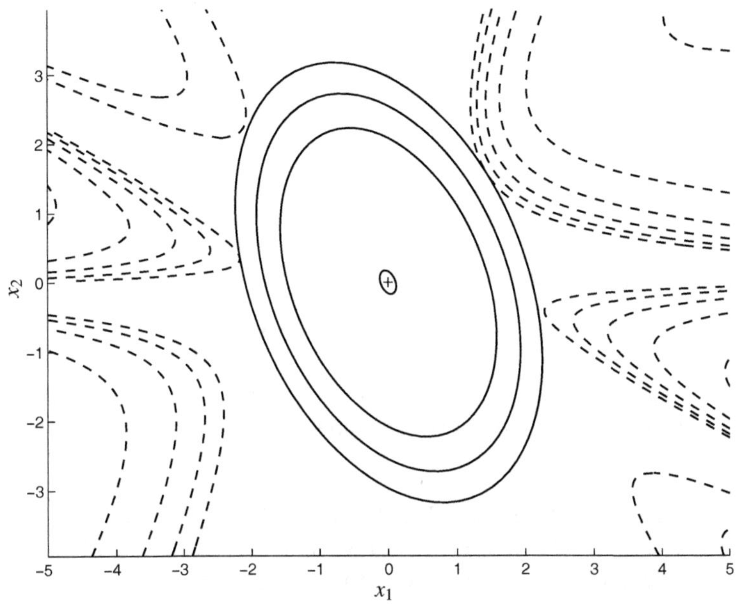

(a)

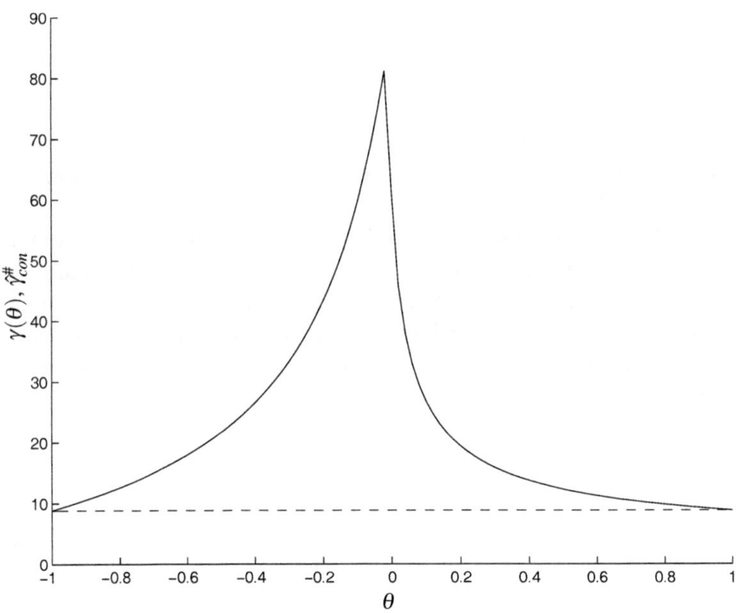

(b)

Fig. 5.14 Example 5.19. (a) Boundary of $\mathscr{V}(\hat{\gamma}^{\#}_{con})$ corresponding to the four found controllers (solid), and curve $\dot{v}(x,\theta) = 0$ corresponding to the controller obtained for $\partial k = 3$ for some values of δ (dashed). (b) Lower bound $\gamma(\theta)$ calculated for some values of θ (solid) and lower bound $\hat{\gamma}^{\#}_{con}$ (dashed) for the controller obtained for $\partial k = 3$.

(observe that this corresponds to an uncontrolled system since $k(y) = 0$ due to the above constraint on k_1). With $\partial k = 1$ we find

$$\hat{\gamma}_{con}^{\#} = 4.338, \;\; k(y) = 0.879y.$$

With $\partial k = 2$ we obtain

$$\hat{\gamma}_{con}^{\#} = 6.507, \;\; k(y) = 0.637y - 0.069y^2$$

while with $\partial k = 3$ we obtain

$$\hat{\gamma}_{con}^{\#} = 8.817, \;\; k(y) = 0.531y - 0.022y^2 + 0.010y^3.$$

Figure 5.14 shows the estimates $\mathscr{V}(\hat{\gamma}_{con}^{\#})$ corresponding to the four found controllers, and the curve $\dot{v}(x,\theta) = 0$ corresponding to the controller obtained for $\partial k = 3$. □

The cases of parameter-dependent LF and variable LF (either common or parameter-dependent) can be similarly addressed, in general via a sequence of convex optimization problems obtained by iteratively fixing some of the variables.

5.6 Notes and References

The estimation of the RDA for uncertain polynomial systems via SOS programming was proposed in [17] in the case of uncertain vectors belonging to the simplex. This chapter has extended the ideas of this paper considering the case of uncertain vectors belonging to semialgebraic sets, and addressing the problem of robust control design for enlarging the RDA.

Relevant references on this topic include [117] which proposes the estimation of the RDA via a suitable decomposition of the system, [114] which addresses local stability analysis for uncertain nonlinear systems, [116] which deals with systems with unmodeled dynamics, [60] which considers state feedback synthesis in the presence of disturbances, and [42] which investigates the RDA in implicit polynomial systems.

Chapter 6
DA and RDA in Non-polynomial Systems

This chapter presents the estimation of the DA of equilibrium for a class of non-polynomial systems, i.e. dynamical systems whose dynamics cannot be described by a polynomial function of the state. In particular, it is supposed that the dynamics is described by an affine combination of non-polynomial functions weighted by polynomials, where each non-polynomial function depends on a single entry of the state. It is shown that a condition for establishing whether a sublevel set of a LF is an estimate of the DA can be obtained in terms of LMI feasibility tests by introducing truncated Taylor expansions with worst-case remainders of the non-polynomial functions. These LMI feasibility tests are independent on each other and correspond to a combination of the bounds of the remainders. Then, the problem of computing the LEDA provided by a LF is addressed, showing that in general a candidate of such an estimate can be found by solving a certain number of GEVPs. A sufficient condition for establishing whether the found candidate is tight is hence provided by looking for power vectors in linear subspaces. Moreover, the computation of optimal estimates of the DA, the problem of establishing global asymptotical stability of the equilibrium point, and the design of polynomial static output controllers for enlarging the DA, are addressed. These problems require to consider simultaneously the systems of LMIs obtained for different combinations of the bounds of the remainders, which consequently have to be determined for an overestimate of the candidate sublevel set. Lastly, the proposed methodology is extended to a class of uncertain non-polynomial systems, in particular non-polynomial systems with coefficients depending polynomially on a time-invariant uncertain vector constrained over a semialgebraic set.

6.1 Non-polynomial Systems

Contrary to the systems considered in Chapter 4 where the functions appearing in the input-state-output representation are polynomial, here we address systems where

G. Chesi: Domain of Attraction, LNCIS 415, pp. 197–234, 2011.
springerlink.com

some of these functions are non-polynomial. In particular, let us assume without loss of generality that the equilibrium point under consideration is the origin. We start by considering the class of non-polynomial systems given by

$$\begin{cases} \dot{x}(t) = f(x(t)) \\ x(0) = x_{init} \\ f(0_n) = 0_n \end{cases} \tag{6.1}$$

where $x \in \mathbb{R}^n$ is the state vector, $t \in \mathbb{R}$ is the time, $\dot{x}(t)$ is the temporal derivative of the state, $x_{init} \in \mathbb{R}^n$ is the initial condition, and $f : \mathbb{R}^n \to \mathbb{R}^n$ is a non-polynomial function (i.e., $f \notin \mathscr{P}_n^n$) of the form

$$f(x) = f^{(0)}(x) + \sum_{i=1}^{n_\phi} f^{(i)}(x)\phi_i(\tau_i) \tag{6.2}$$

where, for some $n_\phi \in \mathbb{N}$ and for all $i = 1, \ldots, n_\phi$:

1. $f^{(0)}, f^{(i)} \in \mathscr{P}_n^n$;
2. $\phi_i : \mathbb{R} \to \mathbb{R}$ is a non-polynomial function (i.e., $\phi_i \notin \mathscr{P}$);
3. τ_i is an entry of x, i.e.

$$\exists j(i) : \tau_i = x_{j(i)}. \tag{6.3}$$

This means that $\phi_i(\tau_i)$ is a non-polynomial function of a scalar variable, and this scalar variable is an entry of x. Hence, $f(x)$ in (6.2) is an affine combination of univariate non-polynomial functions (the functions $\phi_i(\tau_i)$) weighted by polynomials (the functions $f^{(i)}(x)$). In the sequel we will drop the dependence of the state on the time for ease of notation.

System (6.1) has at least one equilibrium point (the origin). Depending on $f(x)$, (6.1) may have other equilibrium points, whose number can be either finite or infinite.

Example 6.1. Let us consider the whirling pendulum in Figure 6.1. By denoting with x_1 the angle ψ of the pole with the vertical and with x_2 the temporal derivative of x_1, the system can be described by

$$\begin{cases} \dot{x}_1 = x_2 \\ \dot{x}_2 = -\dfrac{c}{m}x_2 + \omega^2 \sin x_1 \cos x_1 - \dfrac{g}{l} \sin x_1 \end{cases} \tag{6.4}$$

where c is the friction coefficient, m is the mass of the object attached to the pole, l is the length of the pole, ω is the angular velocity of the arm, and g is the gravity acceleration. In this case $f(x)$ can be expressed as in (6.2) with

$$f^{(0)}(x) = \begin{pmatrix} x_2 \\ -\frac{c}{ml^2}x_2 \end{pmatrix}$$

$$f^{(1)}(x) = \begin{pmatrix} 0 \\ \omega^2 \end{pmatrix}, \quad \phi_1(\tau_1) = \sin\tau_1 \cos\tau_1, \quad \tau_1 = x_1$$

$$f^{(2)}(x) = \begin{pmatrix} 0 \\ -\frac{g}{l} \end{pmatrix}, \quad \phi_2(\tau_2) = \sin\tau_2, \quad \tau_2 = x_1.$$

In the sequel we will consider the numerical values

$$c = 0.2, \quad m = 1, \quad \omega = 0.9, \quad g = 9.8, \quad l = 9.8.$$

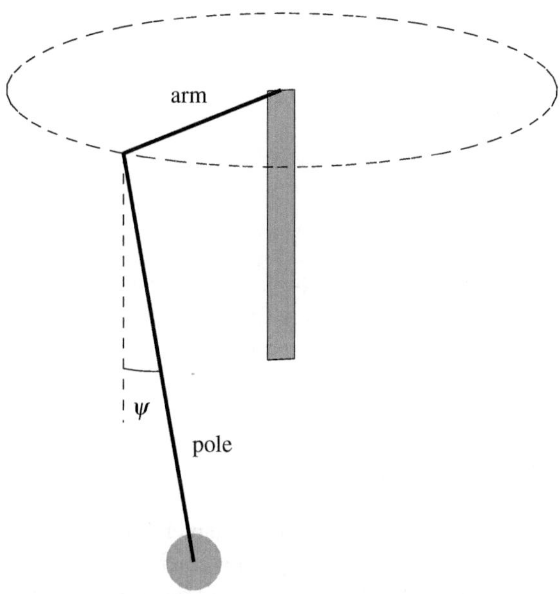

Fig. 6.1 Example 6.1 (whirling pendulum). The rigid arm of length l_a is rotating at angular velocity ω. □

Example 6.2. Let us consider the continuous-flow stirred tank reactor described by [48]

$$\begin{cases} \dot{y}_1 = 700 - 2y_1 + 200y_2 \exp(25 - 10^4/y_1) \\ \dot{y}_2 = 1 - y_2 - y_2 \exp(25 - 10^4/y_1) \end{cases}$$

where y_1 is the temperature and y_2 is the concentration. This system has three equilibrium points as follows:

$$y_{eq1} = \begin{pmatrix} 354 \\ 0.964 \end{pmatrix}, \; y_{eq2} = \begin{pmatrix} 441 \\ 0.089 \end{pmatrix}, \; y_{eq3} = \begin{pmatrix} 400 \\ 0.500 \end{pmatrix}. \tag{6.5}$$

The first two equilibrium points are locally asymptotically stable, while the other is unstable. The equilibrium point y_{eq1} is the equilibrium point of interest. Let us shift this equilibrium point into the origin by defining $x_1 = y_1 - 354$ and $x_2 = y_2 - 0.964$. We obtain:

$$\begin{cases} \dot{x}_1 = -7.486 - 2x_1 + 200(x_2 + 0.964)\exp(25 - 10^4/(x_1 + 354)) \\ \dot{x}_2 = 0.037 - x_2 - (x_2 + 0.964)\exp(25 - 10^4/(x_1 + 354)). \end{cases} \tag{6.6}$$

In this case $f(x)$ can be expressed as in (6.2) with

$$f^{(0)}(x) = \begin{pmatrix} -7.486 - 2x_1 \\ 0.037 - x_2 \end{pmatrix}$$

$$f^{(1)}(x) = \begin{pmatrix} 200(x_2 + 0.964) \\ -(x_2 + 0.964) \end{pmatrix}, \; \phi_1(\tau_1) = \exp(25 - 10^4/(\tau_1 + 354)), \; \tau_1 = x_1.$$

□

6.2 Estimates via LFs

In this section we describe a method for computing estimates of the DA based on polynomial LFs and SOS programming. This method extends to the case of non-polynomial systems the method introduced in Chapter 4 for the case of polynomial systems.

6.2.1 Establishing Estimates

As in the case polynomial systems treated in Chapter 4, the problem of determining estimates of the DA for non-polynomial systems can be addressed through Theorem 3.6, which requires to establish whether a sublevel set of a given function is included in the region where its temporal derivative is negative, i.e. (3.26). This problem can be addressed as proposed in Theorem 4.1 based on the positivity of some functions.

However, in the case of non-polynomial systems, $\dot{v}(x)$ is non-polynomial, and hence Corollaries 4.1 and 4.2 cannot be used. In particular, from (6.2) one has that

$$\dot{v}(x) = \nabla v(x)' \left(f^{(0)}(x) + \sum_{i=1}^{n_\phi} f^{(i)}(x)\phi_i(\tau_i) \right). \tag{6.7}$$

This means that the condition (4.6) requires to investigate positivity of non-polynomial functions.

The idea exploited in this Chapter for coping with this problem is to exploit truncated Taylor expansions of the non-polynomial functions and their worst-case remainders. Specifically, let us denote with the integer m_i the degree of the truncated Taylor expansion of $\phi_i(\tau_i)$. Clearly, we suppose that $\phi_i(\tau_i)$ is m_i-times continuously differentiable on a suitable region, and we will discuss this issue in details in the sequel.

Let us express $\phi_i(\tau_i)$ via a truncated Taylor expansion of degree m_i centered at the origin $\tau_i = 0$ with reminder in the Lagrange form, i.e.

$$\phi_i(\tau_i) = \hat{\phi}_i(\tau_i) + \xi_i \frac{\tau_i^{m_i}}{m_i!} \tag{6.8}$$

where $\hat{\phi}_i(\tau_i)$ contains the monomials of degree lesser than m_i according to

$$\hat{\phi}_i(\tau_i) = \sum_{j=0}^{m_i-1} \left. \frac{d^j \phi_i(\tau_i)}{d\tau_i^j} \right|_{\tau_i=0} \frac{\tau_i^j}{j!} \tag{6.9}$$

and $\xi_i \in \mathbb{R}$ is the coefficient of the reminder. Next, with $\xi = (\xi_1, \ldots, \xi_{n_\phi})'$ let us define

$$\delta(x,\xi) = \nabla v(x)' \left(f^{(0)}(x) + \sum_{i=1}^{n_\phi} f^{(i)}(x) \left(\hat{\phi}_i(\tau_i) + \xi_i \frac{\tau_i^{m_i}}{m_i!} \right) \right). \tag{6.10}$$

Moreover, for some $\xi_1^-, \xi_1^+, \ldots, \xi_{n_\phi}^-, \xi_{n_\phi}^+ \in \mathbb{R}$ let us define the hyper-rectangle

$$\Xi = [\xi_1^-, \xi_1^+] \times \ldots \times [\xi_{n_\phi}^-, \xi_{n_\phi}^+]. \tag{6.11}$$

Lastly, let $\mathcal{V}_i(c)$ is the projection of $\mathcal{V}(c)$ on the τ_i-th axis, i.e.

$$\mathcal{V}_i(c) = \{\tau_i : x \in \mathcal{V}(c)\}. \tag{6.12}$$

The following result provides an extension of the condition in Theorem 3.6 to the case of truncated Taylor expansions with worst-case remainders of the non-polynomial terms in $\dot{v}(x)$.

Theorem 6.1. *Let $v : \mathbb{R}^n \to \mathbb{R}$ be positive definite, continuously differentiable and radially unbounded, and let $c \in \mathbb{R}$ be positive. Define Ξ in* (6.11) *and suppose that*

$$\xi_i^- \leq \frac{d^{m_i} \phi_i(\tau_i)}{d\tau_i^{m_i}} \leq \xi_i^+ \ \ \forall \tau_i \in \mathcal{V}_i(c). \tag{6.13}$$

Suppose that $\delta(x,\xi)$ in (6.10) *satisfies*

$$\delta(x,\xi) < 0 \ \ \forall x \in \mathcal{V}(c) \setminus \{0_n\} \ \forall \xi \in \mathrm{ver}(\Xi). \tag{6.14}$$

Then, $v(x)$ is a LF for the origin of (6.1), *and $\mathcal{V}(c) \subseteq \mathcal{D}$.*

Proof. Let $x \in \mathscr{V}(c) \setminus \{0_n\}$. It follows that there exists $\xi \in \mathbb{R}^{n_\phi}$ such that

$$\dot{v}(x) = \delta(x, \xi)$$

where

$$\xi_i = \frac{d^{m_i} \phi_i(\tilde{\tau}_i)}{d\tau_i^{m_i}}$$

for some $\tilde{\tau}_i$ satisfying

$$\begin{cases} \tilde{\tau}_i \in [0, x_i] \text{ if } x_i \geq 0 \\ \tilde{\tau}_i \in [x_i, 0) \text{ otherwise.} \end{cases}$$

Observe that the last condition is equivalent to $\tilde{\tau}_i \in \mathscr{V}_i(c)$ since $x \in \mathscr{V}(c) \setminus \{0_n\}$. This implies that $\xi \in \Xi$ where Ξ is defined by (6.11) and (6.13). Consequently, given the fact that $\delta(x, \xi)$ is affine linear in ξ, (6.14) implies that

$$\begin{aligned} 0 &> \delta(x, \xi) \\ &= \dot{v}(x). \end{aligned}$$

Hence, it follows that (3.26) holds. Since $c > 0$, this implies that (3.13) holds, i.e. $v(x)$ is a Lyapunov for the origin. From Theorem 3.6 we conclude that $\mathscr{V}(c) \subseteq \mathscr{D}$. □

Theorem 6.1 provides a condition for establishing whether $\mathscr{V}(c)$ is included in $\mathscr{D}$ through a system of inequalities obtained by introducing bounds for the remainders of the truncated Taylor expansions of the functions $\phi_i(\tau_i)$. We will discuss the computation of these bounds in Section 6.2.2.

Let us observe that the inequalities in (6.14) are obtained by evaluating $\delta(x, \xi)$ at the vertices of the hyper-rectangle Ξ. We denote these vertices as $\xi^{(1)}, \ldots, \xi^{(n_\Xi)}$, where

$$n_\Xi = 2^{n_\phi}. \tag{6.15}$$

Let us observe that

$$\begin{aligned} \text{ver}(\Xi) &= \bigcup_{l=1,\ldots,n_\Xi} \xi^{(l)} \\ &= \left\{ \xi \in \mathbb{R}^{n_\phi} : \ \xi_i \in \{\xi_i^-, \xi_i^+\} \right\}. \end{aligned} \tag{6.16}$$

Let us also observe that we could have similarly defined the remainders for the truncated Taylor expansions of $f^{(i)}(x)\phi_i(\tau_i)$ rather than $\phi_i(\tau_i)$, however this would make possibly more difficult to obtain these bounds and possibly more conservative the approach.

The condition provided in Theorem 6.1 does not require a priori knowledge of the fact whether $v(x)$ is LF for the origin of (6.1): indeed, it is easy to see that (6.14) cannot be satisfied for any positive c if $v(x)$ is not a LF for the origin of (6.1).

The condition provided in Theorem 6.1 can be checked by investigating positivity of some functions as explained in the following result, which extends Theorem 4.1.

Theorem 6.2. *Let $v : \mathbb{R}^n \to \mathbb{R}$ be positive definite, continuously differentiable and radially unbounded, and let $c \in \mathbb{R}$ be positive. Define Ξ in* (6.11) *and suppose that* (6.13) *holds. Suppose that, for all $\xi \in \mathrm{ver}(\Xi)$, there exist $q : \mathbb{R}^n \to \mathbb{R}^2$ such that*

$$\left.\begin{array}{r} p(x,\xi) > 0 \\ q(x) > 0 \end{array}\right\} \forall x \in \mathbb{R}_0^n \tag{6.17}$$

where

$$p(x,\xi) = -q(x)' \begin{pmatrix} \delta(x,\xi) \\ c - v(x) \end{pmatrix}. \tag{6.18}$$

Then, $v(x)$ is a LF for the origin of (6.1), *and $\mathscr{V}(c) \subseteq \mathscr{D}$.*

Proof. Suppose that, for all $\xi \in \mathrm{ver}(\Xi)$, (6.17) holds. Let $x \in \mathscr{V}(c) \setminus \{0_n\}$. Then, from the first inequality it follows that

$$\begin{aligned} 0 &< -q_1(x)\delta(x,\xi) - q_2(x)(c - v(x)) \\ &\leq -q_1(x)\delta(x,\xi) \end{aligned}$$

since $q_2(x) > 0$ from the second inequality in (6.17), and since $v(x) \leq c$. Moreover, since $q_1(x) > 0$ from the second inequality in (6.17), this implies that $\delta(x,\xi) < 0$ for all $\xi \in \mathrm{ver}(\Xi)$. From Theorem 6.1 we conclude the proof. □

Theorem 6.2 provides a condition for establishing whether $\mathscr{V}(c)$ is included in $\mathscr{D}$ through the introduction of the auxiliary function $q(x)$.

Let us observe that the systems of inequalities (6.17) obtained for different values of l are independent on each other: in fact, $q(x)$ can change as ξ changes.

In the sequel we will consider the case where the functions $q(x)$ and $v(x)$ are polynomial. In such a case, the condition of Theorem 6.2 can be checked through the SOS index. Now, since $\dot{v}(0_n) = 0$, one has that

$$\begin{gathered} (6.17) \text{ holds} \\ \Downarrow \\ p(x,\xi) \text{ and } q_2(x) \text{ are locally quadratic polynomials in } x \text{ for all } \xi \in \mathbb{R}^{n_\phi}. \end{gathered}$$

This means that, a necessary condition for the SOS index of these polynomials to be positive, is that such a SOS index is defined with respect to the power vector $b_{lin}(x,\cdot)$. The following result summarizes this discussion.

Corollary 6.1. *Let $v \in \mathscr{P}_n$ be positive definite, and let $c \in \mathbb{R}$ be positive. Define Ξ in* (6.11) *and suppose that* (6.13) *holds. Let $p(x,\xi)$ be as in* (6.18). *Suppose that, for all $\xi \in \mathrm{ver}(\Xi)$, there exists $q \in \mathscr{P}_n^2$ such that*

$$\left\{\begin{array}{r} \lambda_{qua}(p(\cdot,\xi)) > 0 \\ \lambda_{pol}(q_1) > 0 \\ \lambda_{qua}(q_2) > 0. \end{array}\right. \tag{6.19}$$

Then, $v(x)$ is a LF for the origin of (6.1), *and $\mathscr{V}(c) \subseteq \mathscr{D}$.*

Proof. Suppose that, for all $\xi \in \mathrm{ver}(\Xi)$, (6.19) holds. From Section 1.3 this implies that, for all $\xi \in \mathrm{ver}(\Xi)$, $p(x,\xi)$ and q_2 are positive definite, and that $q_1(x)$ is positive. Hence, (6.17) holds, and from Theorem 6.2 we conclude the proof. □

Corollary 6.1 provides a condition for establishing whether $\mathscr{V}(c)$ is an inner estimate of $\mathscr{D}$ through the SOS index. This condition can be checked through LMI feasibility tests. Indeed, with $q \in \mathscr{P}_n^2$ and $p(x,\xi)$ is as in (6.18) let us define

$$\begin{aligned} P(Q,\xi)+L(\alpha) &= CSMR_{qua}(p(\cdot,\xi)) \\ Q &= \mathrm{diag}(Q_1,Q_2) \\ Q_1 &= SMR_{pol}(q_1) \\ Q_2 &= SMR_{qua}(q_2). \end{aligned} \tag{6.20}$$

Corollary 6.2. *Let $v \in \mathscr{P}_n$ be positive definite, and let $c \in \mathbb{R}$ be positive. Define Ξ in* (6.11) *and suppose that* (6.13) *holds. Define the matrices in* (6.20)*. Suppose that, for all $\xi \in \mathrm{ver}(\Xi)$, there exist Q and α such that the following LMIs hold:*

$$\begin{cases} P(Q,\xi)+L(\alpha) > 0 \\ Q > 0 \\ \mathrm{trace}(Q_1) = 1. \end{cases} \tag{6.21}$$

Then, $v(x)$ is a LF for the origin of (6.1)*, and $\mathscr{V}(c) \subseteq \mathscr{D}$.*

Proof. Suppose that, for all $\xi \in \mathrm{ver}(\Xi)$, (6.21) holds. This implies that (6.19) holds for all $\xi \in \mathrm{ver}(\Xi)$, and from Corollary (6.1) we conclude the proof. □

Corollary 6.2 shows how the condition (6.19) can be converted into an LMI feasibility test. Let us observe that the constraint $\mathrm{trace}(Q_1) = 1$ normalizes the variables involved in the test: in fact,

$$\text{(6.21) holds for some } Q \text{ and } \alpha \iff \text{(6.21) holds for } \rho Q \text{ and } \rho\alpha \text{ for all } \rho > 0.$$

Let us observe that the systems of inequalities (6.21) obtained for different values of ξ are independent on each other: in fact, the quantities Q and α can change as ξ changes. This means that one can solve one of these systems per time.

Let us consider the selection of the degrees m_i of the truncated Taylor expansions of the functions $\phi_i(\tau_i)$ and the degrees of the polynomials $q_1(x)$ and $q_2(x)$. A possibility is to choose them in order to maximize the degrees of freedom in (6.21) for a fixed degree of $p(x,\xi)$. This is equivalent to require that the degrees of $q_1(x)\delta(x,\xi)$ and $q_2(x)(c-v(x))$, rounded to the smallest following even integers, are equal. This can be achieved as follows:

$$\begin{array}{|l|}\hline \bullet \text{ choose a positive integer } m \text{ satisfying } m \geq \partial f^{(i)} \ \ \forall i = 0,\ldots,n_\phi; \\ \bullet \text{ choose an even degree for } q_1(x) \text{ denoted by } 2k; \\ \bullet \text{ for all } i = 1,\ldots,n_\phi \text{ define } m_i = m - \partial f^{(i)}; \\ \bullet \text{ set the degree of } q_2(x) \text{ as } \partial q_2 = 2\left\lceil \frac{m-1}{2} \right\rceil + 2k; \\ \bullet \text{ hence, the degree of } p(x) \text{ is given by } \partial p = \partial v + \partial q_2. \\ \hline \end{array} \quad (6.22)$$

In (6.22), m is the degree of the truncated Taylor expansion of $f(x)$. For each $\xi \in \mathrm{ver}(\Xi)$, the number of scalar variables in the LMIs (6.21) is given by (4.13). In the sequel we will assume that the degrees of $q_1(x)$ and $q_2(x)$ are chosen according to (6.22) unless specified otherwise.

6.2.2 Bounding the Remainders

The condition provided in Corollary 6.2 exploit the bounds ξ_i^- and ξ_i^+, $i = 1,\ldots,n_\phi$, satisfying (6.13). The computation of these bounds requires the analysis of an univariate function (the m_i-th derivative of $\phi_i(\tau_i)$) over an interval ($\mathscr{V}_i(c)$), which is typically a non-difficult task. Specifically, let us start by considering the computation of $\mathscr{V}_i(c)$. To this end, let us introduce the following result.

Theorem 6.3. *Let $v(x)$ be a locally quadratic polynomial and define $V = SMR_{qua}(v)$. Assume that $V > 0$, and let h_i be the i-th column of the identity matrix with size equal to that of V. Then, $\mathscr{V}_i(c)$ in* (6.12) *satisfies*

$$\mathscr{V}_i(c) \subseteq [-\sigma_i(c), \sigma_i(c)] \tag{6.23}$$

where

$$\sigma_i(c) = \sqrt{c h_{j(i)}' V^{-1} h_{j(i)}} \tag{6.24}$$

and $j(i)$ satisfies (6.3). *Moreover, if $v(x)$ is quadratic, then*

$$\mathscr{V}_i(c) = [-\sigma_i(c), \sigma_i(c)]. \tag{6.25}$$

Proof. Suppose that $x \in \mathscr{V}(c)$. This means that

$$\begin{aligned} v(x) &= b_{lin}(x, \partial v/2)' V b_{lin}(x, \partial v/2) \\ &\leq c. \end{aligned}$$

For any $k = 1,\ldots,d_{lin}(n, \partial v/2)$ let us consider the optimization problem

$$\begin{aligned} z_k(c) = \sup_y \ & |y_k| \\ \text{s.t. } & y' V y \leq c. \end{aligned}$$

By using Lagrangian multipliers, it can be verified that the solution of this optimization problem is given by

$$z_k(c) = \sqrt{c h_k' V^{-1} h_k}.$$

Then, let us observe that

$$\{y:\ y'Vy \le c,\ y = b_{lin}(x, \partial v/2) \text{ for some x}\} \subseteq \{y:\ y'Vy \le c\}.$$

This implies that

$$\mathscr{V}_i(c) \subseteq [-z_{j(i)}(c), z_{j(i)}(c)]$$

and hence (6.23) holds since $z_{j(i)}(c) = \sigma_i(c)$. Lastly, if $v(x)$ is quadratic, one has that

$$\{y:\ y'Vy \le c,\ y = b_{lin}(x, \partial v/2) \text{ for some x}\} = \{y:\ y'Vy \le c\}$$

since $b_{lin}(x, \partial v/2) = x$, and hence (6.25) holds. □

Theorem 6.3 provides an easy way to obtain an outer estimate of the set $\mathscr{V}_i(c)$, which turns out to be exactly $\mathscr{V}_i(c)$ in the case of quadratic LFs. By exploiting this estimate, one looks for bounds ξ_i^- and ξ_i^+ satisfying

$$\xi_i^- \le \frac{d^{m_i}\phi_i(\tau_i)}{d\tau_i^{m_i}} \le \xi_i^+ \ \ \forall \tau_i \in [-\sigma_i(c), \sigma_i(c)]. \tag{6.26}$$

In fact, if (6.26) holds, then also (6.13) holds.

Next, let us consider the computation of ξ_i^- and ξ_i^+ for some typical functions, with $\mathscr{V}_i(c)$ estimated by $[-\sigma_i(c), \sigma_i(c)]$. For instance, in the case where

$$\phi_i(\tau_i) = e^{\tau_i} \tag{6.27}$$

the tightest bounds ξ_i^- and ξ_i^+ satisfying (6.26) are

$$\xi_i^- = e^{-\sigma_i(c)},\ \ \xi_i^+ = e^{\sigma_i(c)}. \tag{6.28}$$

Then, in the case where

$$\phi_i(\tau_i) = \sin \tau_i \tag{6.29}$$

the tightest bounds ξ_i^- and ξ_i^+ satisfying (6.26) are

$$\begin{array}{l} \text{if } \sigma_i(c) \ge \pi : \xi_i^- = -1,\ \xi_i^+ = 1 \\ \text{else if } m_i \text{ is even} : \xi_i^- = -\sin\zeta,\ \xi_i^+ = \sin\zeta,\ \zeta = \min([\sigma_i(c), \pi/2]) \\ \text{else if } (m_i - 1)/2 \text{ is even} : \xi_i^- = \cos\sigma_i(c),\ \xi_i^+ = 1 \\ \text{else} : \xi_i^- = -1,\ \xi_i^+ = -\cos\sigma_i(c). \end{array} \tag{6.30}$$

Also, for the case where

$$\phi_i(\tau_i) = \cos \tau_i \tag{6.31}$$

the tightest bounds ξ_i^- and ξ_i^+ satisfying (6.26) are

$$\begin{array}{l} \text{if } \sigma_i(c) \geq \pi : \xi_i^- = -1,\ \xi_i^+ = 1 \\ \text{else if } m_i \text{ is odd} : \xi_i^- = -\sin\zeta,\ \xi_i^+ = \sin\zeta,\ \zeta = \min([\sigma_i(c), \pi/2]) \\ \text{else if } m_i/2 \text{ is even} : \xi_i^- = \cos\sigma_i(c),\ \xi_i^+ = 1 \\ \text{else} : \xi_i^- = -1,\ \xi_i^+ = -\cos\sigma_i(c). \end{array} \tag{6.32}$$

6.2.3 Choice of the LF

Let us observe that a necessary condition for (6.21) to hold is that (4.14) holds. As in the case of polynomial systems treated in Chapter 4, (4.14) holds if and only if the linearized system is asymptotically stable (i.e., A in (4.14) is Hurwitz) and $v_2(x)$ in (4.15) is a LF for it. This suggests that a LF $v(x)$ for estimating the DA can be chosen as described in (4.17). Moreover, the requirement that the linearized system is asymptotically stable can be removed with a strategy similar to that presented in Section 7.4 for the case of polynomial systems.

Example 6.3. For the whirling pendulum (6.4), let us consider the problem of establishing whether $\mathscr{V}(c)$ is a subset of the DA of the origin for

$$\begin{aligned} v(x) &= x_1^2 + x_1x_2 + 4x_2^2 \\ c &= 0.3. \end{aligned}$$

The temporal derivative is:

$$\begin{aligned} \dot{v}(x) = {} & 1.8x_1x_2 - 0.6x_2^2 + 0.81x_1 \sin x_1 \cos x_1 + 6.48x_2 \sin x_1 \cos x_1 \\ & -x_1 \sin x_1 - 8x_2 \sin x_1. \end{aligned}$$

For this system we have

$$A = \begin{pmatrix} 0 & 1 \\ -0.19 & -0.2 \end{pmatrix}$$

and hence A is Hurwitz. Moreover, $v(x)$ is a LF for the origin. In fact, (4.14) holds with

$$V_2 = \begin{pmatrix} 1 & 0.5 \\ \star & 4 \end{pmatrix}.$$

Let us consider $m = 3$. We have $m_1 = m_2 = 3$. From Theorem 6.3 we have

$$\begin{aligned} \mathscr{V}_i(c) &= [-\sigma_i(c), \sigma_i(c)] \\ \sigma_i(c) &= \sqrt{1.067c}\ \ \forall i = 1,2. \end{aligned}$$

Moreover,

$$\begin{aligned} \frac{d^{m_1}\phi_1(\tau_1)}{d\tau_1^{m_1}} &= -8\cos(2\tau_1) \\ \frac{d^{m_2}\phi_2(\tau_2)}{d\tau_2^{m_2}} &= -\cos\tau_2. \end{aligned}$$

Hence, according to Section 6.2.2, constants ξ_i^- and ξ_i^+ satisfying (6.13) can be chosen as

$$\xi_1^- = -8,\ \xi_1^+ = -3.403,\ \xi_2^- = -1,\ \xi_2^+ = -0.844.$$

The vertices of Ξ are hence given by

$$\xi^{(1)} = \begin{pmatrix} -8 \\ -1 \end{pmatrix},\ \xi^{(2)} = \begin{pmatrix} -3.403 \\ -1 \end{pmatrix},\ \xi^{(3)} = \begin{pmatrix} -8 \\ -0.844 \end{pmatrix},\ \xi^{(4)} = \begin{pmatrix} -3.403 \\ -0.844 \end{pmatrix}.$$

Let us choose $k = 0$ in (6.22). Since $\partial f^{(0)} = 1$, $\partial f^{(1)} = 0$ and $\partial f^{(2)} = 0$, we have

$$\partial q_2 = 2,\ \partial p = 4.$$

The LMIs in (6.21) have 7 scalar variables. We find that these LMIs hold, and hence $\mathscr{V}(c) \subseteq \mathscr{D}$. □

6.2.4 LEDA

The condition provided in Corollary 6.2 is built for a chosen scalar c, which determines the extension of the estimate $\mathscr{V}(c)$. Clearly, it would be helpful if one could search for the largest c for which the LMIs in (6.21) are feasible. In fact, this would allow one to estimate the LEDA introduced in Definition 4.1.

From Corollary 6.2 one can define a natural lower bound of γ in (4.19) by letting c vary. However, this introduces a key difficulty since the hyper-rectangle Ξ depends on c. Indeed, according to (6.13), the quantities ξ_i^- and ξ_i^+ that define Ξ depend on c via $\mathscr{V}_i(c)$, which is the projection of $\mathscr{V}(c)$ on the axis corresponding to τ_i.

Hence, let us denote the dependence on c of the quantities ξ_i^-, ξ_i^+, $\xi^{(l)}$ and Ξ by introducing the corresponding new quantities $\xi_i^-(c)$, $\xi_i^+(c)$, $\xi^{(l)}(c)$ and $\Xi(c)$. The lower bound obtained from Corollary 6.2 is hence given by

$$\begin{aligned} \hat{\gamma} &= \sup_c\ c \\ &\text{s.t. } \forall \xi \in \text{ver}(\Xi(c))\ \exists Q, \alpha : \begin{cases} P(c,Q,\xi) + L(\alpha) > 0 \\ Q > 0 \\ \text{trace}(Q_1) = 1 \end{cases} \end{aligned} \tag{6.33}$$

and $P(c,Q,\xi)$ is the SMR matrix of $p(x,\xi)$ in (6.20) where the dependence on c has been made explicit. Let us observe that $\hat{\gamma}$ can be also obtained as

$$\hat{\gamma} = \min_{l=1,\dots,n_\Xi} \hat{\gamma}_l \tag{6.34}$$

where

$$\hat{\gamma}_l = \sup_{Q,c,\alpha} c$$

$$\text{s.t.} \begin{cases} P(c,Q,\xi^{(l)}(c)) + L(\alpha) > 0 \\ Q > 0 \\ \text{trace}(Q_1) = 1. \end{cases} \tag{6.35}$$

A way to compute each $\hat{\gamma}_l$ is via a one-parameter sweep on c in Corollary 6.2. This search can be conducted in several ways, for instance via a bisection algorithm in order to speed up the convergence. As previously explained, this requires to recalculate the bounds $\xi_i^-(c)$ and $\xi_i^+(c)$ for any different value of c considered in the search, which are used to define the l-th vertex of $\Xi(c)$, i.e. $\xi^{(l)}(c)$.

A simpler strategy consists of not updating the bounds $\xi_i^-(c)$ and $\xi_i^+(c)$ during the search, though this provides in general only a lower bound of $\hat{\gamma}_l$. Specifically, let $c_0 \in \mathbb{R}$ be positive, and define

$$\tilde{\gamma}_l = \sup_{Q,c,\alpha} c$$

$$\text{s.t.} \begin{cases} c_0 - c \geq 0 \\ P(c,Q,\xi^{(l)}(c_0)) + L(\alpha) > 0 \\ Q > 0 \\ \text{trace}(Q_1) = 1. \end{cases} \tag{6.36}$$

Clearly, one has that

$$\tilde{\gamma}_l \leq \hat{\gamma}_l \tag{6.37}$$

i.e. $\tilde{\gamma}_l$ is in general more conservative than $\hat{\gamma}_l$. This happens because:

- the bounds of the remainders are not updated in the search;
- c is bounded by c_0, which defines the initial sublevel set of $v(x)$ where the remainders are bounded.

Let us observe that, since $\xi^{(l)}(c_0)$ is constant, one can obtain $\tilde{\gamma}_l$ via a GEVP. Indeed, let us define the polynomial

$$p_1(x,\xi) = -q(x)' \begin{pmatrix} \delta(x,\xi) \\ -v(x) \end{pmatrix} \tag{6.38}$$

and its SMR matrix as

$$P_1(Q,\xi) + L(\alpha) = CSMR_{qua}(p_1). \tag{6.39}$$

Theorem 6.4. *Let $v \in \mathscr{P}_n$ be positive definite, and let $c_0, \mu \in \mathbb{R}$ be positive. Define $V = SMR_{qua}(v)$ and the matrices in* (6.39). *Assume that $V > 0$, and let $P_2(Q_2)$ be the SMR matrix of $p_2(x)$ in* (4.25) *where the dependence on Q_2 has been made explicit. Then, $\tilde{\gamma}_l$ in* (6.36) *is given by*

$$\tilde{\gamma}_l = -\frac{z^*}{1 + \mu z^*} \tag{6.40}$$

where z^* is the solution of the GEVP

$$z^* = \inf_{Q,\alpha,z} z$$
$$\text{s.t.} \begin{cases} \dfrac{c_0}{1+\mu c_0} + z \geq 0 \\ zP_2(Q_2) + P_1(Q,\xi^{(l)}(c_0)) + L(\alpha) > 0 \\ Q > 0 \\ 1+\mu z > 0 \\ \text{trace}(Q_1) = 1. \end{cases} \tag{6.41}$$

Proof. First of all, observe that $v(x)$ is positive definite since $V > 0$. Now, suppose that the constraints in (6.41) hold. Let us pre- and post-multiply the second inequality in (6.41) by $b_{lin}(x,\partial p/2)'$ and $b_{lin}(x,\partial p/2)$, respectively, where $x \neq 0_n$. By denoting for simplicity $\xi^{(l)}(c_0)$ with ξ, we get:

$$\begin{aligned} 0 &< b_{lin}(x,\partial p/2)'(zP_2(Q_2) + P_1(Q,\xi) + L(\alpha))\, b_{lin}(x,\partial p/2) \\ &= zp_2(x) + p_1(x,\xi) \\ &= zq_2(x)(1+\mu v(x)) - q(x)' \begin{pmatrix} \delta(x,\xi) \\ -v(x) \end{pmatrix} \\ &= -q_1(x)\delta(x,\xi) - q_2(x)(-z - z\mu v(x) - v(x)) \\ &= -q_1(x)\delta(x,\xi) - (1+\mu z)q_2(x)\left(\frac{-z}{1+\mu z} - v(x)\right). \end{aligned}$$

Hence, the second inequality in (6.36) coincides with the second inequality in (6.41) whenever $q_2(x)$ and c are replaced by

$$\begin{aligned} q_2(x) &\rightarrow q_2(x)(1+\mu z) \\ c &\rightarrow \frac{-z}{1+\mu z}. \end{aligned}$$

Since $1+\mu z$ is positive, it follows that the constraints in (6.36) are equivalent to those in (6.41), and hence (6.40) holds. □

Theorem 6.4 states that $\tilde{\gamma}_l$ can be found by solving a GEVP. Let us observe that the first inequality in the constraints of (6.41) ensures that $\tilde{\gamma}_l \leq c_0$.

Example 6.4. For the whirling pendulum (6.4), let us consider the problem of determining $\hat{\gamma}$ in (6.33) with $v(x) = x_1^2 + x_1x_2 + 4x_2^2$. By choosing the constants ξ_i^- and ξ_i^+ similarly to Example 6.3, we obtain the lower bounds shown in Table 6.1. Figure 6.2 shows the estimate $\mathscr{V}(\hat{\gamma})$ for $m = 7$ and the curve $\dot{v}(x) = 0$. As we can see from the figure, the estimate $\mathscr{V}(\hat{\gamma})$ is almost tangent to the curve $\dot{v}(x) = 0$, hence implying that $\hat{\gamma}$ is almost tight. □

Table 6.1 LEDA for the whirling pendulum (6.4) with $v(x) = x_1^2 + x_1x_2 + 4x_2^2$.

m	$\hat{\gamma}_1$	$\hat{\gamma}_2$	$\hat{\gamma}_3$	$\hat{\gamma}_4$	$\hat{\gamma}$
2	0.054	0.027	0.027	0.054	0.027
3	0.601	0.546	0.536	0.598	0.536
4	0.435	0.414	0.414	0.435	0.414
5	0.595	0.712	0.593	0.707	0.593
6	0.673	0.671	0.671	0.673	0.671
7	0.699	0.709	0.699	0.709	0.699

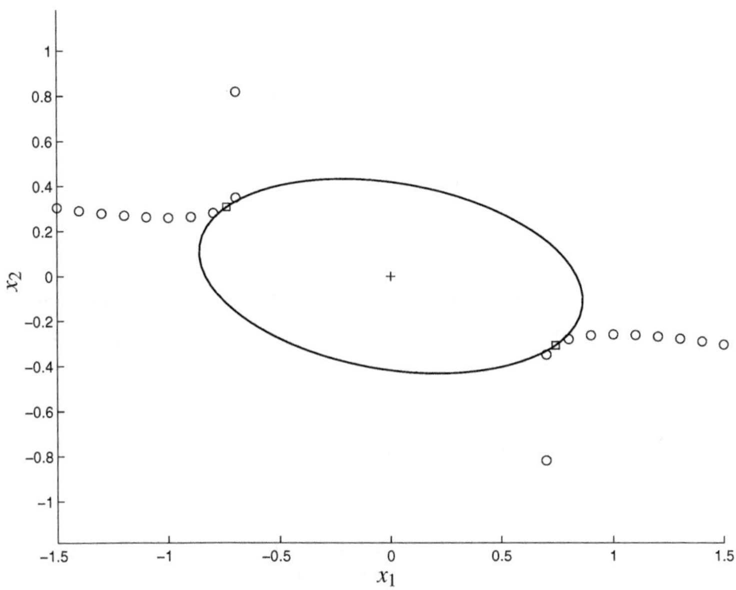

Fig. 6.2 Example 6.4 (whirling pendulum). Boundary of $\mathscr{V}(\hat{\gamma})$ for $m = 7$ (solid) and curve $\dot{v}(x) = 0$ ("o" markers) for $v(x) = x_1^2 + x_1x_2 + 4x_2^2$.

Example 6.5. For the continuous-flow stirred tank reactor (6.6), let us consider the problem of determining $\hat{\gamma}$ in (6.33) with $v(x) = x_1^2 + x_1x_2 + 15x_2^2$. The temporal derivative is:

$$\begin{aligned}\dot{v}(x) &= -4x_1^2 - 2x_1x_2 - 31x_2^2 - 14.972x_1 - 6.339x_2 \\ &\quad +(400x_1x_2 + 169x_2^2 + 385.6x_1 + 162.916x_2)\exp(25 - 10^4/(x_1 + 354)).\end{aligned}$$

For this system we have

$$A = \begin{pmatrix} -1.403 & 7.766 \\ -0.003 & -1.039 \end{pmatrix}$$

and hence A is Hurwitz. Moreover, $v(x)$ is a LF for the origin. In fact, (4.14) holds with

$$V_2 = \begin{pmatrix} 1 & 0.5 \\ \star & 15 \end{pmatrix}.$$

From Theorem 6.3 we have

$$\begin{aligned}\mathscr{V}_1(c) &= [-\sigma_1(c), \sigma_1(c)]\\ \sigma_1(c) &= \sqrt{1.017c}.\end{aligned}$$

Moreover, the derivatives of $\phi_1(\tau_1)$ are monotonically increasing. This implies that constants ξ_i^- and ξ_i^+ satisfying (6.13) can be chosen as

$$\xi_1^- = \left.\frac{d^{m_1}\phi_1(\tau_1)}{d\tau_1^{m_1}}\right|_{\tau_1=-\sigma_1(c)}, \quad \xi_1^+ = \left.\frac{d^{m_1}\phi_1(\tau_1)}{d\tau_1^{m_1}}\right|_{\tau_1=\sigma_1(c)}.$$

The vertices of Ξ are hence given by

$$\xi^{(1)} = \xi_1^-, \quad \xi^{(2)} = \xi_1^+.$$

Let us choose $k = 0$ in (6.22). Since $\partial f^{(0)} = 1$, $\partial f^{(1)} = 0$ and $\partial f^{(2)} = 0$, we have

$$\partial q_2 = 2, \quad \partial p = 4.$$

The LMIs in (6.21) have 7 scalar variables. We obtain the lower bounds shown in Table 6.2. Figure 6.3 shows the estimate $\mathscr{V}(\hat{\gamma})$ for $m = 7$ and the curve $\dot{v}(x) = 0$. As we can see from the figure, $\hat{\gamma}$ is almost tight.

Table 6.2 LEDA for the stirred tank reactor (6.6) with $v(x) = x_1^2 + x_1x_2 + 15x_2^2$.

m	$\hat{\gamma}_1$	$\hat{\gamma}_2$	$\hat{\gamma}$
2	∞	5.891	5.891
3	11.627	10.144	10.144
4	10.772	10.712	10.712
5	10.741	10.739	10.739
6	10.740	10.740	10.740
7	10.740	10.740	10.740

□

6.2.5 Estimate Tightness

Let us study the conservatism of the proposed methodology. The following result investigates the asymptotic relationship between the LEDA level γ introduced in Definition 4.1 and its lower bound $\hat{\gamma}$ obtained in (6.33). In this result we denote with m the degree of the truncated Taylor expansion of $f(x)$ according to (6.22).

Theorem 6.5. *Let γ and $\hat{\gamma}$ be as in (4.19) and (6.33), respectively. Suppose that $0 < \gamma < \infty$, and suppose that there exists $\rho \in \mathbb{R}$ such that the bounds ξ_i^- and ξ_i^+ satisfy*

$$\left.\begin{array}{l}|\xi_i^-| < \rho\\ |\xi_i^+| < \rho\end{array}\right\} \quad \forall i = 1, \ldots, n_\phi \ \forall c \leq \gamma \ \forall m \geq 0. \tag{6.42}$$

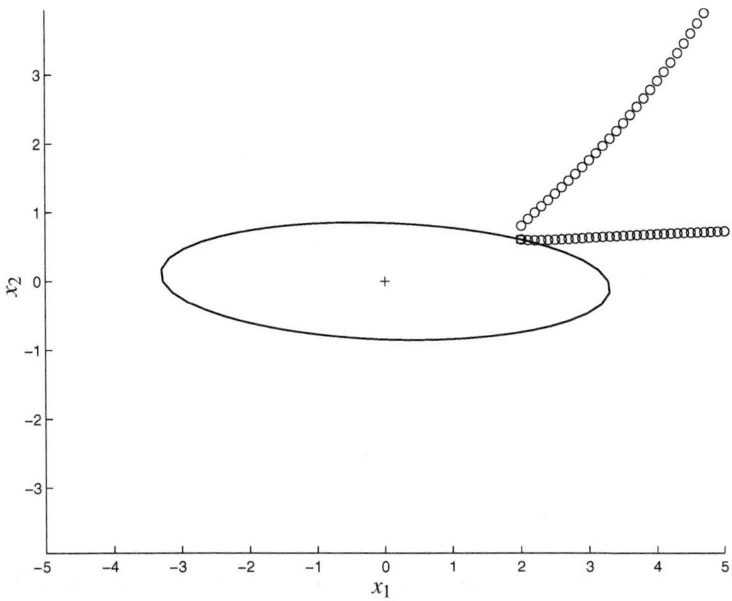

Fig. 6.3 Example 6.5 (stirred tank reactor). Boundary of $\mathscr{V}(\hat{\gamma})$ for $m = 7$ (solid) and curve $\dot{v}(x) = 0$ ("o" markers) for $v(x) = x_1^2 + x_1x_2 + 15x_2^2$.

Then,

$$\lim_{m\to\infty} \hat{\gamma} = \gamma. \tag{6.43}$$

Proof. Let us observe that

$$\dot{v}(x) - \delta(x,\xi) = \nabla v(x)' \sum_{i=1}^{n_\phi} f^{(i)}(x) \left(\phi_i(\tau_i) - \hat{\phi}_i(\tau_i) - \xi_i \frac{\tau_i^{m_i}}{m_i!} \right).$$

Let us define

$$\tilde{f}(x) = \left(\tilde{f}^{(1)}(x), \ldots, \tilde{f}^{(n_\phi)}(x) \right)'$$

where

$$\tilde{f}^{(i)}(x) = \frac{\tau_i^{m_i}}{m_i!} \nabla v(x)' f^{(i)}(x).$$

As explained in the proof of Theorem 6.1, for all $x \in \mathscr{V}(c)$, there exists $\tilde{\xi}(x)$ such that $\tilde{\xi}(x) \in \Xi$ and $\dot{v}(x) = \delta(x, \tilde{\xi})$. This implies that, for all $x \in \mathscr{V}(c)$,

$$\dot{v}(x) - \delta(x,\xi) = (\tilde{\xi}(x) - \xi)' \tilde{f}(x).$$

From (6.42) it follows that, for all $x \in \mathscr{V}(c)$ and for all $\xi \in \Xi$,

$$|\dot{v}(x) - \delta(x,\xi)| \leq 2\rho\sqrt{n_\phi}a^*$$

where

$$a^* = \sup_{x \in \mathscr{V}(c)} \|\tilde{f}(x)\|.$$

Next, let us observe that c is bounded by γ which is finite. This implies that, for all $x \in \mathscr{V}(c)$, $\|x\|$ and each $|\tau_i|$ are bounded by finite quantities, and hence

$$\lim_{m \to \infty} a^* = 0$$

since each m_i tends to ∞ as m does. Consequently,

$$\lim_{m \to \infty} |\dot{v}(x) - \delta(x,\xi)| = 0$$

i.e. the theorem holds. □

Theorem 6.5 provides a condition for asymptotic convergence of the lower bound $\hat{\gamma}$ to γ, which consists of requiring the existence of a constant ρ satisfying (6.42).

Let us observe that, in some cases, the existence of such a constant is easily guaranteed, for instance for trigonometric functions g_i such as sine and cosine. Moreover, this existence can be ensured also for unbounded functions such as $\phi_i(\tau_i) = e^{\tau_i}$: indeed, one has that the m-th derivative is e^{τ_i} and the existence of the constant ρ is automatically implied since $\gamma < \infty$.

Lastly, it would be clearly useful if one could establish whether the found lower bound $\hat{\gamma}$ is tight because the search for tighter values of this lower bound could be immediately terminated. The following result provides a condition to answer to this question.

Theorem 6.6. *Let γ and $\hat{\gamma}$ be as in* (4.19) *and* (6.33), *respectively. For $l = 1,\ldots,n_\Xi$ define the set*

$$\mathscr{M}^{(l)} = \left\{x \in \mathbb{R}^n : \; b_{lin}(x, \partial p/2) \in \ker\left(M^{(l)}\right)\right\} \tag{6.44}$$

where

$$M^{(l)} = P(\hat{\gamma}_l, Q^*, \xi^{(l)}(\hat{\gamma}_l)) + L(\alpha^*) \tag{6.45}$$

and $P(\hat{\gamma}_l, Q^*, \xi^{(l)}(\hat{\gamma}_l)) + L(\alpha^*)$ is the found optimal value of $P(c, Q, \xi^{(l)}(c)) + L(\alpha)$ in (6.35). *Define also*

$$\mathscr{M}_1^{(l)} = \left\{x \in \mathscr{M}^{(l)} : \; v(x) = \hat{\gamma} \text{ and } \dot{v}(x) = 0\right\}. \tag{6.46}$$

Then,

$$\exists l = 1,\ldots,n_\Xi : \; \mathscr{M}_1^{(l)} \neq \emptyset \;\Rightarrow\; \hat{\gamma} = \gamma. \tag{6.47}$$

Proof. Suppose that $\mathscr{M}_1^{(l)} \neq \emptyset$ for some $l = 1,\ldots,n_\Xi$. For one of such values of l, let $x \in \mathscr{M}_1^{(l)}$. It follows that $v(x) = \hat{\gamma}$ and $\dot{v}(x) = 0$. Since $\hat{\gamma}$ is a lower bound of γ, this implies that x is a tangent point between the surface $\dot{v}(x) = 0$ and the sublevel set $\mathscr{V}(\hat{\gamma})$. Hence, $\hat{\gamma} = \gamma$. □

Theorem 6.6 provides a condition for ensuring that the found $\hat{\gamma}$ is the sought γ. In particular, this happens if the set $\mathscr{M}_1^{(l)}$ in (6.46) is nonempty for some l. This set can be found via trivial substitution from the set $\mathscr{M}^{(l)}$ in (6.44) (at least whenever $\mathscr{M}^{(l)}$ is finite), which can be computed as explained in Section 1.5.1.

Example 6.6. For the whirling pendulum (6.4), let us consider the problem of establishing tightness of $\hat{\gamma}$ found in Example 6.4 with $v(x) = x_1^2 + x_1x_2 + 4x_2^2$ and $m = 7$. We have that

$$\mathscr{M}^{(1)} = \left\{(-0.741, 0.308)', (0.741, -0.308)'\right\}.$$

Moreover, $|v(x) - \hat{\gamma}| = 0.000$ and $|\dot{v}(x)| = 0.003$ for all $x \in \mathscr{M}^{(1)}$. This means that the points in $\mathscr{M}^{(1)}$ almost satisfy the conditions for being in $\mathscr{M}_1^{(1)}$. Indeed, these points correspond to the filled dots in Figure 6.2. □

Example 6.7. For the continuous-flow stirred tank reactor (6.6), let us consider the problem of establishing tightness of $\hat{\gamma}$ found in Example 6.5 with $v(x) = x_1^2 + x_1x_2 + 15x_2^2$ and $m = 7$. We have that

$$\mathscr{M}^{(1)} = \left\{(1.990, 0.609)'\right\}.$$

Moreover, $|v(x) - \hat{\gamma}| = 0.007$ and $|\dot{v}(x)| = 0.001$ for $x \in \mathscr{M}^{(1)}$. This means that the point in $\mathscr{M}^{(1)}$ almost satisfy the conditions for being in $\mathscr{M}_1^{(1)}$. Indeed, this point corresponds to the filled dot in Figure 6.3. □

6.3 Optimal Estimates

In the previous sections we have considered the case of a fixed LF. Here we address the problem of obtaining estimates of the DA with a variable LF in order to reduce their conservatism. In particular, we consider the problem of obtaining estimates that are optimal according to specified criteria among the estimates provided by polynomial LFs of a chosen degree.

Let us start by observing that, in the previous sections where the LF was fixed, the optimization problems obtained at different vertices of the hyper-rectangle Ξ were independent on each other: for instance, this is the case of Corollary 6.2 where one can consider separately each of the LMI systems (6.21) obtained for $l = 1,\ldots,n_\Xi$. However, if the LF is variable, these LMI systems are not longer independent since the LF is a common variable.

Moreover, as in the case of polynomial systems treated in Chapter 4, some of the LMIs in the optimization problems derived in the previous sections become BMIs if the LF is allowed to vary. This means that such problems can be solved only locally in general. Hereafter we discuss some strategies for their solution.

6.3.1 Maximizing the Volume of the Estimate

The first criterion that we consider consists of maximizing the volume of the estimate according to (4.62). One way to address this problem is:

- parametrizing the LF $v(x)$;
- for any fixed LF, estimating γ with the lower bound $\hat{\gamma}$ provided by (6.33);
- evaluating the volume of the found estimate, i.e. $\mathrm{vol}(\mathscr{V}(\hat{\gamma}))$;
- updating the LF through a gradient-like method.

By parametrizing $v(x)$ as in (4.64) and defining its complete SMR matrix as in (4.67), the problem is hence replaced by (4.66). The evaluation of the volume of the found estimate can be addressed via (4.69)–(4.70).

Another way to address problem (4.62) in the case of non-polynomial systems consists of alternatively freezing some of the variables in the BMIs obtained by letting $v(x)$ vary, hence solving a sequence of LMI problems. Specifically, let us consider the approximation of the volume of this estimate in (4.70). By exploiting Corollary 6.2, problem (4.62) is replaced by

$$\begin{array}{l} \inf\limits_{c,Q^*,V,\alpha^*} \dfrac{\mathrm{trace}(V)}{c} \\ \text{s.t. } \forall l = 1,\dots,n_\Xi \left\{ \begin{array}{r} P(c,Q^{(l)},V,\xi^{(l)}(c)) + L(\alpha^{(l)}) > 0 \\ Q^{(l)} > 0 \\ V > 0 \\ \mathrm{trace}(Q_1^{(l)}) = 1 \end{array} \right. \end{array} \tag{6.48}$$

where $P(c,Q,V,\xi)$ is the SMR matrix of $p(x,\xi)$ in (6.20) where the dependence on V and C has been made explicit. We have that the first inequality of problem (6.48) is a BMI since $P(c,Q,V,\xi)$ contains products of the entries of V with the elements of Q. Moreover, $\xi^{(l)}(c)$ depends on V and c that are variable.

We hence proceed by alternatively solving two subproblems obtained from (6.48). In particular, for a fixed V, we solve

$$\begin{array}{l} \sup\limits_{c,Q^*,\alpha^*} c \\ \text{s.t. } \forall l = 1,\dots,n_\Xi \left\{ \begin{array}{r} P(c,Q^{(l)},V,\xi^{(l)}(c)) + L(\alpha^{(l)}) > 0 \\ Q^{(l)} > 0 \\ \mathrm{trace}(Q_1^{(l)}) = 1 \end{array} \right. \end{array} \tag{6.49}$$

which is equivalent to solving (6.33). The current estimate of the DA is $\mathscr{V}(c)$ where c is updated with its optimal value in (6.49). Then, we also update $Q^{(l)}$ with its optimal value in (6.49).

Next, we consider $Q^{(l)}$ fixed and V variable. Let us observe that, if the LF is variable, then also the vertices $\xi^{(l)}(c)$ are variable. However, contrary to (6.49) where they are re-calculated for each value of c in the bisection search, they cannot be updated in an SDP if the LF is variable.

In order to cope with this problem, we constraint $\mathscr{V}(c)$ inside a fixed region, and compute the vertices $\xi^{(l)}(c)$ for such a region. Specifically, let $\tilde{v} \in \mathscr{P}_n$ be a locally quadratic polynomial, and let us indicate its sublevel set as

$$\tilde{\mathscr{V}}(c) = \{x \in \mathbb{R}^n : \tilde{v}(x) \leq c\}. \tag{6.50}$$

Let us define an SMR matrix of $\tilde{v}(x)$ as

$$\tilde{V} = SMR_{qua}(\tilde{v}). \tag{6.51}$$

We select $\tilde{v}(x)$ such that

$$\begin{cases} \tilde{V} > 0 \\ V - \tilde{V} > 0 \end{cases} \tag{6.52}$$

where V is the SMR matrix of the LF used in (6.49). Let us observe that the constraint $V - \tilde{V} > 0$ ensures that

$$\mathscr{V}(c) \subset \tilde{\mathscr{V}}(c). \tag{6.53}$$

Let us denote with $\tilde{\xi}^{(l)}(c)$ the vertex $\xi^{(l)}(c)$ obtained by considering $\tilde{\mathscr{V}}(c)$ rather than $\mathscr{V}(c)$ in its definition. We hence solve

$$\begin{aligned} &\inf_{V,\alpha^*} \ \mathrm{trace}(V) \\ &\text{s.t.} \ \begin{cases} 0 < P(c, Q^{(l)}, V, \tilde{\xi}^{(l)}(c)) + L(\alpha^{(l)}) \ \forall l = 1, \ldots, n_{\Xi} \\ 0 < V - \tilde{V} \end{cases} \end{aligned} \tag{6.54}$$

which is an SDP. The current estimate of the DA is $\mathscr{V}(c)$ where $v(x)$ is updated with the optimal value of V in (6.54) according to (4.63). The procedure is hence repeated using in (6.49) the found V. The cost $\mathrm{trace}(V)/c$ is monotonically non-increasing in the procedure, which is terminated when the cost decrease is lesser than a chosen threshold.

6.3.2 Enlarging the Estimate with Fixed Shape Sets

The second criterion that we consider for obtaining less conservative estimates of the DA consists of maximizing the size of a set of fixed shape included in the estimate.

By expressing such a set as the sublevel set of a positive definite polynomial $w(x)$, this strategy amounts to solving the optimization problem (4.74).

Hence, let us define the SMR matrices $P_w(c,d,Q_w,V)+L_w(\alpha_w)$ and Q_w in (4.78). By exploiting Corollary 6.2, a lower bound to the solution of (4.74) in the case of non-polynomial systems is given by

$$\begin{array}{ll} \sup_{c,d,Q^*,Q_w,V,\alpha^*,\alpha_w} & d \\ \text{s.t. } \forall l=1,\ldots,n_\Xi & \left\{ \begin{array}{r} P(c,Q^{(l)},V,\xi^{(l)}(c))+L(\alpha^{(l)}) > 0 \\ Q^{(l)} > 0 \\ P_w(c,d,Q_w,V)+L_w(\alpha_w) > 0 \\ Q_w > 0 \\ \mathrm{trace}(Q_1^{(l)}) = 1 \\ \mathrm{trace}(Q_{w1}) = 1. \end{array} \right. \end{array} \tag{6.55}$$

One way to address (6.55) is as follows:

- solve (6.55) for a fixed V;
- update V through a gradient-like method.

Let us observe that solving (6.55) for a fixed $V > 0$ amounts, first, to determining the largest c that satisfies the constraints of (6.55), and, second, to determining the largest d that satisfies these constraints. The former step is similar to solving (6.33), while the latter step can be solved through a GEVP similar to (4.30). Also, the parameter V can be replaced by the parameter $\mathbf{v}$ in (4.64).

Another way to address problem (6.55) consists of alternatively freezing some of the variables in the BMIs. In particular, let us consider the case of constant Q_{w1}. For fixed V one solves (6.49). The current estimate of the DA is $\mathcal{V}(c)$ where c is updated with its optimal value in (6.49). Then, we also update $Q^{(l)}$ with its optimal value in (6.49). Next, we introduce $\tilde{v}$ and $\tilde{\xi}^{(l)}(c)$ as done in (6.50)–(6.54). We hence solve

$$\begin{array}{ll} \sup_{d,Q_{w2},V,\alpha^*,\alpha_w} & d \\ \text{s.t. } \forall l=1,\ldots,n_\Xi & \left\{ \begin{array}{r} P(c,Q^{(l)},V,\tilde{\xi}^{(l)}(c))+L(\alpha^{(l)}) > 0 \\ P_w(c,d,Q_w,V)+L_w(\alpha_w) > 0 \\ Q_{w2} > 0 \\ V-\tilde{V} > 0 \end{array} \right. \end{array} \tag{6.56}$$

which can be done through a GEVP similarly to (4.30). The current estimate of the DA is $\mathcal{V}(c)$ where $v(x)$ is updated with the optimal value of V in (6.56) according to (4.63). The procedure is hence repeated using in (6.49) the found V. The cost d is monotonically non-decreasing in the procedure, which is terminated when the cost increase is lesser than a chosen threshold.

Example 6.8. Let us consider problem (4.74) with the choice $w(x) = \|x\|^2$ in order to obtain optimal estimates of the DA for the whirling pendulum (6.4). Hence, we address problem (6.55) by solving iteratively (6.49) and (6.56) in the case of polynomial LFs of degree 2 and 4. The found LFs, that we denote with $v_2(x)$ and $v_4(x)$, are obtained by using the following initializations:

1. in the case of $v_2(x)$, the LF used in Example 6.4;
2. in the case of $v_4(x)$, the LF $v_2(x)$.

Table 6.3 and Figure 6.4 show the obtained results.

Table 6.3 Example 6.8 (whirling pendulum). Estimates found with variable LFs by enlarging fixed shape sets.

LF	Cost d	No. of iterations (4.72) and (4.80)
$v_2(x)$	0.299	17
$v_4(x)$	1.647	16

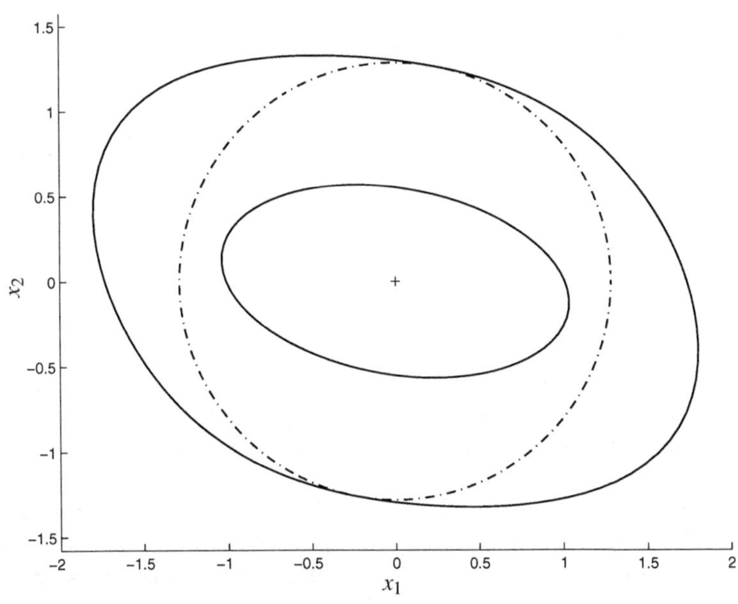

Fig. 6.4 Example 6.8 (whirling pendulum). Boundaries of the estimates of the DA found with variable common LFs of degree 2 and 4 by enlarging fixed shape sets (solid). The "+" denotes the equilibrium point. The dashdot line shows the boundary of the set $\mathscr{W}(d)$ found with $v_4(x)$. □

6.3.3 Establishing Global Asymptotical Stability

Here we consider the specialization of the strategies presented in Sections 6.3.1–6.3.2 to the problem of establishing whether the origin of (6.1) is globally asymptotically stable, i.e. $\mathscr{D} = \mathbb{R}^n$. Analogously to the strategies derived for the case of polynomial systems in Section 4.4.3 and uncertain polynomial systems in Section 5.4.3, such a specialization boils down to an LMI feasibility test as explained in the following result.

Corollary 6.3. *Let $v \in \mathscr{P}_n$ be a locally quadratic polynomial to determine, and define*

$$\begin{aligned} V &= SMR_{qua}(v) \\ P_g(V,\xi) + L(\alpha) &= CSMR_{qua}(p_g) \end{aligned} \tag{6.57}$$

where

$$p_g(x,\xi) = -\delta(x,\xi). \tag{6.58}$$

Define Ξ in (6.11) *and suppose that*

$$\xi_i^- \leq \frac{d^{m_i}\phi_i(\tau_i)}{d\tau_i^{m_i}} \leq \xi_i^+ \ \forall \tau_i \in \mathbb{R}. \tag{6.59}$$

Suppose that there exist V and $\alpha^{(l)}$, $l = 1,\ldots,n_\Xi$, such that the following LMIs hold:

$$\begin{cases} V > 0 \\ P_g(V,\xi^{(l)}) + L(\alpha) > 0 \ \forall l = 1,\ldots,n_\Xi \\ \mathrm{trace}(V) = 1. \end{cases} \tag{6.60}$$

Then, $v(x)$ provided by such a V is a global LF for the origin of (6.1), *and $\mathscr{D} = \mathbb{R}^n$.*

Proof. Suppose that (6.60) holds. This implies that $v(x)$ and $-\delta(x,\xi)$ are positive definite for all $\xi \in \mathrm{ver}(\Xi)$ with $v(x)$ radially unbounded. This implies that $\dot{v}(x)$ is negative definite, and from Theorem 3.2 we conclude the proof. □

Let us observe that, similarly to the LMI conditions derived in this chapter, a necessary condition for (6.60) to hold is that the quadratic part $v_2(x)$ of $v(x)$ is a LF for the system linearized at the origin, i.e. (4.14) holds with V_2 defined by (4.15)–(4.16). Moreover, (6.60) also requires that the highest degree forms of $v(x)$ and $p_g(x,\xi)$ are positive definite. These requirements can be removed with a strategy similar to that presented in Section 7.4 for the case of polynomial systems.

Example 6.9. Let us consider the system

$$\begin{cases} \dot{x}_1 = x_1 - x_2 + x_1 x_2 \sin x_1 - 2x_1^3 \\ \dot{x}_2 = 7x_1 - 2x_2 - 2\sin x_1 - x_2^3. \end{cases} \tag{6.61}$$

For this system we have

$$A = \begin{pmatrix} 1 & -1 \\ 5 & -2 \end{pmatrix}$$

and hence the origin is locally asymptotically stable since A is Hurwitz. Let us investigate global asymptotical stability of the origin through Corollary 6.3. In this case $f(x)$ can be expressed as in (6.2) with

$$f^{(0)}(x) = \begin{pmatrix} x_1 - x_2 - 2x_1^3 \\ 7x_1 - 2x_2 - x_2^3 \end{pmatrix}$$
$$f^{(1)}(x) = \begin{pmatrix} x_1 x_2 \\ -2 \end{pmatrix}, \quad \phi_1(\tau_1) = \sin\tau_1, \quad \tau_1 = x_1.$$

Let us consider $m = 3$. We have $m_1 = 1$. Moreover,

$$\frac{d^{m_1}\phi_1(\tau_1)}{d\tau_1^{m_1}} = -\cos\tau_1.$$

Hence, constants ξ_i^- and ξ_i^+ satisfying (6.59) can be chosen as

$$\xi_1^- = -1, \ \xi_1^+ = 1.$$

By using a variable LF of degree 2 we find that (6.60) can be fulfilled, in particular the found V provides the global LF

$$v(x) = 0.874x_1^2 - 0.368x_1x_2 + 0.126x_2^2.$$

□

6.4 Controller Design

In this section we consider the problem of designing controllers for enlarging the DA in non-polynomial systems. Specifically, we consider the presence of an input and an output in (6.1) as

$$\begin{cases} \dot{x}(t) = f(x(t)) + g(x(t))u(t) \\ y(t) = h(x(t)) \\ x(0) = x_{init} \end{cases} \tag{6.62}$$

where $f(x)$ has the structure in (6.2). Indeed, $u(t) \in \mathbb{R}^{n_u}$ is the input, $y(t) \in \mathbb{R}^{n_y}$ is the output, $g \in \mathscr{P}_n^{n \times n_u}$ and $h \in \mathscr{P}_n^{n_y}$.

Without loss of generality, let us consider that the equilibrium point of interest for $u = 0_{n_u}$ is the origin. We focus on a control strategy of the form $u = k(y)$ where $k \in \mathscr{P}_{n_y}^{n_u}$. Let us parametrize $k(y)$ as in (4.88) where ∂k is the degree of the controller, and $K \in \mathbb{R}^{n_u \times d_{pol}(n_y, \partial k)}$ is a matrix to be determined. Hence, K has to satisfy

$$f^{(0)}(0_n) + \sum_{i=1}^{n_\phi} f^{(i)}(0_n)\phi_i(0) + g(0_n)Kb_{pol}(h(0_n), m_k) = 0_n. \tag{6.63}$$

We consider the presence of constraints on the input to be designed, for simplicity expressed as in (4.90) for some $k_{i,j}^-, k_{i,j}^+ \in \mathbb{R}$.

Let us start by observing that the optimization problems obtained in the case of autonomous systems for different vertices of the hyper-rectangle Ξ are not longer independent in the presence of a controller to be designed.

Indeed, let us consider first the case of a given LF $v(x)$. The temporal derivative of $v(x)$ along the trajectories of (6.62) is given by

$$\dot{v}(x) = \nabla v(x)' \left(f^{(0)}(x) + \sum_{i=1}^{n_\phi} f^{(i)}(x)\phi_i(\tau_i) + g(x)k(h(x)) \right). \tag{6.64}$$

Corollary 6.4. *Let $v \in \mathscr{P}_n$ be positive definite, and let $c \in \mathbb{R}$ be positive. Define the matrices in* (6.20)*, and let $P(K,Q,\xi)$ be the SMR matrix of $p(x,\xi)$ in* (6.20) *with $\dot{v}(x)$ given by* (6.64) *where the dependence on K has been made explicit. Suppose that there exist K, Q^* and α^* such that the following inequalities hold:*

$$\left.\begin{array}{r} P(K,Q^{(l)},\xi^{(l)}) + L(\alpha^{(l)}) > 0 \\ Q^{(l)} > 0 \\ K \in \mathscr{K} \\ \operatorname{trace}(Q_1^{(l)}) = 1 \end{array}\right\} \forall l = 1,\ldots,n_\Xi. \tag{6.65}$$

Then, $v(x)$ is a LF for the origin of (6.62)*, and $\mathscr{V}(c) \subseteq \mathscr{D}$.*

Proof. Analogous to the proof of Corollary 6.2. □

Corollary 6.4 provides a condition for computing a controller $k(y)$ which ensures that $\mathscr{V}(c)$ is an inner estimate of the DA. This condition is in general nonconvex since the first inequality is a BMI due to the product of the entries of $Q_1^{(l)}$ and K. One way to cope of this problem is to fix $Q_1^{(l)}$, for which the optimization problem boils down to an LMI feasibility test.

Similarly to the uncontrolled system (6.1) for which we have defined the LEDA, it is useful to define for the controlled system (6.62) the controlled LEDA, i.e. the maximum LEDA achievable for some admissible controller. Specifically, let us define γ_{con} as in (4.94). From Corollary 6.4 one can define a natural lower bound of γ_{con} as

$$\begin{aligned} \hat{\gamma}_{con} = & \sup_{c,K,Q^*,\alpha^*} \; c \\ & \text{s.t. } \forall l = 1,\ldots,n_\Xi \left\{ \begin{array}{r} P(c,K,Q^{(l)},\xi^{(l)}(c)) + L(\alpha^{(l)}) > 0 \\ Q^{(l)} > 0 \\ K \in \mathscr{K} \\ \operatorname{trace}(Q_1^{(l)}) = 1 \end{array} \right. \end{aligned} \tag{6.66}$$

where $P(c,K,Q,\xi)$ is the SMR matrix of $p(x,\xi)$ in (6.20) where the dependence on K and c has been made explicit. This lower bound can be computed via a one-parameter sweep on the scalar c using the condition provided in Corollary 6.4 for any fixed c. Another way to obtain the lower bound $\hat{\gamma}_{con}$ is via a GEVP similarly to the case of uncontrolled system considered in Theorem 6.4.

Example 6.10. Let us consider the system

$$\begin{cases} \dot{x}_1 = -x_2 + 0.2x_1^2 x_2 + x_2 u \\ \dot{x}_2 = x_1 - x_2 - 0.2x_2^3 + e^{x_1} - 1 \\ y = x_1 + x_2. \end{cases} \tag{6.67}$$

The problem consists of designing a polynomial controller as in (4.86) with $\mathscr{K}$ defined by

$$\mathscr{K} = \{K = (k_1, k_2, \ldots)' : k_i \in (-1,1)\} \tag{6.68}$$

with $v(x) = 2x_1^2 - x_1x_2 + x_2^2$.

We have that (6.63) is satisfied for all $K \in \mathscr{K}$. We compute $\sigma_1(c)$ as in Theorem 6.3, and the constants ξ_i^- and ξ_i^+ as in Section 6.2.2. We obtain the lower bounds shown in Table 6.4. Figure 6.5 shows the estimates $\mathscr{V}(\hat{\gamma}_{con})$ corresponding to the three found controllers for $m = 7$, and the curve $\dot{v}(x) = 0$ corresponding to the controller obtained for $\partial k = 2$.

Table 6.4 Example 6.10. Controller design with $v(x) = 2x_1^2 - x_1x_2 + x_2^2$.

∂k	m	$\hat{\gamma}_{con}$	Controller $k(y)$
0	3	2.893	-0.947
0	4	3.759	-0.947
0	5	4.368	-0.947
0	6	4.675	-0.947
0	7	4.778	-0.947
1	3	3.815	$-0.903 - 0.182y$
1	4	5.514	$-0.739 - 0.281y$
1	5	6.551	$-0.909 - 0.207y$
1	6	7.892	$-0.899 - 0.237y$
1	7	8.185	$-0.907 - 0.215y$
2	3	12.583	$0.026 - 0.039y - 0.458y^2$
2	4	5.522	$-0.258 - 0.301y - 0.165y^2$
2	5	18.653	$0.170 + 0.006y - 0.450y^2$
2	6	10.335	$-0.183 - 0.199y - 0.223y^2$
2	7	26.219	$0.168 + 0.008y - 0.429y^2$

□

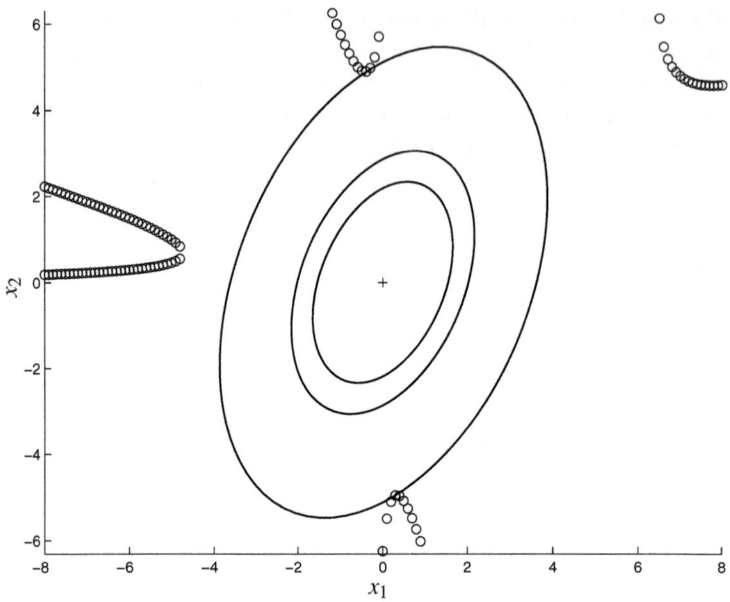

Fig. 6.5 Example 6.10. Boundary of $\mathscr{V}(\hat{\gamma}_{con})$ corresponding to the three found controllers for $m = 7$ (solid), and curve $\dot{v}(x) = 0$ corresponding to the controller obtained for $\partial k = 2$ and $m = 7$ ("o" markers).

6.5 Uncertain Non-polynomial Systems

In the remaining part of this chapter we consider a class of uncertain non-polynomial systems, in particular systems where the temporal derivative of the state is a non-polynomial function of the state and a polynomial function of the uncertainty. For these systems the problem is to investigate the RDA of a common equilibrium point of interest, that without loss of generality we assume to be the origin. We start by considering the class of non-polynomial systems given by

$$\begin{cases} \dot{x}(t) = f(x(t), \theta) \\ x(0) = x_{init} \\ \theta \in \Theta \\ f(0_n, \theta) = 0_n \ \ \forall \theta \in \Theta \end{cases} \tag{6.69}$$

where $\Theta \subseteq \mathbb{R}^{n_\theta}$ is as in (5.2) for some $a \in \mathscr{P}^{n_a}_{n_\theta}$, $b \in \mathscr{P}^{n_b}_{n_\theta}$, and

$$f(x, \theta) = f^{(0)}(x, \theta) + \sum_{i=1}^{n_\phi} f^{(i)}(x, \theta)\phi_i(\tau_i) \tag{6.70}$$

where $f^{(i)} \mathscr{P}^n_{n,n_\theta}$ for all $i = 0, \ldots, n_\phi$, $\phi_i : \mathbb{R} \to \mathbb{R}$ is non-polynomial (i.e., $\phi_i \notin \mathscr{P}$) for all $i = 1, \ldots, n_\phi$, and τ_i is an entry of x for all $i = 1, \ldots, n_\phi$, i.e. (6.3) holds.

System (6.69) has at least one common equilibrium point (the origin). Depending on $f(x,\theta)$, (6.69) may have other equilibrium points, either common or dependent on the uncertainty. The RDA of the origin is limited by these equilibria: in fact, by denoting with $\mathscr{X}_0(\theta)$ the set of equilibrium points of (6.69) except the origin, one has that (5.3)–(5.4) hold.

Example 6.11. Let us consider the synchronous generator described by

$$\begin{cases} \dot{y}_1 = y_2 \\ \dot{y}_2 = -Dy_2 - \sin y_1 + \sin \delta_0 \end{cases}$$

where y_1 is the power angle, y_2 is the corresponding speed deviation, and D and δ_0 are parameters [48]. The equilibrium point of interest is $(\delta_0, 0)'$.

Let us choose $\delta_0 = 0.412$, and let us suppose that D is uncertain in $[0.5, 2]$. Defining $x_1 = y_1 - 0.412$, $x_2 = y_2$ and $\theta = D$, we can express this system as

$$\begin{cases} \dot{x}_1 = x_2 \\ \dot{x}_2 = -\theta x_2 - \sin(x_1 + 0.412) + \sin 0.412 \\ \theta \in \Theta = [0.5, 2]. \end{cases} \tag{6.71}$$

This is an uncertain non-polynomial system, moreover the origin is a common equilibrium point. The function $f(x)$ can be expressed as in (6.70) with

$$f^{(0)}(x,\theta) = \begin{pmatrix} x_2 \\ -\theta x_2 + \sin 0.412 \end{pmatrix}$$

$$f^{(1)}(x,\theta) = \begin{pmatrix} 0 \\ \omega^2 \end{pmatrix}, \quad \phi_1(\tau_1) = \sin(\tau_1 + 0.412), \quad \tau_1 = x_1.$$

The set Θ can be expressed as in (5.2) with

$$a(\theta) = -1 + 2.5\theta - \theta^2.$$

□

6.6 Estimates for Uncertain Non-polynomial Systems

In this section we describe a method for computing estimates of the RDA based on polynomial LFs and SOS programming. This method extends to the case of uncertain non-polynomial systems the methods introduced in Chapter 5 for the case of uncertain polynomial systems and in the first part of this chapter for the case of non-polynomial systems.

6.6.1 Establishing Estimates

Let $v(x)$ be a candidate common LF, and define

$$\delta(x,\theta,\xi) = \nabla v(x)' \left(f^{(0)}(x,\theta) + \sum_{i=1}^{n_\phi} f^{(i)}(x,\theta) \left(\hat{\phi}_i(\tau_i) + \xi_i \frac{\tau_i^{m_i}}{m_i!} \right) \right). \tag{6.72}$$

The following result provides a condition for establishing whether a sublevel set of $v(x)$ is an inner estimate of the RDA.

Theorem 6.7. *Let $v : \mathbb{R}^n \to \mathbb{R}$ be positive definite, continuously differentiable and radially unbounded, and let $c \in \mathbb{R}$ be positive. Define Ξ in* (6.11) *and suppose that* (6.13) *holds. Suppose that $\delta(x,\theta,\xi)$ in* (6.72) *satisfies*

$$\delta(x,\theta,\xi) < 0 \;\; \forall x \in \mathscr{V}(c) \setminus \{0_n\} \; \forall \theta \in \Theta \; \forall \xi \in \mathrm{ver}(\Xi). \tag{6.73}$$

Then, $v(x)$ is a common LF for the origin of (6.69), *and $\mathscr{V}(c) \subseteq \mathscr{R}$.*

Proof. Let $x \in \mathscr{V}(c) \setminus \{0_n\}$ and $\theta \in \Theta$. It follows that there exists $\xi \in \Xi$ such that

$$\dot{v}(x,\theta) = \delta(x,\theta,\xi).$$

Consequently, given the fact that $\delta(x,\theta,\xi)$ is affine linear in ξ, (6.73) implies that

$$\begin{aligned} 0 &> \delta(x,\theta,\xi) \\ &= \dot{v}(x,\theta). \end{aligned}$$

Hence, it follows that (3.60) holds. Since $c > 0$, this implies that (3.46) holds, i.e. $v(x)$ is a common Lyapunov for the origin. From Theorem 3.13 we conclude that $\mathscr{V}(c) \subseteq \mathscr{R}$. □

The condition provided in Theorem 6.7 can be checked by investigating positivity of some functions as explained in the following result, which extends Theorem 6.2.

Theorem 6.8. *Let $v : \mathbb{R}^n \to \mathbb{R}$ be positive definite, continuously differentiable and radially unbounded, and let $c \in \mathbb{R}$ be positive. Define Ξ in* (6.11) *and suppose that* (6.13) *holds. Suppose that, for all $\xi \in \mathrm{ver}(\Xi)$, there exist $q : \mathbb{R}^n \times \mathbb{R}^{n_\theta} \to \mathbb{R}^2$, $r : \mathbb{R}^n \times \mathbb{R}^{n_\theta} \to \mathbb{R}^{n_a}$ and $s : \mathbb{R}^n \times \mathbb{R}^{n_\theta} \to \mathbb{R}^{n_b}$ such that*

$$\left. \begin{array}{r} p(x,\theta,\xi) > 0 \\ q(x,\theta) > 0 \\ r(x,\theta) \geq 0 \end{array} \right\} \;\; \forall x \in \mathbb{R}_0^n \; \forall \theta \in \mathbb{R}^{n_\theta} \tag{6.74}$$

where

$$p(x,\theta,\xi) = -q(x,\theta)' \begin{pmatrix} \delta(x,\theta,\xi) \\ c - v(x) \end{pmatrix} - r(x,\theta)' a(\theta) - s(x,\theta)' b(\theta). \tag{6.75}$$

Then, $v(x)$ is a common LF for the origin of (6.69), *and $\mathscr{V}(c) \subseteq \mathscr{R}$.*

Proof. Suppose that, for all $\xi \in \text{ver}(\Xi)$, (6.74) holds. Let $x \in \mathscr{V}(c) \setminus \{0_n\}$ and $\theta \in \Theta$. Then, from the first inequality it follows that

$$\begin{aligned} 0 &< -q_1(x,\theta)\delta(x,\theta,\xi) - q_2(x,\theta)(c - v(x)) - r(x,\theta)'a(\theta) - s(x,\theta)'b(\theta) \\ &\le -q_1(x)\delta(x,\xi) \end{aligned}$$

since $q_2(x,\theta) > 0$ from the second inequality in (6.74), $v(x) \le c$, $r(x,\theta) \ge 0$ from the third inequality in (6.74), and $a(\theta) \ge 0$ and $b(\theta) = 0$. Moreover, since $q_1(x,\theta) > 0$ from the second inequality in (6.74), this implies that $\delta(x,\theta,\xi) < 0$ for all $\xi \in \text{ver}(\Xi)$. From Theorem 6.7 we conclude the proof. □

In the sequel we will consider the case where the functions $q(x,\theta)$, $r(x,\theta)$, $s(x,\theta)$ and $v(x)$ are polynomial. In such a case, the condition of Theorem 6.8 can be checked through the SOS index as explained hereafter.

Corollary 6.5. *Let $v \in \mathscr{P}_n$ be positive definite, and let $c \in \mathbb{R}$ be positive. Define Ξ in* (6.11) *and suppose that* (6.13) *holds. Suppose that, for all $\xi \in \text{ver}(\Xi)$, there exist $q \in \mathscr{P}^2_{n,n_\theta}$, $r \in \mathscr{P}^{n_a}_{n,n_\theta}$ and $s \in \mathscr{P}^{n_b}_{n,n_\theta}$ such that*

$$\begin{cases} \lambda_{quapol}(p(\cdot,\cdot,\xi)) > 0 \\ \lambda_{polpol}(q_1) > 0 \\ \lambda_{quapol}(q_2) > 0 \\ \lambda_{quapol}(r_i) \ge 0 \;\; \forall i = 0,\ldots,n_a. \end{cases} \tag{6.76}$$

Then, $v(x)$ is a common LF for the origin of (6.69)*, and $\mathscr{V}(c) \subseteq \mathscr{R}$.*

Proof. Suppose that, for all $\xi \in \text{ver}(\Xi)$, (6.76) holds. This implies that

$$\begin{cases} p(x,\theta,\xi) > 0 \;\; \forall x \in \mathbb{R}^n_0 \; \forall \theta \in \mathbb{R}^{n_\theta} \\ q_1(x,\theta) > 0 \;\; \forall x \in \mathbb{R}^n \; \forall \theta \in \mathbb{R}^{n_\theta} \\ q_2(x,\theta) > 0 \;\; \forall x \in \mathbb{R}^n_0 \; \forall \theta \in \mathbb{R}^{n_\theta} \\ r_i(x,\theta) \ge 0 \;\; \forall x \in \mathbb{R}^n_0 \; \forall \theta \in \mathbb{R}^{n_\theta} \; \forall i = 1,\ldots,n_a. \end{cases}$$

Consequently, (6.74) holds for all $\xi \in \text{ver}(\Xi)$, and from Theorem 6.8 we conclude the proof. □

The condition provided in Corollary 6.5 can be checked through LMI feasibility tests. Indeed, let us define SMR matrices

$$\begin{aligned} P(Q,R,\mathbf{s},\xi) + L(\alpha) &= CSMR_{quapol}(p(\cdot,\cdot,\xi)) \\ Q &= \text{diag}(Q_1,Q_2) \\ Q_1 &= SMR_{polpol}(q_1) \\ Q_2 &= SMR_{quapol}(q_2) \\ R &= \text{diag}(R_1,\ldots,R_{n_a}) \\ R_i &= SMR_{quapol}(r_i) \;\; \forall i = 1,\ldots,n_a \\ \mathbf{s} &= (\mathbf{s}_1',\ldots,\mathbf{s}_{n_b}')' \\ \mathbf{s}_i &= COE_{quapol}(s_i) \;\; \forall i = 1,\ldots,n_b. \end{aligned} \tag{6.77}$$

Corollary 6.6. *Let $v \in \mathscr{P}_n$ be positive definite, and let $c \in \mathbb{R}$ be positive. Define Ξ in* (6.11) *and suppose that* (6.13) *holds. Define the matrices in* (6.77)*. Suppose that, for all $\xi \in \mathrm{ver}(\Xi)$, there exist $Q, R, \mathbf{s}$ and α such that the following LMIs hold:*

$$\begin{cases} P(Q,R,\mathbf{s},\xi)+L(\alpha) > 0 \\ Q > 0 \\ R \geq 0 \\ \mathrm{trace}(Q_1) = 1. \end{cases} \tag{6.78}$$

Then, $v(x)$ is a common LF for the origin of (6.69)*, and $\mathscr{V}(c) \subseteq \mathscr{R}$.*

Proof. Suppose that, for all $\xi \in \mathrm{ver}(\Xi)$, (6.78) holds. This implies that (6.76) holds for all $\xi \in \mathrm{ver}(\Xi)$, and from Corollary (6.5) we conclude the proof. □

Let us consider the selection of the degrees m_i of the truncated Taylor expansions of the functions $\phi_i(\tau_i)$ and the degrees of the polynomials in $q(x,\theta)$, $r(x,\theta)$ and $s(x,\theta)$. A possibility is to choose them in order to maximize the degrees of freedom in (6.78) for fixed degrees of $p(x,\theta,\xi)$. This is equivalent to require that the degrees of $q_1(x,\theta)\delta(x,\theta)$, $q_2(x,\theta)(c-v(x))$, $r(x,\theta)'a(\theta)$ and $s(x,\theta)'b(\theta)$, rounded to the smallest following even integers, are equal. This can be achieved as follows:

$$\boxed{\begin{array}{l}
\bullet \text{ choose a positive integer } m \text{ satisfying} \\
\quad m \geq \partial^x f^{(i)} \;\; \forall i = 0,\ldots,n_\phi; \\
\bullet \text{ choose even degrees for } q_1(x,\theta) \text{ denoted by} \\
\quad 2k_x = \partial^x q_1 \text{ and } 2k_\theta = \partial^\theta q_1; \\
\bullet \text{ for all } i = 1,\ldots,n_\phi \text{ define} \\
\quad m_i = m - \partial^x f^{(i)}; \\
\bullet \text{ define} \\
\quad m_\theta = \max\limits_{i=1,\ldots,n_\phi} \partial^\theta f^{(i)}; \\
\bullet \text{ set the degrees of } q_2(x,\theta) \text{ as} \\
\quad \partial^x q_2 = 2\left\lceil \dfrac{m-1}{2} \right\rceil + 2k_x \text{ and } \partial^\theta q_2 = 2\left\lceil \dfrac{m_\theta}{2} \right\rceil + 2k_\theta; \\
\bullet \text{ set the degrees of } r_i(x,\theta) \text{ as} \\
\quad \partial^x r_i = \partial v + \partial^x q_2 \text{ and } \partial^\theta r_i = \partial^\theta q_2 - 2\left\lceil \dfrac{\partial a_i}{2} \right\rceil; \\
\bullet \text{ set the degrees of } s_i(x,\theta) \text{ as} \\
\quad \partial^x s_i = \partial^x r_i \text{ and } \partial^\theta s_i = \partial^\theta q_2 - \partial b_i; \\
\bullet \text{ hence, the degrees of } p(x,\theta) \text{ are given by} \\
\quad \partial^x p = \partial v + \partial^x q_2 \text{ and } \partial^\theta p = \partial^\theta q_2.
\end{array}} \tag{6.79}$$

In (6.79), m is the degree of the truncated Taylor expansion of $f(x)$. For each $\xi \in \mathrm{ver}(\Xi)$, the number of scalar variables in the LMIs (6.78) is given by (5.14). In the sequel we will assume that the degrees of the polynomials in $q(x,\theta)$, $r(x,\theta)$ and $s(x,\theta)$ are chosen according to (6.79) unless specified otherwise.

6.6.2 Choice of the LF

It is useful to observe that a necessary condition for (6.78) to hold is that (5.15) holds. As in the case of uncertain polynomial systems treated in Chapter 5, (5.15) holds if and only if the linearized system is robustly asymptotically stable (i.e., $A(\theta)$ in (5.15) is Hurwitz for all $\theta \in \Theta$) and $v_2(x)$ in (4.15) is a common LF for it. This suggests that the function $v(x)$ for estimating the RDA can be chosen as described in (5.16).

Similarly to the case of uncertain polynomial systems, (4.14) guarantees that (6.78) holds for some $c > 0$, but in this case under the condition that $m_i \geq 2$ for all $i = 1,\dots,n_{phi}$.

Example 6.12. For the synchronous generator (6.71), let us consider the problem of establishing whether $\mathscr{V}(c)$ is a subset of the RDA of the origin for

$$\begin{aligned} v(x) &= 2x_1^2 + x_1x_2 + 2x_2^2 \\ c &= 1. \end{aligned}$$

The temporal derivative is:

$$\dot{v}(x,\theta) = x_1x_2(4-\theta) + x_2^2(1-4\theta) + (x_1+4x_2)(\sin(x_1+0.412) - \sin 0.412).$$

For this system we have

$$A(\theta) = \begin{pmatrix} 0 & 1 \\ -\cos 0.412 & -\theta \end{pmatrix}$$

and hence $A(\theta)$ is Hurwitz for all $\theta \in \Theta$. Moreover, $v(x)$ is a common LF for the origin. In fact, (5.15) holds with

$$V_2 = \begin{pmatrix} 2 & 0.5 \\ \star & 2 \end{pmatrix}.$$

Let us consider $m = 3$. We have $m_1 = 3$. From Theorem 6.3 we have

$$\mathscr{V}_1(c) = [-\sigma_1(c), \sigma_1(c)],\ \sigma_1(c) = \sqrt{0.533c}.$$

Moreover,

$$\frac{d^{m_1}\phi_1(\tau_1)}{d\tau_1^{m_1}} = -\cos(\tau_1 + 0.412).$$

Hence, according to Section 6.2.2, constants ξ_1^- and ξ_1^+ satisfying (6.13) can be chosen as

$$\xi_1^- = -1,\ \xi_1^+ = -0.416.$$

The vertices of Ξ are hence given by

$$\xi^{(1)} = -1,\ \xi^{(2)} = -0.416.$$

Let us choose $k_x = k_\theta = 0$ in (6.79). Since $\partial^x f^{(0)} = 1$, $\partial^\theta f^{(0)} = 1$, $\partial^x f^{(1)} = 0$ and $\partial^\theta f^{(1)} = 0$, we have

$$\partial^x q_2 = 2,\ \partial^\theta q_2 = 2,\ \partial^x p = 4,\ \partial^\theta p = 2.$$

The LMIs in (6.21) have 44 scalar variables. We find that these LMIs hold, and hence $\mathscr{V}(c) \subseteq \mathscr{R}$. □

6.6.3 LERDA

Here we address the problem of estimating the largest sublevel of the RDA that can be obtained through the common LF $v(x)$, i.e. the LERDA introduced in Definition 5.2.

Clearly, the LERDA coincides with the intersection of all the LEDAs provided by the common LF $v(x)$ for any admissible value of the uncertainty, i.e.

$$\mathscr{V}(\hat{\gamma}^{\#}) = \bigcap_{\theta \in \Theta} \mathscr{V}(\gamma(\theta)) \tag{6.80}$$

where $\gamma(\theta)$ is defined analogously to Definition 5.1. Hence, one has that

$$\hat{\gamma}^{\#} = \inf_{\theta \in \Theta} \gamma(\theta). \tag{6.81}$$

From Corollary 6.6 one can define a natural lower bound of $\hat{\gamma}^{\#}$ by letting c vary. However, let us observe that the hyper-rectangle Ξ depends on c. Hence, let us denote the dependence on c of the quantities ξ_i^-, ξ_i^+, $\xi^{(l)}$ and Ξ by introducing the corresponding new quantities $\xi_i^-(c)$, $\xi_i^+(c)$, $\xi^{(l)}(c)$ and $\Xi(c)$. The lower bound obtained from Corollary 6.6 is hence given by

$$\begin{aligned} \hat{\gamma}^{\#} &= \sup_{c}\ c \\ &\text{s.t. } \forall \xi \in \text{ver}(\Xi(c))\ \exists Q,R,\mathbf{s},\alpha : \begin{cases} P(c,Q,R,\mathbf{s},\xi) + L(\alpha) > 0 \\ Q > 0 \\ R \geq 0 \\ \text{trace}(Q_1) = 1 \end{cases} \end{aligned} \tag{6.82}$$

where $P(c,Q,R,\mathbf{s},\xi)$ is the SMR matrix of $p(x,\theta,\xi)$ in (6.77) where the dependence on c has been made explicit.

Let us observe that $\hat{\gamma}^{\#}$ can be also obtained as

$$\hat{\gamma}^{\#} = \min_{l=1,\ldots,n_\Xi} \hat{\gamma}_l^{\#} \tag{6.83}$$

where

$$\hat{\gamma}_l^{\#} = \sup_{c,Q,R,\mathbf{s},\alpha} c \quad \text{s.t.} \begin{cases} P(c,Q,R,\mathbf{s},\xi^{(l)}(c)) + L(\alpha) > 0 \\ Q > 0 \\ R \geq 0 \\ \text{trace}(Q_1) = 1. \end{cases} \tag{6.84}$$

A way to compute each $\hat{\gamma}_l^{\#}$ is via a one-parameter sweep on c in Corollary 6.6. This search can be conducted in several ways, for instance via a bisection algorithm in order to speed up the convergence. As previously explained, this requires to recalculate the bounds $\xi_i^-(c)$ and $\xi_i^+(c)$ for any different value of c considered in the search, which are used to define the l-th vertex of $\Xi(c)$, i.e. $\xi^{(l)}(c)$.

A simpler strategy consists of not updating the bounds $\xi_i^-(c)$ and $\xi_i^+(c)$ during the search, though this provides in general only an upper bound of $\hat{\gamma}_l$. Specifically, let $c_0 \in \mathbb{R}$ be positive, and define

$$\tilde{\gamma}_l^{\#} = \sup_{c,Q,R,\mathbf{s},\alpha} c \quad \text{s.t.} \begin{cases} c_0 - c \geq 0 \\ P(c,Q,R,\mathbf{s},\xi^{(l)}(c_0)) + L(\alpha) > 0 \\ Q > 0 \\ R \geq 0 \\ \text{trace}(Q_1) = 1. \end{cases} \tag{6.85}$$

Clearly, one has that

$$\tilde{\gamma}_l^{\#} \leq \hat{\gamma}_l^{\#} \tag{6.86}$$

i.e. $\tilde{\gamma}_l^{\#}$ is in general more conservative than $\hat{\gamma}_l$. However, since $\xi^{(l)}(c_0)$ is constant, one can obtain $\tilde{\gamma}_l^{\#}$ via a GEVP.

Indeed, let us define the polynomial

$$p_1(x,\theta,\xi) = -q(x,\theta)' \begin{pmatrix} \delta(x,\theta,\xi) \\ -v(x) \end{pmatrix} - r(x,\theta)'a(\theta) - s(x,\theta)'b(\theta) \tag{6.87}$$

and its SMR matrix as

$$P_1(Q,R,\mathbf{s},\xi) + L(\alpha) = CSMR_{quapol}(p_1). \tag{6.88}$$

Theorem 6.9. *Let $v \in \mathscr{P}_n$ be positive definite, and let $c_0, \mu \in \mathbb{R}$ be positive. Define $V = SMR_{qua}(v)$ and the matrices in (6.88). Assume that $V > 0$, and let $P_2(Q_2)$ be the SMR matrix of $p_2(x,\theta)$ in (5.28) where the dependence on Q_2 has been made explicit. Then, $\tilde{\gamma}_l^{\#}$ in (6.85) is given by*

$$\tilde{\gamma}_l^{\#} = -\frac{z^*}{1 + \mu z^*} \tag{6.89}$$

where z^* is the solution of the GEVP

$$z^* = \inf_{Q,R,\mathbf{s},z,\alpha} z$$
$$\text{s.t.} \begin{cases} \dfrac{c_0}{1+\mu c_0} + z \geq 0 \\ zP_2(Q_2) + P_1(Q,R,\mathbf{s},\xi^{(l)}(c_0)) + L(\alpha) > 0 \\ Q > 0 \\ R \geq 0 \\ 1+\mu z > 0 \\ \mathrm{trace}(Q_1) = 1. \end{cases} \tag{6.90}$$

Proof. First of all, observe that $v(x)$ is positive definite since $V > 0$. Now, suppose that the constraints in (6.90) hold. For $x \in \mathbb{R}_0^n$ and $\theta \in \mathbb{R}^{n_\theta}$ define $y = b_{pol}(\theta, \partial^\theta p_2/2) \otimes b_{lin}(x, \partial^x p_2/2)$, and let us pre- and post-multiply the second inequality in (6.90) by y' and y, respectively. By denoting for simplicity $\xi^{(l)}(c_0)$ with ξ, we get:

$$\begin{aligned} 0 &< y'\left(zP_2(Q_2) + P_1(Q,R,\mathbf{s},\xi) + L(\alpha)\right) y \\ &= zp_2(x,\theta) + p_1(x,\theta,\xi) \\ &= zq_2(x,\theta)(1+\mu v(x)) - q(x,\theta)' \begin{pmatrix} \delta(x,\theta,\xi) \\ -v(x) \end{pmatrix} - r(x,\theta)'a(\theta) - s(x,\theta)'b(\theta) \\ &= -q_1(x,\theta)\delta(x,\theta,\xi) - q_2(x,\theta)(-z - z\mu v(x) - v(x)) - r(x,\theta)'a(\theta) \\ &\quad -s(x,\theta)'b(\theta) \\ &= -q_1(x,\theta)\delta(x,\theta,\xi) - (1+\mu z)q_2(x,\theta)\left(\frac{-z}{1+\mu z} - v(x)\right) - r(x,\theta)'a(\theta) \\ &\quad -s(x,\theta)'b(\theta). \end{aligned}$$

Hence, the second inequality in (6.85) coincides with the second inequality in (6.90) whenever $q_2(x,\theta)$ and c are replaced by

$$\begin{aligned} q_2(x,\theta) &\to q_2(x,\theta)(1+\mu z) \\ c &\to \frac{-z}{1+\mu z}. \end{aligned}$$

Since $1+\mu z$ is positive, it follows that the constraints in (6.85) are equivalent to those in (6.90), and hence (6.89) holds. □

Theorem 6.9 states that $\tilde{\gamma}_l^\#$ can be found by solving a GEVP. Let us observe that the first inequality in the constraints of (6.90) ensures that $\tilde{\gamma}_l \leq c_0$.

In the case where Θ is a polytope and $\partial^\theta f = 1$, the LERDA can be further characterized as explained in the following result.

Corollary 6.7. *Let $v \in \mathscr{P}_n$ be positive definite. Suppose that Θ is a bounded convex polytope and $\partial^\theta f = 1$. Then,*

$$\hat{\gamma}^\# = \min_{\theta \in \mathrm{ver}(\Theta)} \gamma(\theta). \tag{6.91}$$

Proof. Analogous to the proof of Theorem 5.5. □

From Corollary 6.7 a lower bound of $\hat{\gamma}^{\#}$ can be directly computed as

$$\min_{\theta \in \text{ver}(\Theta)} \hat{\gamma}(\theta) \tag{6.92}$$

where $\hat{\gamma}(\theta)$ is the lower bound in (6.33) parametrized by θ.

Example 6.13. For the synchronous generator (6.71), let us consider the problem of determining $\hat{\gamma}^{\#}$ in (6.81) with $v(x) = 2x_1^2 + x_1x_2 + 2x_2^2$. By choosing the constants ξ_i^- and ξ_i^+ similarly to Example 6.12, we obtain the lower bounds shown in Table 6.5. Figure 6.6 shows the estimate $\mathcal{V}(\hat{\gamma}^{\#})$ and the curve $\dot{v}(x,\theta) = 0$. As we can see from the figure, the estimate $\mathcal{V}(\hat{\gamma}^{\#})$ is almost tangent to the curve $\dot{v} = 0$, hence implying that $\hat{\gamma}^{\#}$ is almost tight. □

Table 6.5 LERDA for the synchronous generator (6.71) with $v(x) = 2x_1^2 + x_1x_2 + 2x_2^2$.

m	$\hat{\gamma}_1$	$\hat{\gamma}_2$	$\hat{\gamma}$
2	2.094	3.301	2.094
3	3.477	6.833	3.477
4	3.207	4.809	3.207
5	3.955	4.295	3.955
6	4.209	4.328	4.209
7	4.234	4.248	4.234

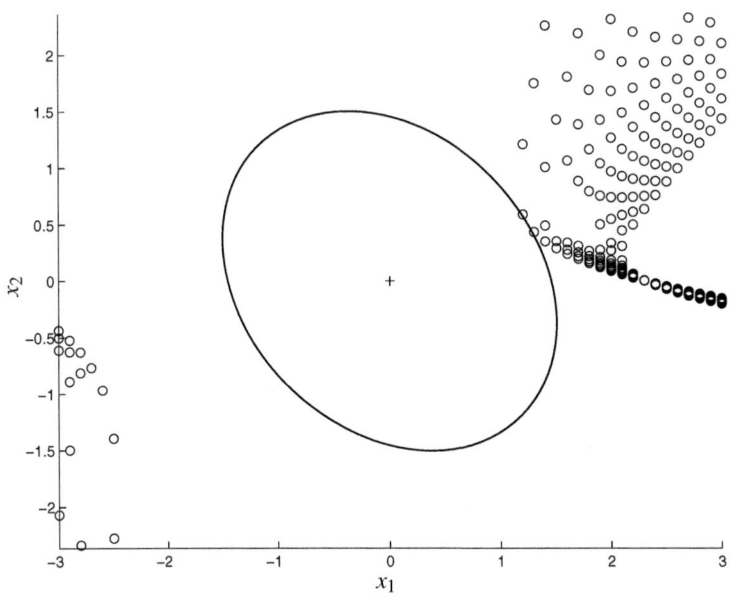

Fig. 6.6 Example 6.13 (synchronous generator). Boundary of $\mathcal{V}(\hat{\gamma}^{\#})$ (solid) and curve $\dot{v}(x,\theta) = 0$ for some values of θ ("o" markers).

6.6.4 Extensions

Similarly to the case of common LFs, one can estimate the RDA for uncertain non-polynomial system by using parameter-dependent LFs. This can be done by extending the ideas introduced in Section 5.3 with those presented in this chapter for non-polynomial systems.

One thing to observe is that, bounding the reminders of the non-polynomial functions according to (6.13) becomes more difficult in the case of parameter-dependent LFs, since the sublevel set is parameter-dependent. One way to cope with this problem is to bound the remainders in a simpler set that contains such a parameter-dependent sublevel set, for instance according to (6.50)–(6.54).

Also, one can analogously consider the case of variable LFs in the search for optimal estimates, and the design of polynomial static output controllers for enlarging the RDA. The details of these extensions are omitted for conciseness.

6.7 Notes and References

The estimation of the DA for non-polynomial systems via SOS programming described in this chapter was introduced in [18, 23] for the case of uncertainty-free polynomial systems.

Relevant references on this topic include [81] which addresses stability of an equilibrium point of a non-polynomial system through suitable changes of coordinate.

Chapter 7
Miscellaneous

This chapter addresses some miscellaneous problems that can be addressed with the framework described in the previous chapters. First, the problem of estimating the DA via union of a continuous family of estimates is considered for polynomial and non-polynomial systems. The family of estimates corresponds to the LEDA provided by a continuous family of LFs, and hence the problem amounts to computing a parameter-dependent LEDA. The problem can be solved by exploiting the results introduced in Chapters 5 and 6 and by constructing a suitable parameter-dependent LF. The second problem addressed in this chapter considers the estimation of the set of admissible equilibrium points for an uncertain nonlinear system, either polynomial or non-polynomial. Specifically, it is shown how outer estimates with fixed shape can be computed by solving an SDP by exploiting SOS polynomials. A sufficient condition for establishing whether the found candidate is tight is hence provided by looking for power vectors in linear subspaces, which turns out to be also necessary in the case of uncertain polynomial systems. Then, the methodology is exteded to address the computation of the minimum volume outer estimate and the construction of the smallest convex outer estimate. The third problem addressed in this chapter considers the computation and the minimization via constrained control design of the extremes of the trajectories of a polynomial or non-polynomial system over a given set of initial conditions. It is shown how upper bounds of the sought extremal values as well as candidates of the sought controllers can be computed by solving optimization problems with BMI constraints by looking for suitable invariant sets through SOS polynomials. Moreover, a necessary and sufficient condition is proposed to establish the tightness of the found upper bound by exploiting a reverse trajectory system. Lastly, the chapter provides some remarks on the estimation of the DA and construction of global LFs for degenerate polynomial systems, in particular polynomial systems whose linearized system at the equilibrium point is marginally stable and polynomial systems where the global LF or its temporal derivative have not positive definite highest degree forms.

G. Chesi: Domain of Attraction, LNCIS 415, pp. 235–263, 2011.
springerlink.com

7.1 Union of Estimates

A locally asymptotically stable equilibrium point admits different LFs. This means that the DA of such an equilibrium point can be estimated through different LFs, which provide estimates with different shapes. This suggests that the DA can be estimated by taking the union of all these estimates, in particular the union of all the LEDAs corresponding to these LFs. However, this would require to repeat the computation of the LEDA for each of these functions, operation that is clearly intractable if the number of these functions is infinite as it happens in the case of continuous families of Lypaunov functions.

A possibility to cope with this problem could be to parametrize the family of LFs, and then to obtain a parameter-dependent LEDA for such a family. Specifically, let us consider either the polynomial system (4.1) or the non-polynomial system (6.1). For these systems, let us denote the parametrized LF of the origin as $v(x,\theta)$ where $\theta \in \mathbb{R}^{n\theta}$ is the parameter, that we assume constrained into a semialgebraic set expressed as

$$\Theta = \{\theta \in \mathbb{R}^{n\theta} : a(\theta) \geq 0,\ b(\theta) = 0\} \tag{7.1}$$

for some $a \in \mathscr{P}_{n\theta}^{n_a}$ and $b \in \mathscr{P}_{n\theta}^{n_b}$. Let us define the parameter-dependent sublevel set of $v(x,\theta)$ as $\mathscr{V}(c(\theta),\theta)$ in (3.63) for some $c : \mathbb{R}^{n\theta} \to \mathbb{R}$. The estimate of the DA provided by this family of LFs is given by

$$\mathscr{U} = \bigcup_{\theta \in \Theta} \mathscr{V}(\gamma(\theta),\theta) \tag{7.2}$$

where $\gamma(\theta)$ is given by (5.57) for the case of polynomial $f(x)$, and similarly defined for the case of non-polynomial $f(x)$.

Hence, the problem is to estimate $\gamma(\theta)$. Let us observe that a lower bound $\hat{\gamma}_w(\theta)$ of $\gamma(\theta)$ can be obtained through either Corollary 5.5 for the case of polynomial $f(x)$ or a similar one for the case of non-polynomial $f(x)$, where w specifies the criterion used to maximize the lower bound.

In order to choose the LF, let us observe that a necessary condition for the LMIs in the theorems to hold is given by (5.49), which in the present case boils down to

$$\left.\begin{array}{rcl} V_2(\theta) & > & 0 \\ V_2(\theta)A + A'V_2(\theta) & < & 0 \\ A & = & \mathrm{lp}(f) \end{array}\right\} \forall \theta \in \Theta \tag{7.3}$$

where $V_2(\theta) = V_2(\theta)'$ defines the quadratic part of $v(x,\theta)$ according to (5.50)–(5.51).

Condition (7.3) holds if and only if the linearized system is asymptotically stable (i.e., A is Hurwitz) and $v_2(x,\theta) = x'V_2(\theta)x$ is a parameter-dependent LF for it.

A function $V_2(\theta)$ satisfying (7.3) can be obtained as the solution of the Lyapunov equation

$$V_2(\theta)A + A'V_2(\theta) + Q(\theta) = 0 \tag{7.4}$$

where $Q(\theta) = Q(\theta)'$ is such that

$$Q(\theta) > 0 \ \forall \theta \in \Theta. \tag{7.5}$$

This solution has the property

$$\partial V_2 = \partial Q. \tag{7.6}$$

Also, this solution is linear in $Q(\theta)$, which means that $Q(\theta)$ can be normalized, for instance by setting one entry equal to a constant. For instance, in the case $n = 3$, one can select

$$Q(\theta) = \begin{pmatrix} 1 & \theta_1 & \theta_2 \\ \star & \theta_3 & \theta_4 \\ \star & \star & \theta_5 \end{pmatrix}.$$

Let us consider now the choice of the set Θ. This set has to be chosen in order to guarantee the fulfillment of (7.5). Let us observe that (7.5) holds if and only if

$$\det(Q_i(\theta)) > 0 \ \forall i = 1, \ldots, n \ \forall \theta \in \Theta \tag{7.7}$$

where $Q_i(\theta)$ is the $i \times i$ top-left submatrix of $Q(\theta)$ (in other words, $\det(Q_i(\theta))$ is the i-th principal minor of $Q(\theta)$). Hence, a simple way to guarantee the fulfillment of (7.5) is to define Θ as in (7.1) with $n_b = 0$ and

$$a_i(\theta) = \det(Q_i(\theta)) - \varepsilon \ \forall i = 1, \ldots, n \tag{7.8}$$

for some $\varepsilon > 0$. This suggests that a function $v(x, \theta)$ can be chosen as follows:

- choose $Q(\theta)$ and Θ satisfying (7.5);
- find $V_2(\theta)$ satisfying (7.4);
- define $v_2(x, \theta) = x'V_2(\theta)x$;
- define $v(x, \theta) = v_2(x, \theta) + \sum_{i=3}^{\partial v} v_i(x, \theta)$ where $v_3(x, \theta), v_4(x, \theta), \ldots$ are arbitrarily chosen under the constraint that $v(x, \theta)$ is positive definite in x for all $\theta \in \Theta$.

(7.9)

Let us observe that one can establish whether $v(x, \theta)$ is positive definite in x for all $\theta \in \Theta$ with the technique in (5.86)–(5.87).

Example 7.1. Let us consider the problem of determining $\mathscr{U}$ for the Van der Pol system

$$\begin{cases} \dot{x}_1 = x_2 \\ \dot{x}_2 = -2x_1 - 3x_2 + x_1^2 x_2. \end{cases}$$

Let us consider a family of quadratic LFs depending on one free parameter. Let us select the parameter-dependent LF as $v(x, \theta) = x'V(\theta)x$ where $V(\theta) = F(Q(\theta))$ and

$$Q(\theta) = \begin{pmatrix} 1 & 0 \\ \star & \theta \end{pmatrix}.$$

We have

$$V_2(\theta) = \frac{1}{12}\begin{pmatrix} 11 & 3 \\ \star & 1 \end{pmatrix} + \frac{1}{6}\begin{pmatrix} 2 & 0 \\ \star & 1 \end{pmatrix}.$$

Let us observe that $Q(\theta)$ is positive definite whenever $\theta > 0$. In order to avoid to consider matrices $Q(\theta)$ that are too close to singular matrices, we choose

$$\Theta = [1/5, 5]$$

through the choice

$$a(\theta) = -1 + 26/5\theta - \theta^2.$$

By using polynomials $c(\theta)$ of degree ranging from 0 to 4 and by choosing w according to (5.62), we obtain the following lower bounds of $\gamma(\theta)$:

$$\begin{aligned}
\hat{\gamma}_w(\theta) &= 1.4908 \\
\hat{\gamma}_w(\theta) &= 1.6794 + 1.5296\theta \\
\hat{\gamma}_w(\theta) &= 1.5717 + 2.1211\theta - 0.1249\theta^2 \\
\hat{\gamma}_w(\theta) &= 1.4729 + 2.6393\theta - 0.3944\theta^2 + 0.0361\theta^3 \\
\hat{\gamma}_w(\theta) &= 1.3561 + 3.3377\theta - 0.9863\theta^2 + 0.2085\theta^3 - 0.0163\theta^4.
\end{aligned}$$

Figure 7.1a shows the set $\mathscr{U}$ and the boundary of the DA, which has been found with the trajectory reverse method [48] and which represents an unstable limit cycle. Figure 7.1b shows the found lower bounds $\hat{\gamma}_w(\theta)$ and the quantity $\gamma(\theta)$ calculated for some values of θ. □

7.2 Equilibrium Points in Uncertain Systems

Let us consider the uncertain polynomial system described by

$$\begin{cases} \dot{x}(t) = f(x(t), \theta) \\ \theta \in \Theta \end{cases} \tag{7.10}$$

where $x \in \mathbb{R}^n$ is the state, $t \in \mathbb{R}$ is the time, $\theta \in \mathbb{R}^{n_\theta}$ is the uncertainty, $\Theta \subset \mathbb{R}^{n_\theta}$ is the set of admissible uncertainties, and $f : \mathbb{R}^n \times \mathbb{R}^{n_\theta} \to \mathbb{R}^n$ is a non-polynomial function of the form (6.70). We consider the case where Θ is a semialgebraic set expressed as

$$\Theta = \{\theta \in \mathbb{R}^{n_\theta} : \ a(\theta) \geq 0, \ b(\theta) = 0\} \tag{7.11}$$

for some $a \in \mathscr{P}_{n\theta}^{n_a}$ and $b \in \mathscr{P}_{n\theta}^{n_b}$. The set of admissible equilibrium points of (7.10) is given by

$$\mathscr{O} = \{x \in \mathbb{R}^n : \ f(x,\theta) = 0_n \ \text{ for some } \theta \in \Theta\}. \tag{7.12}$$

The first problem that we address consists of determining outer estimates of $\mathscr{O}$ of the form

$$\mathscr{V}(c) = \{x \in \mathbb{R}^n : \ v(x) \leq c\} \tag{7.13}$$

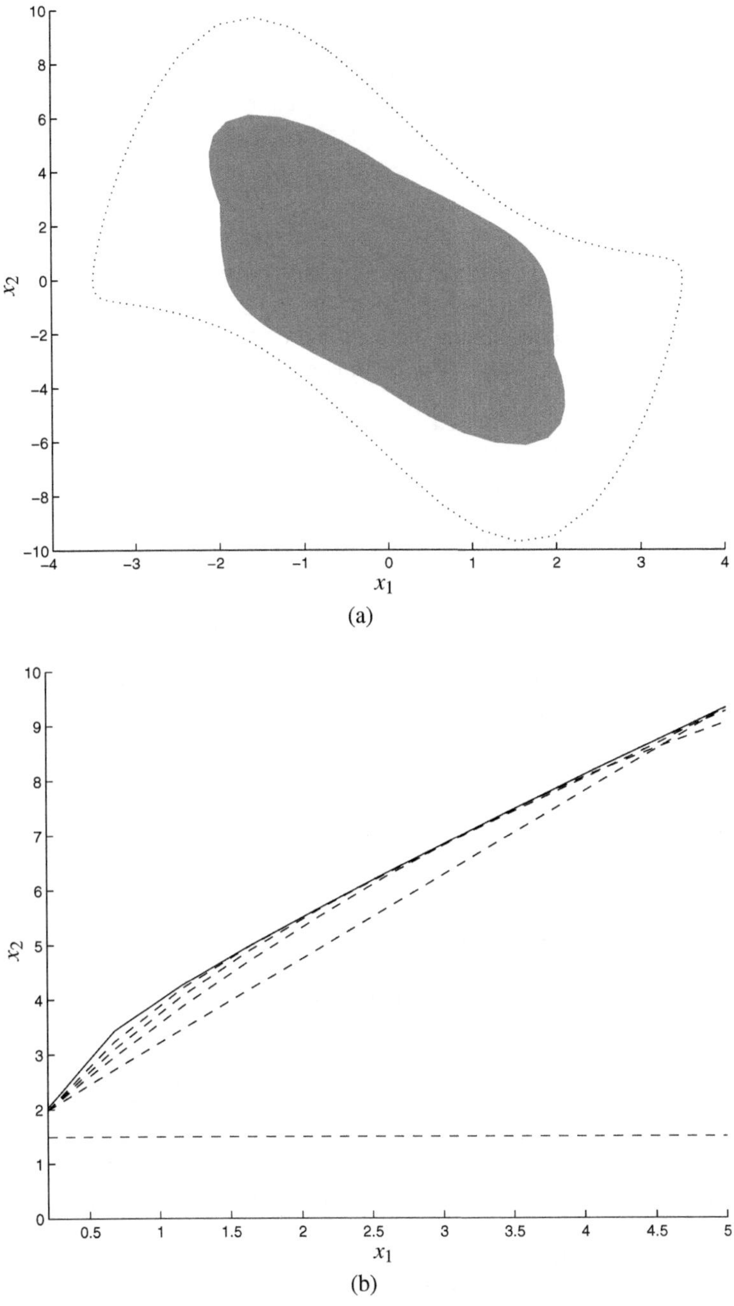

Fig. 7.1 Example 7.1 (Van der Pol system). (a) Estimate $\mathscr{U}$ found with a family of quadratic LFs depending on one free parameter (gray area). (b) Lower bound $\gamma(\theta)$ calculated for some values of θ (solid) and lower bounds $\hat{\gamma}_w(\theta)$ for $\partial c = 0, \ldots, 4$ (dashed). In (a), the dotted line is the boundary of the DA.

where $v \in \mathscr{P}_n$ and $c \in \mathbb{R}$. In particular, the problem consists of estimating the smallest outer estimate of $\mathscr{O}$ with fixed shape defined by $v(x)$, which is denoted by $\mathscr{V}(\gamma)$ where γ is the solution of the optimization problem

$$\begin{aligned} \gamma = & \inf_{c} c \\ & \text{s.t. } \begin{cases} \mathscr{O} \subseteq \mathscr{V}(c) \\ c \geq 0. \end{cases} \end{aligned} \tag{7.14}$$

The second problem consists of determining less conservative outer estimates of $\mathscr{O}$ by using a variable shape. This is firstly addressed by considering the problem of determining the minimum volume outer estimate of $\mathscr{O}$ among the sublevel sets of polynomials of a given degree, i.e. the estimate

$$\mathscr{V}^* = \{x \in \mathbb{R}^n : v^*(x) \leq 1\} \tag{7.15}$$

where $v^*(x)$ is the solution of the optimization problem

$$\begin{aligned} v^*(x) = & \arg\inf_{v} \text{vol}(\mathscr{V}(1)) \\ & \text{s.t. } \begin{cases} \mathscr{O} \subseteq \mathscr{V}(1) \\ \partial v = constant \end{cases} \end{aligned} \tag{7.16}$$

where vol$(\cdot)$ denotes the volume. Lastly, we address the problem of constructing the smallest convex outer estimate of $\mathscr{O}$, i.e. the set

$$\begin{aligned} \mathscr{H}^* = & \arg\inf_{\mathscr{H}} \text{vol}(\mathscr{H}) \\ & \text{s.t. } \begin{cases} \mathscr{O} \subseteq \mathscr{H} \\ \mathscr{H} \text{ is convex} \end{cases} \end{aligned} \tag{7.17}$$

7.2.1 *Estimate Computation*

With $\xi = (\xi_1, \ldots, \xi_{n_\phi})'$ and for some integers m_i let us define

$$z(x,\theta,\xi) = f^{(0)}(x,\theta) + \sum_{i=1}^{n_\phi} f^{(i)}(x,\theta)\left(\hat{\phi}_i(\tau_i) + \xi_i \frac{\tau_i^{m_i}}{m_i!}\right). \tag{7.18}$$

Theorem 7.1. *Define Ξ in* (6.11) *and suppose that* (6.13) *holds. Suppose there exists $g \in \mathscr{P}^n_{n,n_\theta}$, $q \in \mathscr{P}_{n,n_\theta}$, $r \in \mathscr{P}^{n_a}_{n,n_\theta}$ and $s \in \mathscr{P}^{n_b}_{n,n_\theta}$ such that*

$$\left.\begin{array}{r} p(x,\theta,\xi) \geq 0 \\ q(x,\theta) > 0 \\ r(x,\theta) \geq 0 \end{array}\right\} \forall x \in \mathbb{R}^n \; \forall \theta \in \mathbb{R}^{n_\theta} \; \forall \xi \in \text{ver}(\Xi) \tag{7.19}$$

where

$$p(x,\theta,\xi) = g(x,\theta)'z(x,\theta,\xi) + (c - v(x))q(x,\theta) - r(x,\theta)'a(\theta) - s(x,\theta)'b(\theta). \tag{7.20}$$

Then, $\gamma \leq c$.

Proof. Let us consider any $\bar{x} \in \mathscr{O}$, and let $\bar{\theta} \in \Theta$ be such that $f(\bar{x},\bar{\theta}) = 0$. Let $\bar{\xi}$ be the coefficient of the reminder corresponding to x. One has hence that $z(\bar{x},\bar{\theta},\bar{\xi}) = 0_n$. Let us suppose that the constraints in (7.19) are fulfilled. Since $\bar{\xi} \in \Xi$ and $p(x,\theta,\xi)$ depends affine linearly on ξ, from the second constraint we obtain

$$\begin{aligned} 0 &\leq p(\bar{x},\bar{\theta},\bar{\xi}) \\ &= g(\bar{x},\bar{\theta})'z(\bar{x},\bar{\theta},\bar{\xi}) + (c - v(\bar{x}))q(\bar{x},\bar{\theta}) - r(\bar{x},\bar{\theta})'a(\bar{\theta}) - s(\bar{x},\bar{\theta})'b(\bar{\theta}) \\ &\leq g(\bar{x},\bar{\theta})'z(\bar{x},\bar{\theta},\bar{\xi}) + (c - v(\bar{x}))q(\bar{x},\bar{\theta}) \\ &= (c - v(\bar{x}))q(\bar{x},\bar{\theta}). \end{aligned}$$

Finally, let us observe that $q(\bar{x},\bar{\theta}) > 0$ from the second constraint in (7.19), which implies that $v(\bar{x}) \leq c$. Therefore, $\gamma \leq c$. □

From Theorem 7.1 one can define a natural upper bound of γ by exploiting the SOS index as follows. Specifically, let us denote the dependence on c of the quantities ξ_i^-, ξ_i^+, $\xi^{(l)}$ and Ξ by introducing the corresponding new quantities $\xi_i^-(c)$, $\xi_i^+(c)$, $\xi^{(l)}(c)$ and $\Xi(c)$. Define

$$\hat{\gamma} = \inf_{c,g,q,r,s} c$$
$$\text{s.t.} \begin{cases} \lambda_{polpol}(p(\cdot,\cdot,\xi^{(l)}(c))) \geq 0 \ \forall l = 1,\ldots,n_\Xi \\ \lambda_{polpol}(q) > 0 \\ \lambda_{polpol}(r_i) \geq 0 \ \forall i = 1,\ldots,n_a. \end{cases} \tag{7.21}$$

Then,

$$\gamma \leq \hat{\gamma}. \tag{7.22}$$

The upper bound $\hat{\gamma}$ can be found via a bisection search where at each iteration the condition of Theorem 7.1 is investigated through the SOS index of some polynomials, i.e. via LMI feasibility tests.

7.2.2 Estimate Tightness

At this point, a question that naturally arises concerns the tightness of the found upper bound: is $\hat{\gamma} = \gamma$? The following theorem provides an answer to this question.

Theorem 7.2. *Let* γ *and* $\hat{\gamma}$ *be as in* (7.14) *and* (7.21), *respectively. Suppose that* $0 < \gamma < \infty$, *and for* $l = 1,\ldots,n_\Xi$ *define the set*

$$\mathscr{M}^{(l)} = \left\{(x,\theta) : b_{pol}(\theta, \partial^\theta p/2) \otimes b_{pol}(x, \partial^x p/2) \in \ker(M^{(l)})\right\} \tag{7.23}$$

where $M^{(l)}$ is the found optimal value of the SMR matrix of $p(x,\theta,\xi^{(l)}(\hat{\gamma}))$ in (7.21). Define also the set

$$\mathscr{M}_1^{(l)} = \left\{(x,\theta) \in \mathscr{M}^{(l)} : \theta \in \Theta \text{ and } v(x) = \hat{\gamma} \text{ and } f(x,\theta) = 0_n\right\}. \tag{7.24}$$

Then,

$$\exists l = 1,\ldots,n_{\Xi} : \mathscr{M}_1^{(l)} \neq \emptyset \Rightarrow \hat{\gamma} = \gamma. \tag{7.25}$$

Moreover, in the case that $f(x,\theta)$ is polynomial, one has that

$$\exists l = 1,\ldots,n_{\Xi} : \mathscr{M}_1^{(l)} \neq \emptyset \iff \hat{\gamma} = \gamma. \tag{7.26}$$

Proof. "$\Rightarrow$" Let us suppose that $\mathscr{M}_1^{(l)} \neq \emptyset$ for some l, and let $(x,\theta) \in \mathscr{M}_1^{(l)}$. We have that x is an admissible equilibrium point of the system since $f(x,\theta) = 0_n$ and $\theta \in \Theta$. Moreover, x satisfies $v(x) = \hat{\gamma}$, i.e. x lies on the boundary of $\mathscr{V}(\hat{\gamma})$. This implies that $\hat{\gamma}$ is a lower bound of γ from the definition of γ. Moreover, $\hat{\gamma}$ is an upper bound of γ from (7.22). Therefore, we conclude that $\hat{\gamma} = \gamma$.

"$\Leftarrow$" Consider the case that $f(x,\theta)$ is polynomial. Let us suppose that $\hat{\gamma} = \gamma$. Let $x \in \mathscr{O}$ be a tangent point between $\mathscr{O}$ and $\mathscr{V}(\hat{\gamma})$, and let $\theta \in \Theta$ be such that $f(x,\theta) = 0_n$. Since $M^{(l)}$ is positive semidefinite it follows that

$$\begin{aligned}
0 &\leq \left(b_{pol}(\theta,\partial^\theta p/2) \otimes b_{pol}(x,\partial^x p/2)\right)' M^{(l)} \left(b_{pol}(\theta,\partial^\theta p/2) \otimes b_{pol}(x,\partial^x p/2)\right) \\
&= p(x,\theta) \\
&= g(x,\theta)'z(x,\theta) + (\hat{\gamma} - v(x))q(x,\theta) - r(x,\theta)'a(\theta) - s(x,\theta)'b(\theta) \\
&= -r(x,\theta)'a(\theta)
\end{aligned}$$

since $z(x,\theta) = 0_n$, $\hat{\gamma} - v(x) = 0$ and $b(\theta) = 0$. Moreover, $a(\theta) \geq 0$ and $r(x,\theta) \geq 0$, which implies that

$$-r(x,\theta)'a(\theta) = 0.$$

Hence, $b_{pol}(\theta,\partial^\theta p/2) \otimes b_{pol}(x,\partial^x p/2) \in \ker(M^{(l)})$ since $M^{(l)}$ is positive semidefinite, i.e. $(x,\theta) \in \mathscr{M}_1^{(l)}$ and the theorem holds. □

Theorem 7.2 provides a condition for establishing whether the upper bound provided by (7.21) is tight. In particular, this happens if the set $\mathscr{M}_1^{(l)}$ in (7.24) is nonempty. This set can be found via trivial substitution from the set $\mathscr{M}^{(l)}$ in (7.23) (at least whenever $\mathscr{M}^{(l)}$ is finite), which can be computed as explained in Section 1.5.2 after (1.202).

7.2.3 Variable Shape Estimates

Let us start by showing how the methodology proposed in the previous sections can further be elaborated in order to search for the minimum volume estimate of $\mathscr{O}$. If $v(x)$ is quadratic, then

$$\begin{cases} \mathrm{vol}(\mathscr{V}(1)) \propto \det(V)^{-1/2} \\ V : V > 0, \;\; v(x) = (x-x_0)'V(x-x_0) \end{cases} \tag{7.27}$$

and hence the problem amounts to maximizing $\det(V)$. Let us observe that, with $w \in \mathbb{R}$ and with a suitable positive integer η depending on n, the condition

$$\begin{cases} \sqrt[2\eta]{\det(V)} \geq w \\ V \geq 0 \end{cases} \tag{7.28}$$

can equivalently be written through a suitable LMI in V, w and some additional variables, see Appendix B. Hence, problem (7.21) is replaced by

$$\hat{\gamma} = \sup_{V,d,g,r,s} \; d \quad \text{s.t.} \begin{cases} \lambda_{polpol}(p(\cdot,\cdot,\xi)) \geq 0 \;\; \forall \xi \in \mathrm{ver}(\Xi) \\ \lambda_{polpol}(r_i) \geq 0 \;\; \forall i = 1,\dots,n_a \\ \sqrt[2\eta]{\det(V)} - d \geq 0 \end{cases} \tag{7.29}$$

where $p(x,\theta,\xi)$ is defined with $c = 1$ and a chosen positive parameter-dependent polynomial $q(x,\theta)$. If $v(x)$ has degree larger than 2, one possibility is to use the same procedure by maximizing the determinant of an SMR matrix of $v(x)$ (this will approximately minimize $\mathrm{vol}(\mathscr{V}(1))$).

Next, we consider the construction of the smallest convex estimate of $\mathscr{O}$, i.e. the set $\mathscr{H}^*$ in (7.17). The basic idea consists of generating a convex polytopic estimate via intersection of a sequence of half-spaces. Specifically, we start by computing an outer estimate of $\mathscr{O}$ with spherical shape. This can be done by choosing $v(x) = \|x - x_0\|^2$ for some $x_0 \in \mathbb{R}^n$ by exploiting (7.21). We hence obtain a sphere $\mathscr{V}(\hat{\gamma})$ which is guaranteed to contain $\mathscr{O}$. Then, we sample the boundary of $\mathscr{V}(\hat{\gamma})$ at n_s points $x^{(1)},\dots,x^{(n_s)}$, for example by using equally distributed points. For each of these points, we compute an outer estimate of $\mathscr{O}$ by choosing in place of $v(x)$ the functions

$$v_i(x) = \left(\frac{(x_0 - x^{(i)})'(x - x^{(i)})}{\|x_0 - x^{(i)}\|} \right)^2. \tag{7.30}$$

The level sets of $v_i(x)$ are planes orthogonal to the ray $x_0 - x^{(i)}$ of the sphere, and $v_i(x)$ is the square of the distance of the plane containing x from $x^{(i)}$, see Figure 7.2a. Let $\hat{\gamma}_i$ be the upper bound found for $v_i(x)$. We hence obtain that the set $\mathscr{V}_i(\hat{\gamma}_i)$, given by

$$\mathscr{V}_i(\gamma_i) = \{x \in \mathbb{R}^n : \; v_i(x) \leq \hat{\gamma}_i\} \tag{7.31}$$

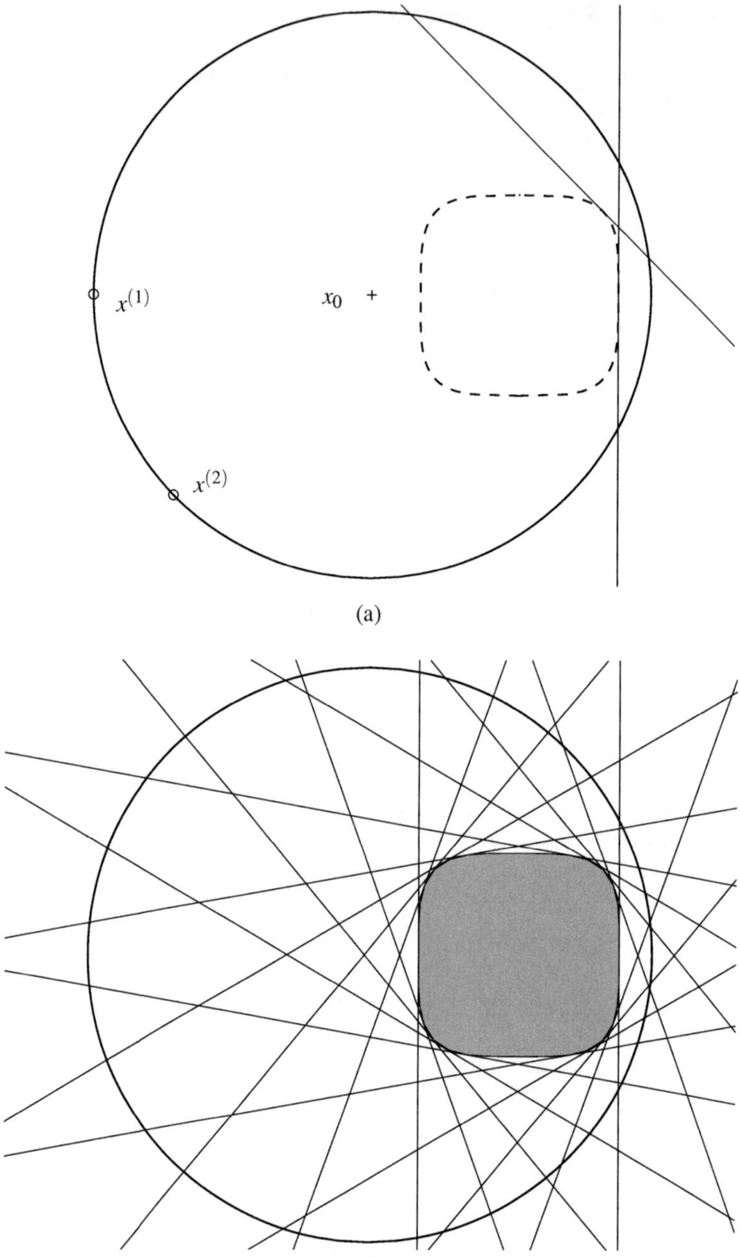

Fig. 7.2 Procedure for estimating the convex hull of $\mathscr{O}$. (a) For each point $x^{(i)}$ on a spherical outer estimate centered at x_0, an outer estimate with planar level sets perpendicular to $x_0 - x^{(i)}$ is computed via (7.21). (b) A convex estimate is obtained as intersection of the half-spaces identified by the previous computations.

is an outer estimate of $\mathscr{O}$. Lastly, we define

$$\mathscr{H}_i = \left\{ x \in \mathbb{R}^n : \frac{(x_0 - x^{(i)})'(x - x^{(i)})}{\|x_0 - x^{(i)}\|} \le \sqrt{\hat{\gamma}_i} \right\}. \tag{7.32}$$

We have that $\mathscr{H}_i$ is a half-space delimited by one of the two planes delimiting $\mathscr{V}_i(\hat{\gamma}_i)$ and containing $\mathscr{O}$. From $\mathscr{H}_1, \ldots, \mathscr{H}_{n_s}$ we define the final estimate

$$\mathscr{H} = \bigcap_{i=1}^{n_s} \mathscr{H}_i \tag{7.33}$$

which is a convex polytope containing $\mathscr{O}$. See Figure 7.2b for details. It turns out that, for any chosen x_0 and n_s,

$$\mathscr{O} \subseteq \mathscr{H}^* \subseteq \mathscr{H}. \tag{7.34}$$

Example 7.2. Let us consider the uncertain polynomial system

$$\begin{cases} \dot{x}_1 = x_1^2 + (\theta - 2)x_1x_2 + x_2^2 + 3\theta - 4 \\ \dot{x}_2 = x_1^2 + (1 - 4\theta)x_1x_2 + x_2^2 + (1 - 2\theta)x_1 - 2\theta \\ \theta \in \Theta = [0, 1]. \end{cases}$$

The set Θ can be expressed as in (7.11) with $a(\theta) = \theta - \theta^2$. Let us select the shape function $v(x) = \|x\|^2$. By solving (7.21) we find the upper bound $\hat{\gamma} = 6.811$ of γ. Figure 7.3a shows the boundary of the found estimate $\mathscr{V}(\hat{\gamma})$ and the equilibrium points of the system computed for 101 values of θ equally distributed in $[0, 1]$.

In order to establish whether the found upper bound is tight, we use Theorem 7.2. We find that $\mathscr{M}_1 \neq \emptyset$, in particular $(x, \theta) \in \mathscr{M}_1$ with

$$x = (1.857, 1.834)', \quad \theta = 0.624.$$

This implies that $\hat{\gamma} = \gamma$. Figure 7.3b shows the equilibrium points of the system achieved for $\theta = 0.624$.

It is worth observing that the computation of the equilibrium points shown in Figure 7.3a for 101 values of θ is unable to find the extreme equilibrium points that delimit the outer estimate of $\mathscr{O}$ (in fact, no equilibrium point lies on the boundary of $\mathscr{V}(\hat{\gamma})$ in Figure 7.3a). Such an extreme point is obtained via Theorem 7.2 and is shown in Figure 7.3b.

Figure 7.4a shows the smallest outer estimate with ellipsoidal shape $\mathscr{V}^{\#}$ centered in the origin obtained as described in Section 7.2.3. In particular, the optimal quadratic function $v(x)$ found by solving (7.29) is

$$v^*(x) = 0.299x_1^2 - 0.354x_1x_2 + 0.346x_2^2.$$

Lastly, Figure 7.4b shows the polytopic convex estimate $\mathscr{H}$ obtained from Section 7.2.3 by using 18 points equally distributed on the boundary of $\mathscr{V}(\hat{\gamma})$. □

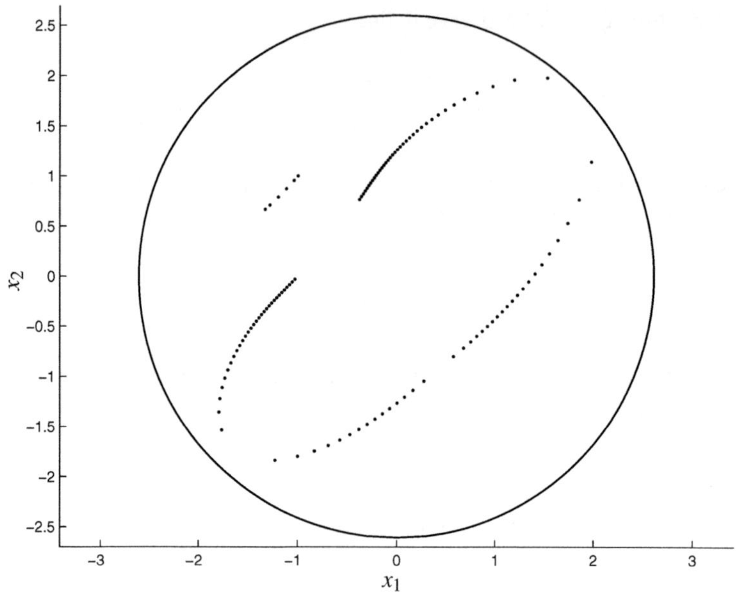

(a)

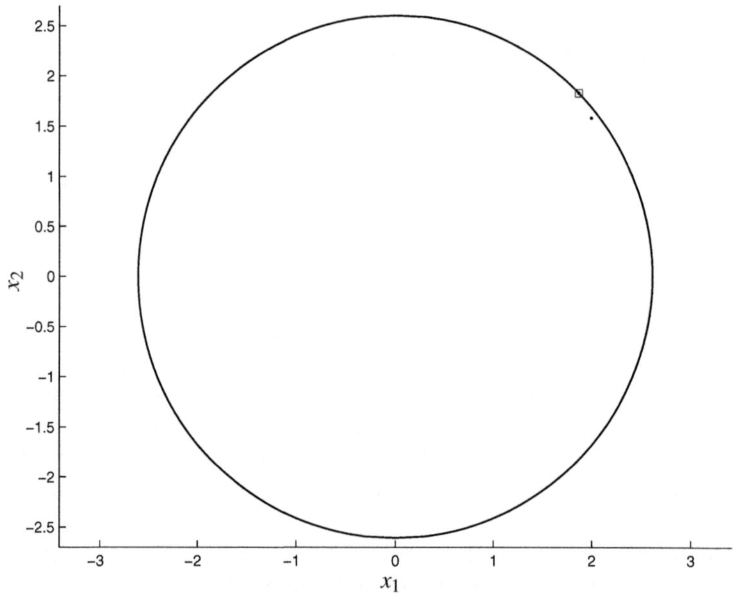

(b)

Fig. 7.3 Example 7.2. (a) Boundary of $\mathscr{V}(\hat{\gamma})$ for $v(x) = \|x\|^2$ (solid) and equilibrium points for 101 values of θ equally distributed in $[0,1]$ (dots). (b) Equilibrium points of the system for the uncertain parameter found with Theorem 7.2 (dots). The "□" marker is the x in $\mathscr{M}_1$.

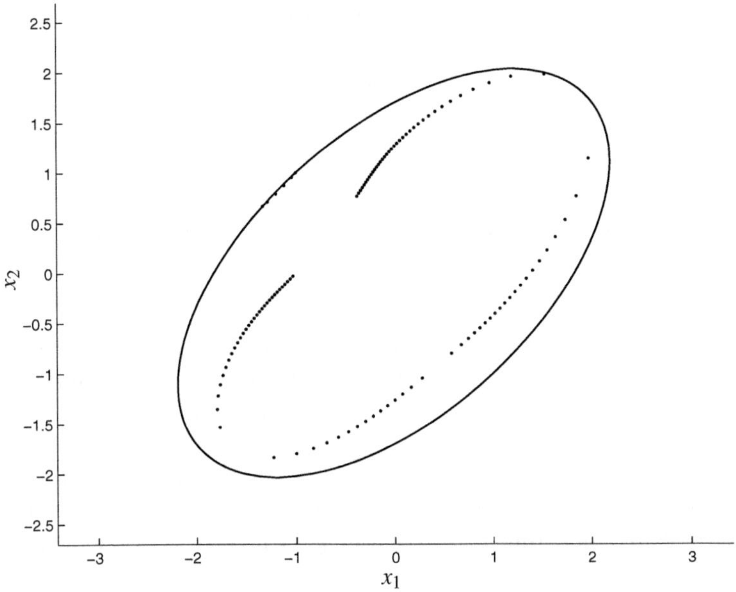

(a)

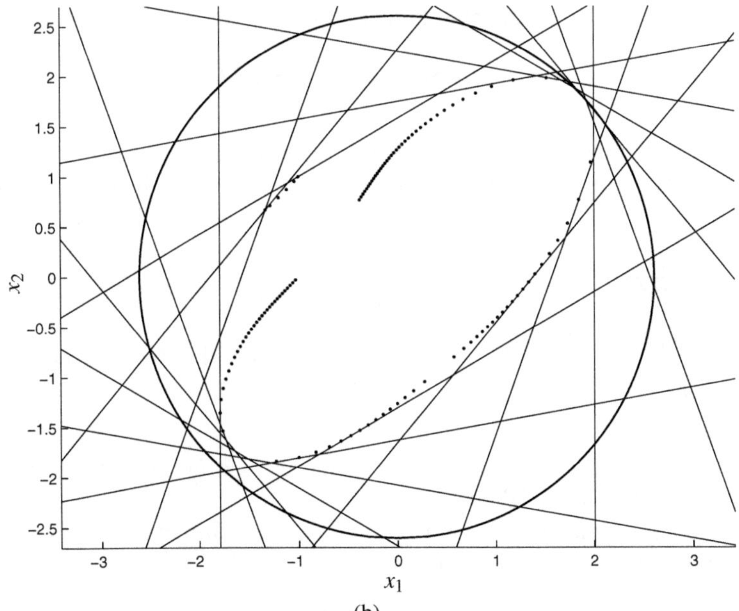

(b)

Fig. 7.4 Example 7.2. (a) Minimum volume estimate with ellipsoidal shape (solid). (b) Convex estimate constructed with 18 half-spaces (straight lines).

Example 7.3. Let us consider the uncertain non-polynomial system

$$\begin{cases} \dot{x}_1 = x_1^2 - x_1x_2 + 3x_2^2 + 2\theta x_1 - 1 \\ \dot{x}_2 = x_2^2 + 2x_1x_2 - \theta - \sin x_1 \\ \theta \in \Theta = [0,1]. \end{cases}$$

The set Θ can be expressed as in (7.11) with $a(\theta) = \theta - \theta^2$. Let us select the shape function $v(x) = \|x\|^2$. Condition (6.13) is simply satisfied by choosing

$$\xi_1^- = -1, \ \xi_1^+ = 1$$

for any truncation order m_1 of $\sin x_1$ and for any c. By using $m_1 = 1,3,5$ we find the upper bounds $\hat{\gamma} = 6.993$, $\hat{\gamma} = 6.799$ and $\hat{\gamma} = 6.289$, respectively. Figure 7.5a shows the boundary of $\mathscr{V}(\hat{\gamma})$ for $k = 5$ and the equilibrium points of the system computed for 101 values of θ equally distributed in $[0,1]$.

As we can see, the upper bound $\hat{\gamma}$ for $m_1 = 5$ is nearly tight as there are equilibrium points close to the boundary of $\mathscr{V}(\hat{\gamma})$. Moreover, tighter upper bounds can be obtained either by increasing k or by repeating the procedure with tighter quantities ξ_1^- and ξ_1^+.

Figure 7.5b shows the smallest outer estimate with ellipsoidal shape $\mathscr{V}^\#$ centered in the origin obtained as described in Section 7.2.3. In particular, the optimal quadratic function $v(x)$ found by solving (7.29) is

$$v(x)^* = 0.169x_1^2 - 0.296x_1x_2 + 2.053x_2^2. \qquad \square$$

Example 7.4. Let us consider the electrical circuit in Figure 7.6a with three tunnel diodes and two variable resistors. By indicating with x_i the voltage of the capacitor C_i, $i = 1,2$, and selecting x_2 as output, the system can be described by

$$\begin{cases} R_0C_1\dot{x}_1 = E - x_1 - x_2 - \dfrac{R_0}{R_1(\beta_1)}x_1 - R_0h(x_1) - R_0h(x_1 + x_2) \\ R_0C_2\dot{x}_2 = E - x_1 - x_2 - \dfrac{R_0}{R_2(\beta_2)}x_1 - R_0h(x_2) - R_0h(x_1 + x_2) \\ y = x_2 \end{cases}$$

where $R_1(\beta_1)$ and $R_2(\beta_2)$ are variable resistances, and $h(\cdot)$ is the voltage-current characteristic of the tunnel diodes.

Let us consider the problem of determining the maximum absolute value of the output of the system in steady-state, i.e.

$$y^* = \sup\{|x_2| : x \in \mathscr{O}\}$$

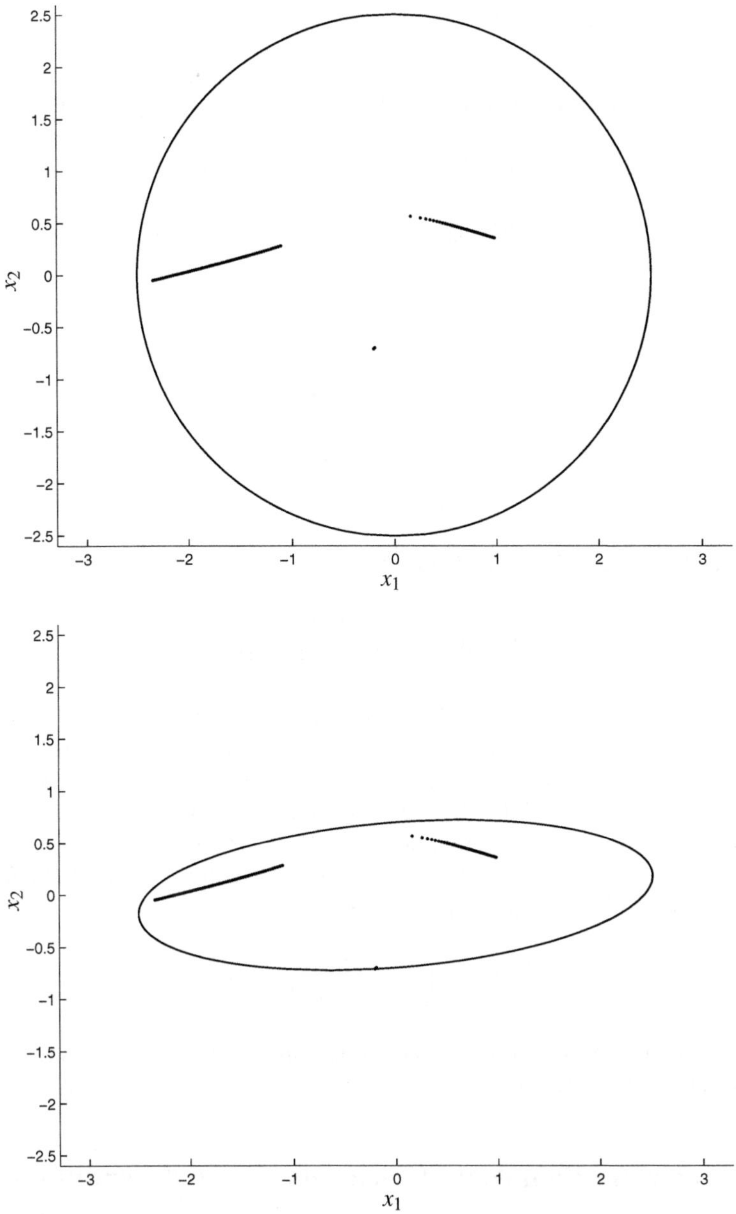

Fig. 7.5 Example 7.3. (a) Boundary of $\mathscr{V}(\hat{\gamma})$ for $m_1 = 5$ (solid) and equilibrium points for 101 values of θ equally distributed in $[0, 1]$ (dots). (b) Minimum volume estimate with ellipsoidal shape (solid).

where $\mathscr{O}$ is the set of equilibrium points of the circuit. We select the plausible values

$$C_1 = 2\cdot 10^{-9},\ C_2 = 2\cdot 10^{-9},\ E = 1.2,\ R_0 = 500$$
$$R_1(\beta_1) = 500 + 3000\beta_1,\ R_2(\beta_2) = 1500 + 3000\beta_2,\ \beta \in [0,1]^2.$$

Moreover, we adopt the expression of $h(\cdot)$ in [41, 65] given by

$$h(z) = \left(17.76z - 103.79z^2 + 229.62z^3 - 226.31z^4 + 83.72z^5\right) 10^{-3}$$

For this system one can select

$$\theta = \left(\frac{R_0}{R_1(\beta_1)}, \frac{R_0}{R_2(\beta_2)}\right)'.$$

Hence, the set of admissible uncertainty θ is given by

$$\Theta = [1/7, 1] \times [1/9, 1/3]$$

which can be expressed as in (7.11) with

$$a(\theta) = \begin{pmatrix} -1/7 + 8/7\theta_1 - \theta_1^2 \\ -1/27 + 4/9\theta_2 - \theta_2^2. \end{pmatrix}.$$

Let us select the shape function $v(x) = x_2^2$, hence implying $\gamma = (y^*)^2$. We find the upper bound $\hat{\gamma} = 0.762$. In order to establish whether $\hat{\gamma}$ is tight, we use Theorem 7.2, and we find that $\mathscr{M}_1 \neq \emptyset$, in particular $(x,\theta) \in \mathscr{M}_1$ with $x = (0.021, 0.873)'$. This implies that $\hat{\gamma} = \gamma$. Figure 7.6b shows the boundary of $\mathscr{V}(\hat{\gamma})$, i.e. y^*.

Figure 7.6b also shows the minimum volume ellipsoidal estimate obtained as described in Section 7.2.3 for $x_0 = (0.450, 0.450)'$ (solid line). In particular, the optimal matrix V found by solving (7.29) is

$$V^* = \begin{pmatrix} 14.228 & 9.886 \\ \star & 10.921 \end{pmatrix}$$

Lastly, Figure 7.6b also shows the polytopic convex estimate $\mathscr{H}$ constructed with 18 half-planes (dashed area). □

7.3 Trajectory Bounds for Given Sets of Initial Conditions

Let us consider the non-polynomial system

$$\begin{cases} \dot{x}(t) = f(x(t)) + g(x(t))u(t) \\ y(t) = h(x(t)) \\ x(0) = x_{init} \end{cases} \tag{7.35}$$

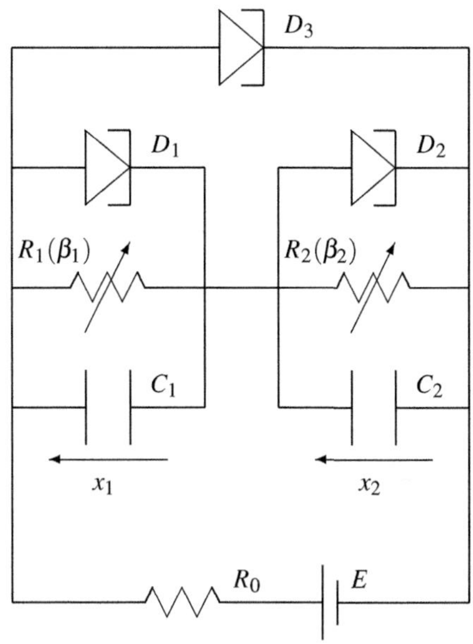

(a)

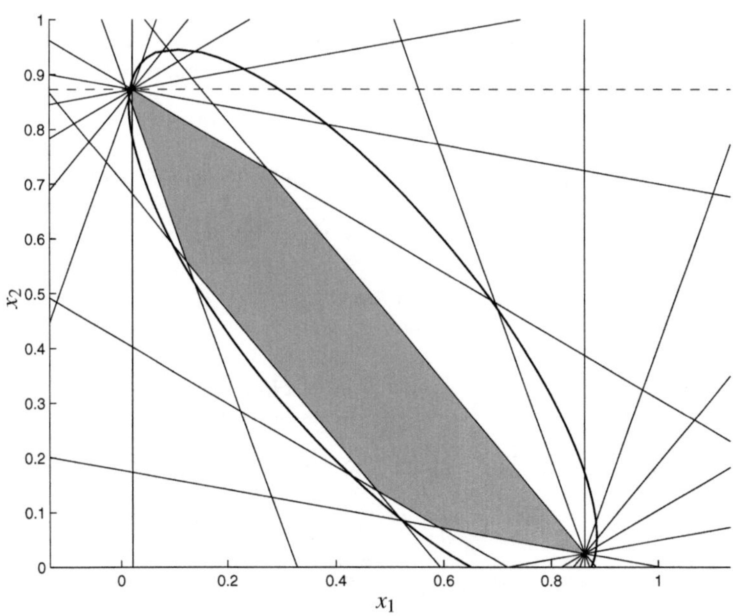

(b)

Fig. 7.6 Example 7.4 (Tunnel diodes circuit). (a) Electrical circuit with three tunnel diodes and two variable resistors. (b) Solution y^* (dashed), minimum volume ellipsoidal estimate (solid), and polytopic convex estimate (dashed area).

where $x \in \mathbb{R}^n$ is the state, $u(t) \in \mathbb{R}^{n_u}$ is the input, $y(t) \in \mathbb{R}^{n_y}$ is the output, $f : \mathbb{R}^n \to \mathbb{R}^n$ has the structure in (6.2), $g \in \mathscr{P}_n^{n \times n_u}$ is the input weighting function, and $h \in \mathscr{P}_n^{n_y}$ is the output function. We consider that x_{init} is constrained according to

$$\begin{aligned} x_{init} &\in \Theta \\ \Theta &= \{x \in \mathbb{R}^n : \ a(x) \geq 0, \ b(x) = 0\} \end{aligned} \tag{7.36}$$

where Θ is the set of admissible initial conditions defined for some $a \in \mathscr{P}_n^{n_a}$ and $b \in \mathscr{P}_n^{n_b}$ such that Θ is compact. It is assumed that the origin is the equilibrium point of interest and the output vanishes at the origin, i.e. $h(0_n) = 0_{n_y}$.

For some $z_p \in \mathscr{P}_n^{n_{zp}}$, $z_l \in \mathscr{P}_n^{n_{zl}}$, $w_p \in \mathscr{P}_n^{n_{wp}}$ and $w_l \in \mathscr{P}_n^{n_{wl}}$ with $\partial z_l = \partial w_l = 1$, let us define

$$\begin{aligned} z(t) &= (z_p(x(t))', z_l(u(t))')' \\ w(t) &= (w_p(x(t))', w_l(u(t))')'. \end{aligned} \tag{7.37}$$

The first problem that we consider consists of characterizing the extremes of the admissible trajectories of the autonomous system, in particular computing

$$\begin{aligned} \gamma = \sup_{t, x_{init}, u(t)} \ & \max(z(t)) \\ \text{s.t.} \ & \begin{cases} t \geq 0 \\ x_{init} \in \Theta \\ u(t) = 0. \end{cases} \end{aligned} \tag{7.38}$$

The second problem is the design of a polynomial static output controller minimizing the extremes of the admissible trajectories of the system, which is stated as

$$\begin{aligned} \gamma^* &= \inf_k \gamma(k) \\ & \text{s.t.} \ \begin{cases} \text{the origin is locally asymptotically stable with } u(t) = k(y(t)) \\ \max(w(t)) \leq w_0 \ \ \forall t \geq 0 \end{cases} \end{aligned} \tag{7.39}$$

where

$$\begin{aligned} \gamma(k) = \sup_{t, x_{init}, u} \ & \max(z(t)) \\ \text{s.t.} \ & \begin{cases} t \geq 0 \\ x_{init} \in \Theta \\ u(t) = k(y(t)). \end{cases} \end{aligned} \tag{7.40}$$

Let us observe that, depending on the choice of z, one can select several costs such as:

$$\begin{gathered} z = (h(x)' B_1' B_1 h(x), u' B_2', -u' B_2', B_3 x)' \\ \Downarrow \\ \max(z) = \max\{\|B_1 y\|_2^2, \|B_2 u\|_\infty, b_{3,1} x, \ldots, b_{3,m} x\} \end{gathered}$$

where B_1, B_2, B_3 are weighting matrices of suitable dimensions, and $b_{3,1}, \ldots, b_{3,m}$ are the rows of B_3. Moreover, one can analogously define several constraints by similarly defining w and w_0.

7.3.1 Analysis

Let us consider the analysis problem with $u = 0$ (and, consequently, with $z = z_p(x)$). The basic idea is to look for a LF whose unitary sublevel set is an invariant set and contain Θ. Then, we also require that this sublevel set is contained in the region where $\max(z_p(x))$ is bounded by a certain quantity d. If we can find this LF, it is guaranteed that $d \geq \gamma$.

More formally, let us denote with $v(x)$ the LF candidate, and define

$$\begin{array}{rl} \mathscr{V}(c) = & \{x \in \mathbb{R}^n : \ v(x) \leq c\} \\ \mathscr{Q}(d) = & \left\{x \in \mathbb{R}^n : \ \max(z_p(x)) \leq d\right\}. \end{array} \tag{7.41}$$

With $\xi = (\xi_1, \ldots, \xi_{n_\phi})'$ and for some integers m_i let us define

$$\delta(x,\xi) = \nabla v(x)' \left(f^{(0)}(x) + \sum_{i=1}^{n_\phi} f^{(i)}(x) \left(\hat{\phi}_i(\tau_i) + \xi_i \frac{\tau_i^{m_i}}{m_i!} \right) \right). \tag{7.42}$$

Theorem 7.3. *Let $v \in \mathscr{P}_n$ be given. Define Ξ in* (6.11) *and suppose that* (6.13) *holds with $c = 1$. Suppose there exists $d \in \mathbb{R}$, $q \in \mathscr{P}_n$, $r_c \in \mathscr{P}_n^{n_a}$, $s_c \in \mathscr{P}_n^{n_b}$, and $q_{zp} \in \mathscr{P}_n^{n_{zp}}$ such that*

$$\left.\begin{array}{r} p(x,\xi) > 0 \\ q(x) > 0 \\ p_c(x) > 0 \\ r_c(x) > 0 \\ p_{zp}(x) > 0 \\ q_{zp}(x) > 0 \end{array}\right\} \quad \forall x \in \mathbb{R}_0^n \ \forall \xi \in \mathrm{ver}(\Xi) \tag{7.43}$$

where

$$\begin{array}{rl} p(x,\xi) = & -\delta(x,\xi) - q(x)(1 - v(x)) \\ p_c(x) = & 1 - v(x) - r_c(x)'a(x) - s_c(x)'b(x) \\ p_{zpi}(x) = & v(x) - 1 - q_{zpi}(x)(z_{pi}(x) - d) \ \ \forall i = 1, \ldots, n_{zp}. \end{array} \tag{7.44}$$

Then, $d \geq \gamma$.

Proof. Suppose that (7.43) holds. Then, the first two inequalities imply that $\delta(x,\xi) < 0$ for all $x \in \mathscr{V}(1) \setminus \{0_n\}$ for all $\xi \in \mathrm{ver}(\Xi)$, and hence that $\dot{v}(x) < 0$ for all $x \in \mathscr{V}(1) \setminus \{0_n\}$. Similarly, the third and fourth inequalities imply that $1 - v(x) < 0$ for all $x \in \Theta$, and hence that $\Theta \subset \mathscr{V}(1)$. Lastly, the last two inequalities imply that $z_{pi}(x) - d < 0$ for all $x \in \mathscr{V}(1)$, and hence that $\mathscr{V}(1) \subset \mathscr{Q}(d)$. □

Theorem 7.3 provides a sufficient condition for establishing an upper bound of γ. For chosen $v(x)$ and d, this condition amounts to solving an LMI feasibility test by exploiting the SMR.

For a chosen $v(x)$, the best upper bound of γ provided by Theorem 7.3 can be computed via a bisection search on d where (7.43) is investigated at each iteration.

For a variable $v(x)$, the condition (7.43) contains a BMI. In this case, the best upper bound of γ can be computed by solving a sequence of LMI feasibility tests where $v(x)$ and the other polynomials are alternatively frozen.

7.3.2 Synthesis and Tightness

Let us consider the synthesis problem. To this end, let us define

$$p_{zli} = q_{zli}(x)(v(x)-1) - (z_{li}(x)-d) \ \forall i = 1,\ldots,n_{zl}. \tag{7.45}$$

for some $q_{zl} \in \mathscr{P}_n^{n_{zl}}$. Similarly, we introduce the polynomials $p_{wpi}(x)$, $q_{wpi}(x)$, $p_{wli}(x)$, $q_{wli}(x)$.

Corollary 7.1. *Let $v \in \mathscr{P}_n$ be given. Define Ξ in* (6.11) *and suppose that* (6.13) *holds with $c = 1$. Suppose there exists $d \in \mathbb{R}$, $k \in \mathscr{P}_{n_y}^{n_u}$, $q \in \mathscr{P}_n$, $r_c \in \mathscr{P}_n^{n_a}$, $s_c \in \mathscr{P}_n^{n_b}$ and $q_{zp} \in \mathscr{P}_n^{n_{zp}}$ such that*

$$\left.\begin{array}{r} p(x,\xi) > 0 \\ q(x) > 0 \\ p_c(x) > 0 \\ r_c(x) > 0 \\ p_*(x) > 0 \\ q_*(x) > 0 \end{array}\right\} \ \forall x \in \mathbb{R}_0^n \ \forall \xi \in \mathrm{ver}(\Xi) \ \forall * \in \{zp, zl, wp, wl\}. \tag{7.46}$$

Then, $d \geq \gamma^*$.

Proof. Analogous to the proof of Theorem 7.3. □

By exploiting the SMR, the condition of Corollary 7.1 amounts to solving an LMI feasibility test for chosen $v(x)$ and d.

For a chosen $v(x)$, the best upper bound of γ provided by Corollary 7.1 can be computed via a bisection search on d where (7.46) is investigated at each iteration.

For a variable $v(x)$, the condition (7.46) contains a BMI. In this case, the best upper bound of γ can be computed by solving a sequence of LMI feasibility tests where $v(x)$ and the other polynomials are alternatively frozen.

The next result explains how one can establish tightness of the found upper bound in the case of autonomous systems.

Theorem 7.4. *Let $v(x)$ be the LF corresponding to the found upper bound d. Then,*

$$d = \gamma \iff \exists \bar{t} \geq 0, x_{R,init} \in \mathscr{T} : x_R(\bar{t}) \in \Theta \tag{7.47}$$

where

$$\mathscr{T} = \partial\mathscr{V}(1) \cap \partial\mathscr{Q}(d) \tag{7.48}$$

and $x_R(t)$ is the solution of the system

$$\begin{cases} \dot{x}_R(t) = -f(x(t)) \\ x_R(0) = x_{R,init}. \end{cases} \tag{7.49}$$

Proof. "$\Leftarrow$" Suppose that there exist $\bar{t} \geq 0$ and $x_{R,init} \in \mathscr{T}$ such that $x_R(\bar{t}) \in \Theta$. Then, since the system (7.49) evolves reversely with respect to the system (7.35), it follows that by initializing (7.35) with $x_{init} = x_R(\bar{t})$ we have that $x(\bar{t}) = x_{R,init}$. Since $x_{init} \in \Theta$, it follows from the definition of γ that

$$\begin{aligned} d &= \max(z(\bar{t})) \\ &\leq \gamma. \end{aligned}$$

On the other hand, d is an upper bound of γ. Therefore, $d = \gamma$.

"$\Rightarrow$" Suppose that $d = \gamma$. From the definition of γ and since Θ is compact, it follows that $\exists x_{init} \in \Theta$ and $\bar{t} \geq 0$ such that $\max(z(\bar{t})) = \gamma$. Observe that $x(\bar{t}) \in \partial\mathscr{Q}(d)$ because $d = \gamma$. Suppose now for contradiction that $x(\bar{t})$ does not belong to $\partial\mathscr{V}(1)$. Then, this implies either that $v(x(\bar{t})) > 1$ or that $v(x(\bar{t})) < 1$. But the former implies that $x(\bar{t})$ lies outside an invariant set containing the initial condition $x_{init} \in \Theta$, and the second implies that $\mathscr{V}(1)$ is not included in $\mathscr{Q}(d)$ as ensured by the last two inequalities in (7.43). Hence, both hypotheses are impossible and consequently $x(\bar{t}) \in \partial\mathscr{V}(1)$. Therefore, $x(\bar{t}) \in \mathscr{T}$. Then, by initializing the reverse system (7.49) with $x_{R,init} = x(\bar{t})$ we obtain that $x_R(\bar{t}) = x_{init} \in \Theta$ which concludes the proof. □

The set $\mathscr{T}$ can be found by solving

$$\begin{cases} v(x) = 1 \\ \max(z_p(x)) = d. \end{cases} \tag{7.50}$$

However, solving the system (7.50) can be a difficult task because it is nonlinear system. The following result provides an alternative technique for finding $\mathscr{T}$.

Theorem 7.5. *Let M_i be the SMR matrix of $p_{zpi}(x)$ for the optimal values of the variables in* (7.3). *Then,*

$$\mathscr{T} = \left\{ x \in \bigcup_{i=1}^{n_{zp}} \mathscr{M}_i : \ x \text{ satisfies } (7.50) \right\} \tag{7.51}$$

where

$$\mathscr{M}_i = \left\{ x \in \mathbb{R}^n : \ b_{pol}(x, \partial p_{zpi}/2) \in \ker(M_i) \right\}. \tag{7.52}$$

Proof. Consider $x \in \mathscr{T}$. Since x satisfies (7.50) there exists i such that $p_{zpi}(x) = 0$. It follows that:

$$\begin{aligned} 0 &= p_{zpi}(x) \\ &= b_{pol}(x, \partial p_{zpi}/2)' M_i b_{pol}(x, \partial p_{zpi}/2). \end{aligned}$$

Since $M_i \geq 0$, it follows that $b_{pol}(x, \partial p_{zpi}/2) \in \ker(M_i)$, i.e. $x \in \mathcal{M}_i$. □

Let us observe that the set $\mathcal{T}$ can be found via trivial substitution from the set $\mathcal{M}_i$ in (7.52) (at least whenever $\mathcal{M}_i$ is finite), which can be computed as explained in Section 1.5.1.

Example 7.5. Let us consider the non-polynomial system

$$\begin{cases} \dot{x}_1 = x_2 + 0.2x_1^2 x_2 + 0.2\,(1 - e^{x_1}) \\ \dot{x}_2 = -1.5x_1 - 2x_2 - 1.1x_2^2 + u \\ y = x_1 \\ x_{init} \in \Theta = \{x \in \mathbb{R}^2 : \ \|x\| \leq 2\}. \end{cases}$$

The set Θ can be expressed as in (7.36) with $a(x) = 4 - \|x\|^2$. Figure 7.7a shows the trajectories of the system for $u = 0$. We consider the following synthesis problem:

$$\begin{aligned} &\inf_{k} \ \sup_{t, x_{init}, u} \ \|x(t)\|_\infty \\ &\text{s.t.} \ \begin{cases} t \geq 0 \\ x_{init} \in \Theta \\ \text{the origin is locally asymptotically stable with } u(t) = k(y(t)) \\ |u(t)| < 3. \end{cases} \end{aligned}$$

This problem can be reformulated as in (7.39) with

$$z = (x', -x')', \ \ w = (u, -u)', \ \ w_0 = (3,3)'.$$

Let us use Corollary 7.1. Table 7.1 shows the upper bound d computed with a variable LF by alternatively freezing the variables involved in the BMI condition. The controller found for $(\partial v, \partial k) = (6,3)$ is

$$k_{6,3}(y) = 0.900y + 0.918y^2 - 0.184y^3.$$

Let us consider the problem of establishing if the upper bound $d_{6,3}$ found for this controller is tight. This can be done by using Theorems 7.4–7.5. We have that

$$\mathcal{T} = \mathcal{M}_2 = \{(-0.665, -2.191)'\}.$$

The trajectory $x_R(t)$ of (7.49) with initial condition $x_{R,init} \in \mathcal{M}_2$ intersects Θ in $x_R(\bar{t})$ as shown in Figure 7.7b. Therefore, the upper bound is tight.

Table 7.1 Example 7.5. Upper bound d.

$\partial v \setminus \partial k$	1	2	3
2	∞	∞	∞
4	∞	2.727	2.687
6	∞	2.463	2.191

□

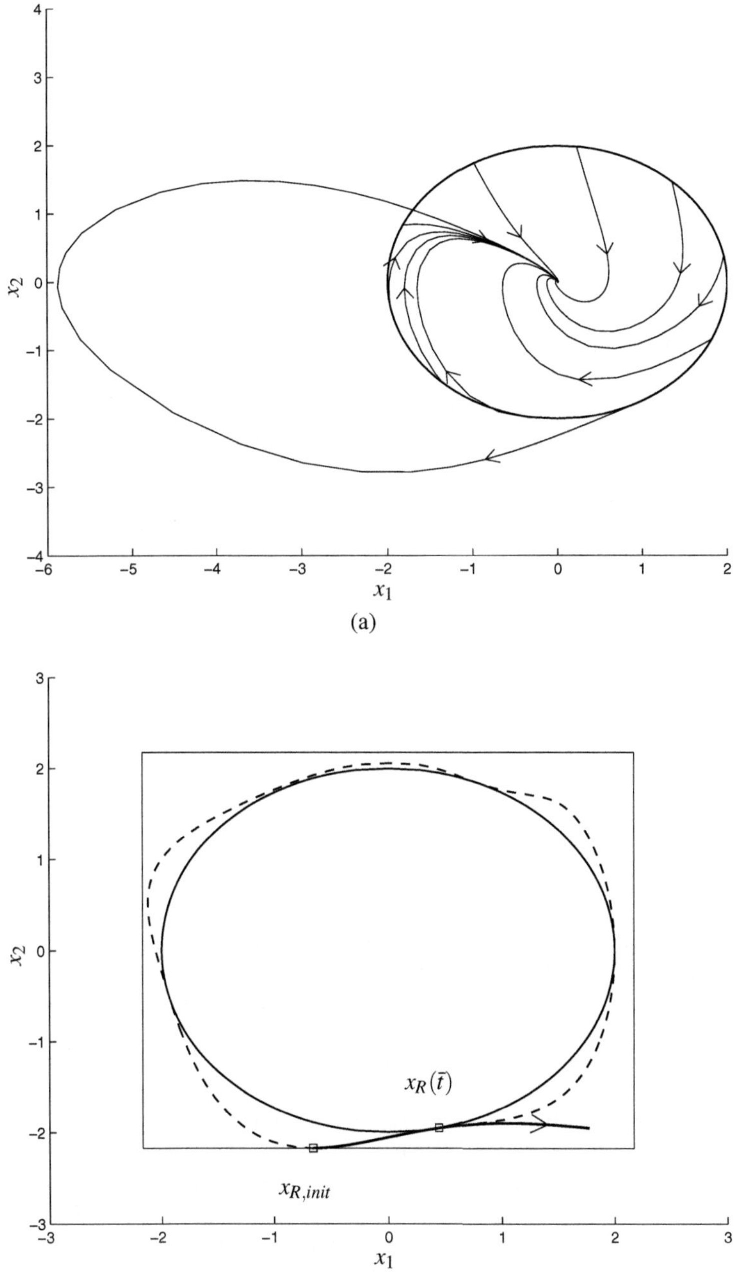

(b)

Fig. 7.7 Example 7.5. (a) Uncontrolled system trajectories (oriented curves) starting in Θ (solid). (b) System controlled with $u = k_{6,3}(y)$: the trajectories are confined in the unitary sublevel set $\mathcal{V}(1)$ (dashed) included in the region $\{x: \|x\|_\infty < d_{6,3}\}$ (box). The trajectory $x_R(t)$ (oriented curve) intersects Θ (solid) which means that $d_{6,3}$ is tight.

Example 7.6. Consider the inverted pendulum controlled with a DC motor in Figure 7.8a. By denoting with x_1 the angle ψ of the pole with the vertical, with x_2 its temporal derivative, and with x_3 the armature current, the system can be described by

$$\begin{cases} \dot{x}_1 = x_2 \\ \dot{x}_2 = \dfrac{-cx_2 + k_t x_3 + mgl\sin x_1}{I + ml^2} \\ \dot{x}_3 = \dfrac{-k_e x_2 - Rx_3 + u}{L} \\ y = x_1 \end{cases}$$

where u is the input voltage, c is the friction coefficient, I is the moment of inertia of the pole, m is the mass of the pole, l is the distance from the center of gravity of the pole to the pivot, k_t is the torque constant, k_e is the electric constant, L is the inductance, R is the resistance, and g is the gravitational constant. We consider the following numerical values:

$$c = 0.5,\ g = 9.8,\ I = 0.03,\ k_t = 1,\ k_e = 0.01,\ L = 0.001,\ l = 0.5,\ m = 0.1,\ R = 0.1.$$

The synthesis problem we address is

$$\begin{array}{ll} \inf\limits_{k} \sup\limits_{t, x_{init}, u} & |u(t)| \\ \text{s.t.} & \begin{cases} t \geq 0 \\ x_{init} \in \Theta \\ \text{the origin is locally asymptotically stable with } u(t) = k(y(t)) \end{cases} \end{array}$$

where

$$\Theta = \left\{ x \in \mathbb{R}^3 : \left(\frac{x_1}{\pi/2}\right)^2 + \left(\frac{x_2}{0.1}\right)^2 + \left(\frac{x_3}{0.1}\right)^2 \leq 1 \right\}.$$

In order to initialize our procedure we observe that the linearized system can be stabilized by $u = k_{lin}y$ with $k_{lin} \in (-6.540, -0.049)$ and select for example $K = -1$. Table 7.2a shows the found upper bound d. The controller found for $(\partial v, \partial k) = (4,3)$ is

$$k_{4,3}(y) = -0.049y + 0.007y^3$$

whose performance is illustrated in Figures 7.8b and 7.9a.

Let us observe from Figure 7.9a that the convergence can be quite slow. This happens because the linearized controlled system is almost marginally stable only. In order to cope with this problem we can repeat the procedure by imposing that the maximum real part of the eigenvalues of the linearized controlled system is less than a negative value, for example -0.4. From standard root locus investigation this is equivalent to adding the LMIs $k_1 > -5.977$ and $k_1 < -0.072$ in the optimization, where k_1 is the coefficient of the linear term of the controller. Table 7.2b shows the obtained upper bounds, and Figure 7.9b the corresponding performance. Observe that the convergence is much faster, clearly at the expense of a larger amplitude of the control signal.

Table 7.2 Example 7.6. Upper bound d obtained with unconstrained k_1 (a) and constrained k_1 (b).

(a)

$\partial v \setminus \partial k$	1	2	3
2	0.078	0.078	0.050
4	0.077	0.077	0.050

(b)

$\partial v \setminus \partial k$	1	2	3
2	0.114	0.114	0.058
4	0.113	0.113	0.058

□

7.4 A Note on Degenerate Polynomial Systems

For degenerate polynomial systems we mean polynomial systems that are singular according to some criterion among all possible polynomial systems. In particular, a case of interest concerns the set of polynomial systems in the form (4.1) for which the linearized system at the origin is marginally stable, i.e. the matrix $A = \mathrm{lp}(f)$ is marginally Hurwitz according to the definition

$$\begin{cases} \mathrm{Re}(\lambda) \leq 0 \ \ \forall \lambda \in \mathrm{spc}(A) \\ \mathrm{Re}(\lambda) < 0 \ \ \forall \lambda \in \mathrm{spc}(A): \ \mu_g(\lambda) < \mu_a(\lambda) \end{cases} \tag{7.53}$$

where $\mu_a(\lambda)$ and $\mu_g(\lambda)$ denote the algebraic and geometric multiplicity of the eigenvalue λ in A, respectively.

In the previous chapters we have derived techniques for analysis and control of the DA based on SOS programming. As pointed out, these techniques require that the linearized system is asymptotically stable, and hence cannot be used when A is marginally Hurwitz.

One way to cope with this problem consists of adopting the strategy presented in (2.19)–(2.21) for establishing whether a locally quadratic polynomial (with second degree and highest degree forms possibly not positive definite) is positive definite.

Specifically, let us consider for simplicity the first condition we have derived in this book for establishing estimates of the DA via SOS programming, in particular the condition provided in Corollary 4.2. As explained in Section 4.2.2, a necessary requirement for this condition to hold is that A is Hurwitz. The following result provides a modified version of Corollary 4.2 that does not need such a requirement.

Corollary 7.2. *Let $v \in \mathscr{P}_n$ be positive definite, and let $c \in \mathbb{R}$ be positive. For $\varepsilon \in \mathbb{R}$, $\varepsilon > 0$, suppose that there exist Q, Ψ and α satisfying the LMIs*

$$\begin{cases} \tilde{P}(Q,\Psi) + L(\alpha) \geq 0 \\ Q > 0 \\ \mathrm{trace}(Q_1) = 1 \\ \sum_{j=1}^{m} \Psi_{i,j} \geq \varepsilon \ \ \forall i = 1,\ldots,n \\ \Psi_{i,j} \geq 0 \ \ \forall i = 1,\ldots,n \ \forall j = 1,\ldots,m \end{cases} \tag{7.54}$$

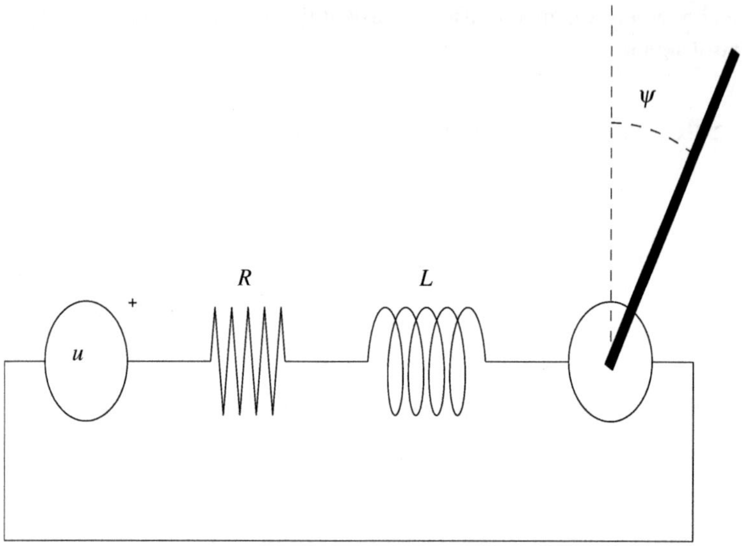

(a)

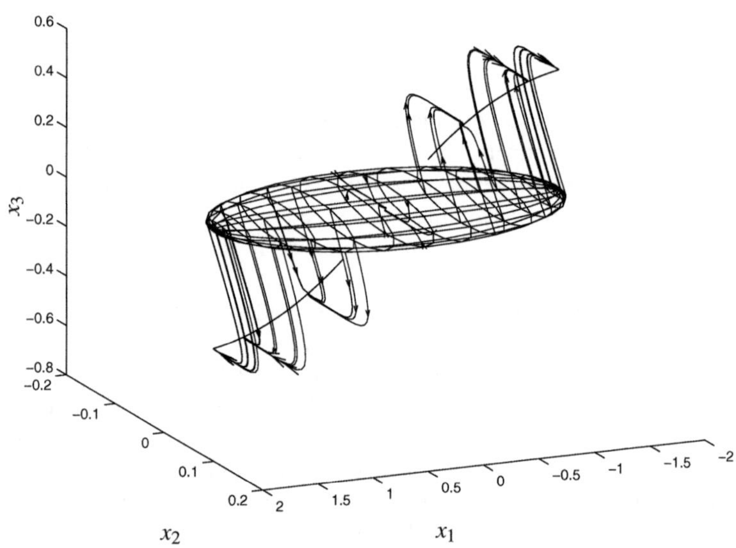

(b)

Fig. 7.8 Example 7.6 (Inverted pendulum). (a) Inverted pendulum controlled with a DC motor. (b) Trajectories starting in Θ (ellipsoid) with $u = k_{4,3}(y)$ (oriented curves).

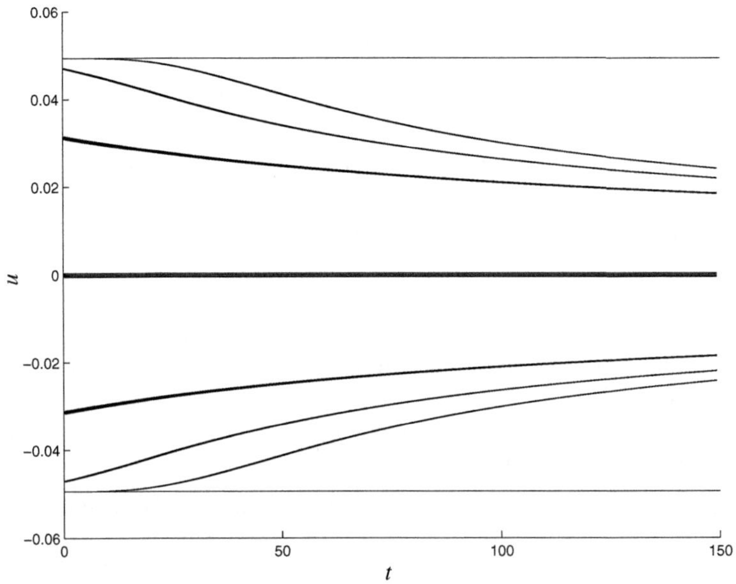

(a)

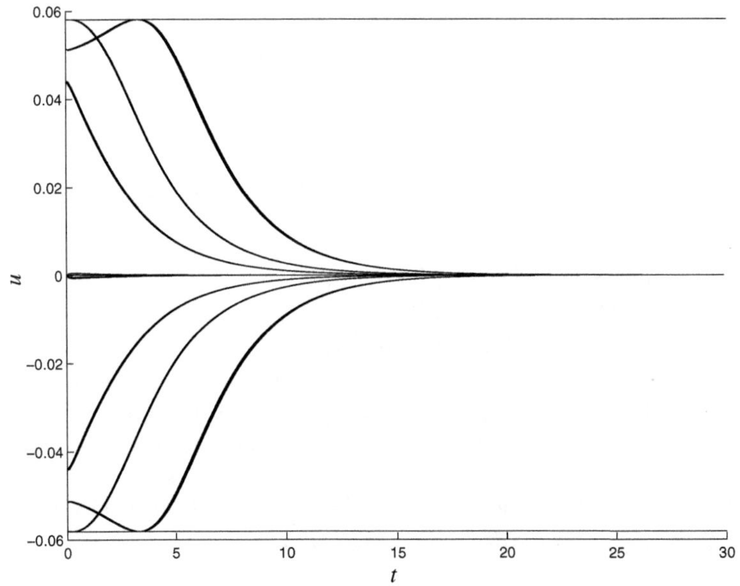

(b)

Fig. 7.9 Example 7.6. (a) Control input with $u = k_{4,3}(y)$. (b) Control input achieved by constraining k_1.

where

$$\begin{aligned}\tilde{P}(Q,\Psi)+L(\alpha) &= CSMR_{qua}(\tilde{p})\\ \tilde{p}(x) &= p(x)-\Sigma_{(i,j)=(1,1)}^{(n,m)}\Psi_{i,j}x_i^{2j}\\ m &= \lceil \partial p/2 \rceil\end{aligned} \tag{7.55}$$

with $p(x)$ and Q as in (4.7) *and* (4.10). *Then, $v(x)$ is a LF for the origin of* (4.1), *and $\mathscr{V}(c) \subseteq \mathscr{D}$.*

Proof. Direct consequence of Corollary 4.2 and (2.21). □

It is interesting to observe that the same strategy can be used for reducing the conservatism when attempting to establish global asymptotical stability of the origin (i.e., $\mathscr{D} = \mathbb{R}^n$). Indeed, as pointed out in Section 4.4.3, a necessary requirement for the condition provided by Corollary 4.5 to hold is that the highest degree forms of $v(x)$ and $-\dot{v}(x)$ are positive definite. The following result provides a modified version of this corollary that does not need such a requirement.

Corollary 7.3. *Let $v \in \mathscr{P}_n$ be a locally quadratic polynomial to determine, and define*

$$\begin{aligned}\mathbf{v} &= COE_{qua}(v)\\ \tilde{V}(\mathbf{v},\Omega)+L(\alpha_v) &= CSMR_{qua}(\tilde{v})\\ \tilde{P}(\mathbf{v},\Psi)+L(\alpha_p) &= CSMR_{qua}(\tilde{p})\end{aligned} \tag{7.56}$$

where

$$\begin{aligned}\tilde{v}(x) &= v(x)-\Sigma_{(i,j)=(1,1)}^{(n,\partial v/2)}\Omega_{i,j}x_i^{2j}\\ \tilde{p}(x) &= -\dot{v}(x)-\Sigma_{(i,j)=(1,1)}^{(n,m)}\Psi_{i,j}x_i^{2j}\\ m &= \lceil \partial \dot{v}/2 \rceil .\end{aligned} \tag{7.57}$$

For $\varepsilon \in \mathbb{R}$, $\varepsilon > 0$, suppose that there exist $\mathbf{v}$, Ω, Ψ, α_p and α_v such that the following LMIs hold:

$$\begin{cases}\tilde{V}(\mathbf{v},\Omega)+L(\alpha_v) \geq 0\\ \tilde{P}(\mathbf{v},\Psi)+L(\alpha_p) \geq 0\\ \Sigma_{j=1}^{\partial v/2}\Omega_{i,j} \geq \varepsilon \ \forall i=1,\ldots,n\\ \Sigma_{j=1}^{m}\Psi_{i,j} \geq \varepsilon \ \forall i=1,\ldots,n\\ \Omega_{i,j} \geq 0 \ \forall i=1,\ldots,n \ \forall j=1,\ldots,\partial v/2\\ \Psi_{i,j} \geq 0 \ \forall i=1,\ldots,n \ \forall j=1,\ldots,m.\end{cases} \tag{7.58}$$

Then, $v(x)$ provided by such a $\mathbf{v}$ is a global LF for the origin of (4.1), *and $\mathscr{D} = \mathbb{R}^n$.*

Proof. Direct consequence of Corollary 4.5 and (2.21). □

7.5 Notes and References

The estimation of the DA via union of a continuous family of estimates was proposed in [20]. Related works include [3] where the logical compositions of LFs is proposed.

The determination of outer estimates of the set of admissible equilibrium points for uncertain nonlinear systems was considered in [25]. See also the related work [52].

The computation and the minimization via constrained control design of the extremes of the trajectories for a given set of initial conditions was introduced in [35].

Lastly, the construction of LFs for degenerate polynomial systems derived in Section 7.4 exploits the idea proposed in [81].

Appendix A
LMI Problems

In this section we provide some basic notions about LMIs and optimization problems with LMI constraints. The interested reader is referred to [10, 11] for an extensive treatment.

Definition A.1 (Linear Matrix Function). The function $A : \mathbb{R}^{n \times m} \to \mathbb{R}^{p \times q}$ is a *linear matrix function* if

$$A(\alpha X + \beta Y) = \alpha A(X) + \beta A(Y) \;\; \forall X, Y \in \mathbb{R}^{n \times m} \, \forall \alpha, \beta \in \mathbb{R}.$$

Moreover, if $A(X) = A(X)'$, then $A(X)$ is said a *symmetric linear matrix function.* □

Definition A.2 (LMI). Let $A : \mathbb{R}^{n \times m} \to \mathbb{S}^p$ be a symmetric linear matrix function, and let $A_0 \in \mathbb{S}^p$. Then, the constraint

$$A(X) + A_0 \, * \, 0 \;\; \text{for any} * \in \{>, <, \geq, \leq\} \tag{A.1}$$

is an *LMI*. □

Definition A.3 (LMI Feasibility Set). For $i = 1, \ldots, k$ let $A_i : \mathbb{R}^{n \times m} \to \mathbb{S}^{p_i}$ be a symmetric linear matrix function, and let $A_{0,i} \in \mathbb{S}^{p_i}$. Let us define the set of LMIs

$$A_i(X) + A_{0,i} \, * \, 0 \;\; \text{for any} * \in \{>, <, \geq, \leq\} \; \forall i = 1, \ldots, k. \tag{A.2}$$

Then, the set

$$\mathscr{X} = \left\{ X \in \mathbb{R}^{n \times m} : \text{ (A.2) holds} \right\} \tag{A.3}$$

is the *feasibility set* of the set of LMIs (A.2). □

The feasibility set of a set of LMIs is a convex set [10].

Definition A.4 (LMI Feasibility Test). For $i = 1, \ldots, k$ let $A_i : \mathbb{R}^{n \times m} \to \mathbb{S}^{p_i}$ be a symmetric linear matrix function, and let $A_{0,i} \in \mathbb{S}^{p_i}$. The problem of establishing whether there exists $X \in \mathbb{R}^{n \times m}$ such that (A.2) holds is an *LMI feasibility test*. □

Solving an LMI feasibility test amounts to solving a convex optimization problem [10].

Definition A.5 (SDP). For $i = 1,\ldots,k$ let $A_i : \mathbb{R}^{n\times m} \to \mathbb{S}^{p_i}$ be a symmetric linear matrix function, and let $A_{0,i} \in \mathbb{S}^{p_i}$. Let $c(X) : \mathbb{R}^{n\times m} \to \mathbb{R}$ be a linear function. The optimization problem

$$\begin{aligned} &\inf_X \; c(X) \\ &\text{s.t. } A_i(X) + A_{0,i} \ast 0 \;\text{ for any } \ast \in \{>,<,\geq,\leq\}\; \forall i = 1,\ldots,k \end{aligned}$$

is an *SDP*. □

An SDP is a convex optimization problem [10]. SDPs are also known as eigenvalue problems (EVPs) [118].

Definition A.6 (GEVP). For $i = 1,\ldots,k$ let $A_i, B_i : \mathbb{R}^{n\times m} \to \mathbb{S}^{p_i}$ and $C_i : \mathbb{R}^{n\times m} \to \mathbb{S}^{q_i}$ be symmetric linear matrix functions, and let $A_{0,i}, B_{0,i} \in \mathbb{S}^{p_i}$ and $C_{0,i} \in \mathbb{S}^{q_i}$. The optimization problem

$$\begin{aligned} &\inf_{t,X} \; t \\ &\text{s.t. } \begin{cases} t\,(B_i(X) + B_{0,i}) - A_i(X) - A_{0,i} > 0 \\ \qquad\qquad\qquad B_i(X) + B_{0,i} > 0 \\ \qquad\qquad\qquad C_i(X) + C_{0,i} > 0 \\ \qquad\qquad\qquad\qquad\qquad i = 1,\ldots,k \end{cases} \end{aligned}$$

is a *GEVP*. □

A GEVP is a quasiconvex optimization problem [10].

Appendix B
Determinant and Rank Constraints via LMIs

Consider $V \in \mathbb{S}^n$ and $w \in \mathbb{R}$, and define

$$\eta = \min\{2^k : \; n \le 2^k, \;\; k \in \mathbb{N}\}.$$

As explained in [76], the condition

$$\begin{cases} \sqrt[2\eta]{\det(V)} \ge w \\ \qquad\quad V \ge 0 \end{cases}$$

can be equivalently expressed via a system of LMIs

$$F_i(V, w, x) \ge 0 \;\; \forall i = 1, \ldots, k$$

where x is an auxiliary vector variable, and $F_i(V, w, x)$ is affine linear in V, w and x. For instance, in the case $n = 2$, one has $\eta = 2$ and the LMIs $F_i(V, w, x)$ are given by

$$\begin{aligned} F_1(V, w, x) &= \begin{pmatrix} x_1 & x_4 \\ \star & x_3 \end{pmatrix} \\ F_2(V, w, x) &= x_4 \\ F_3(V, w, x) &= x_4 - w \\ F_4(V, w, x) &= \begin{pmatrix} I_2 & \begin{pmatrix} x_1 & x_2 \\ 0 & x_3 \end{pmatrix} \\ \star & V \end{pmatrix}. \end{aligned}$$

This also allows one to express some rank constraints via LMIs. Specifically, consider $V \in \mathbb{S}^n$ with $V \ge 0$ and $m \in \mathbb{N}$ with $m \in [0, n]$. The constraint

$$\mathrm{rank}(V) \ge m$$

can be equivalently rewritten as

$$\exists j \in [1, p] : \; \det(V_j) > 0$$

where $V_1,\ldots,V_p$ are all the principal submatrices of V with size $m \times m$. Hence, $\operatorname{rank}(V) \geq m$ if and only if

$$\begin{array}{l} t_1 + \ldots + t_k > 0 \\ F_i(V_j, t_j, x_j) \geq 0 \ \forall i = 1, \ldots, k_j \ \forall j = 1, \ldots, p \end{array}$$

where $t_1,\ldots,t_p \in \mathbb{R}$ and $x_1,\ldots,x_p$ are auxiliary vector variables.

Appendix C
MATLAB Code: SMRSOFT

The techniques presented in this book for SOS programming (Chapters 1 and 2) and for the analysis and control of the DA (Chapters 4–7) are available in the freely downloadable MATLAB toolbox SMRSOFT [27].

For instance, the computation of the generalized SOS index in Example 2.2 and the non-positivity test can be simply done as follows:

```
>> f='3.3-2x1+3x2-x1x2+x1^2x2-x1x2^3+x1^3x2^2+x1^6+x2^6';
>> k=0;
>> smrsos(f,k)

SOS index (k=0): -1.8300e-005

Points in set M:
x=[-1.0386,-1.1439] -> f(x)=-0.0015
```

Then, the constrained minimization in Example 2.11 is done as follows:

```
>> f='7+x1x2^2+x1^4-x2^3';
>> g='1+x2+x2^2';
>> a='4-x1^2-2x2^2';
>> b='-1+x1^2+x1+x1x2+x2+x2^3';
>> k=0;
>> smropt(f,k,g,a,b)

Function lower bound (k=0): 1.9127

Set M1 is nonempty -> Lower bound is tight

Points in set M1:
x=[-0.8144,0.9891] -> f(x)/g(x)=1.9127
```

Lastly, the computation of the LEDA in Example 4.7 is obtained with the following:

```
>> f{1}='x2';
>> f{2}='-x1-x2(1-x1^2)';
>> v='x1^2+x1x2+x2^2';
>> smrda(f,v)

LEDA lower bound (k=0): 0.9152

Set M1 is nonempty -> Lower bound is tight

Points in set M1:
x=[-0.8586,1.0313] -> dotv(x)=0.0000, v(x)=0.9152
x=[0.8586,-1.0313] -> dotv(x)=0.0000, v(x)=0.9152
```

We conclude this section providing a simplified version of the code in this toolbox for the construction of the matrices involved in the complete SMR (1.34). Specifically, Algorithms 1 and 2 in Tables C.1 and C.2 construct an SMR matrix F and an SMR parametrization $L(\alpha)$ in (1.34), respectively, while Algorithm 3 in Table C.3 constructs the power vector $b_{pol}(x,d)$ in (1.7) required by Algorithms 1 and 2. Indeed, the complete SMR in Example 1.4 can be found as follows:

```
>> f=[4,-3,0,0,1];
>> F=smrF(f,1,2);
>> L=smrL(1,2);

F =
     4.0000   -1.5000         0
    -1.5000         0         0
          0         0    1.0000

L{1} =
     0     0    -1
     0     2     0
    -1     0     0
```

Table C.1 Algorithm 1: computation of an SMR matrix F in (1.34)

```
function F=smrF(f,n,m)

[B1,b1,l1]=powvec(n,m);
[B2,b2,l2]=powvec(n,2*m);
F=zeros(l1);
for i=1:l1
    for j=i:l1
        p=[1:l2]*((B2*b2)==((B1(i,:)+B1(j,:))*b2));
        F(i,j)=f(p)/2;
        F(j,i)=F(j,i)+f(p)/2;
        f(p)=0;
    end
end
```

Table C.2 Algorithm 2: computation of an SMR parametrization $L(\alpha)$ in (1.34)

```
function Lb=smrL(n,m)

[B1,b1,l1]=powvec(n,m);
[B2,b2,l2]=powvec(n,2*m);
Lb=[];
w=zeros(l2,3);
k=0;
for i=1:l1
    for j=i:l1
        p=[1:l2]*((B2*b2)==((B1(i,:)+B1(j,:))*b2));
        w(p,1)=w(p,1)+1;
        if w(p,1)==1
            w(p,2:3)=[i,j];
        else
            L=zeros(l1);
            ii=w(p,2);
            jj=w(p,3);
            L(ii,jj)=L(ii,jj)-1;
            L(jj,ii)=L(jj,ii)-1;
            L(i,j)=L(i,j)+1;
            L(j,i)=L(j,i)+1;
            k=k+1;
            Lb{k}=L;
        end
    end
end
```

Table C.3 Algorithm 3: computation of a power vector $b_{pol}(x,m)$ in (1.7)

```
function [B,b,l]=powvec(n,d)

l=prod((n+1:n+d))/prod((2:d));
B=zeros(l,n+1);
B(1,1)=d;
j=1;
for i=2:l
    B(i,:)=B(i-1,:);
    if B(i,j)>0
        B(i,j)=B(i,j)-1;
        B(i,j+1)=B(i,j+1)+1;
        if j<n
            j=j+1;
        end
    else
        k=B(i,j+1);
        B(i,j+1)=0;
        while B(i,j)==0
            j=j-1;
        end
        B(i,j)=B(i,j)-1;
        B(i,j+1)=k+1;
        j=j+1;
    end
end
B=B(:,2:end);
b=ones(n,1);
k=1;
for i=2:n
    k=k*(d+1);
    b(n+1-i)=k;
end
```

References

1. Alberto, L.F.C., Chiang, H.-D.: Characterization of stability region for general autonomous nonlinear dynamical systems. IEEE Trans. on Automatic Control (to appear, 2011)
2. Baier, R., Gerdts, M.: A computational method for non-convex reachable sets using optimal control. In: European Control Conference, Budapest, Hungary, pp. 97–102 (2009)
3. Balestrino, A., Caiti, A., Crisostomi, E.: Logical composition of lyapunov functions. Int. Journal of Control 84(3), 563–573 (2011)
4. Barkin, A., Zelentsovsky, A.: Method of power transformations for analysis and stability of nonlinear control systems. Systems and Control Letters 3, 303–310 (1983)
5. Blanchini, F.: Set invariance in control - a survey. Automatica 35(11), 1747–1768 (1999)
6. Blanchini, F., Miani, S.: Set-Theoretic Methods in Control. In: Systems and Control: Foundations and Applications, Birkhauser (2008)
7. Blekherman, G.: There are significantly more nonnegative polynomials than sums of squares. Israel Journal of Mathematics 153(1), 355–380 (2006)
8. Bochnak, J., Coste, M., Roy, M.-F.: Real Algebraic Geometry. Springer, Heidelberg (1998)
9. Bose, N., Li, C.: A quadratic form representation of polynomials of several variables and its applications. IEEE Trans. on Automatic Control 13(8), 447–448 (1968)
10. Boyd, S., El Ghaoui, L., Feron, E., Balakrishnan, V.: Linear Matrix Inequalities in System and Control Theory. SIAM (1994)
11. Boyd, S., Vandenberghe, L.: Convex Optimization. Cambridge University Press, Cambridge (2004)
12. Brockett, R.W.: Lie algebra and Lie groups in control theory. In: Mayne, D.Q., Brockett, R.W. (eds.) Geometric Methods in Systems Theory, Dordrecht, pp. 43–82 (1973)
13. Camilli, F., Gruene, L., Wirth, F.: A regularization of zubov's equation for robust domains of attraction. In: Isidori, A., Lamnabhi-Lagarrigue, F., Respondek, W. (eds.) Nonlinear Control in the Year 2000, pp. 277–290. Springer, Heidelberg (2000)
14. Camilli, F., Gruene, L., Wirth, F.: Domains of attraction of interconnected systems: A zubov method approach. In: European Control Conference, Budapest, Hungary, pp. 91–96 (2009)
15. Chesi, G.: New convexification techniques and their applications to the analysis of dynamical systems and vision problems. PhD thesis, University of Bologna (2001)
16. Chesi, G.: Computing output feedback controllers to enlarge the domain of attraction in polynomial systems. IEEE Trans. on Automatic Control 49(10), 1846–1850 (2004)

17. Chesi, G.: Estimating the domain of attraction for uncertain polynomial systems. Automatica 40(11), 1981–1986 (2004)
18. Chesi, G.: Domain of attraction: estimates for non-polynomial systems via LMIs. In: IFAC World Congress on Automatic Control, Prague, Czech Republic (2005)
19. Chesi, G.: Establishing tightness in robust H-infinity analysis via homogeneous parameter-dependent Lyapunov functions. Automatica 43(11), 1992–1995 (2007)
20. Chesi, G.: Estimating the domain of attraction via union of continuous families of Lyapunov estimates. Systems and Control Letters 56(4), 326–333 (2007)
21. Chesi, G.: On the gap between positive polynomials and SOS of polynomials. IEEE Trans. on Automatic Control 52(6), 1066–1072 (2007)
22. Chesi, G.: Optimal representation matrices for solving polynomial systems via LMI. Int. Journal of Pure and Applied Mathematics 45(3), 397–412 (2008)
23. Chesi, G.: Estimating the domain of attraction for non-polynomial systems via LMI optimizations. Automatica 45(6), 1536–1541 (2009)
24. Chesi, G.: LMI techniques for optimization over polynomials in control: a survey. IEEE Trans. on Automatic Control 55(11), 2500–2510 (2010)
25. Chesi, G.: On the admissible equilibrium points of nonlinear dynamical systems affected by parametric uncertainty: Characterization via LMIs. In: IEEE Multi-Conf. on Systems and Control, Yokohama, Japan, pp. 351–356 (2010)
26. Chesi, G.: Robustness analysis of genetic regulatory networks affected by model uncertainty. Automatica 47(6), 1131–1138 (2011)
27. Chesi, G.: SMRSOFT: Optimization over polynomials and dynamical systems study via SOS programming (2011), `http://www.eee.hku.hk/~chesi`
28. Chesi, G., Garulli, A., Tesi, A., Vicino, A.: An LMI-based approach for characterizing the solution set of polynomial systems. In: IEEE Conf. on Decision and Control, Sydney, Australia, pp. 1501–1506 (2000)
29. Chesi, G., Garulli, A., Tesi, A., Vicino, A.: Characterizing the solution set of polynomial systems in terms of homogeneous forms: an LMI approach. Int. Journal of Robust and Nonlinear Control 13(13), 1239–1257 (2003)
30. Chesi, G., Garulli, A., Tesi, A., Vicino, A.: Robust stability for polytopic systems via polynomially parameter-dependent Lyapunov functions. In: IEEE Conf. on Decision and Control, Maui, Hawaii, pp. 4670–4675 (2003)
31. Chesi, G., Garulli, A., Tesi, A., Vicino, A.: Solving quadratic distance problems: an LMI-based approach. IEEE Trans. on Automatic Control 48(2), 200–212 (2003)
32. Chesi, G., Garulli, A., Tesi, A., Vicino, A.: Polynomially parameter-dependent Lyapunov functions for robust stability of polytopic systems: an LMI approach. IEEE Trans. on Automatic Control 50(3), 365–370 (2005)
33. Chesi, G., Garulli, A., Tesi, A., Vicino, A.: Homogeneous Polynomial Forms for Robustness Analysis of Uncertain Systems. Springer, Heidelberg (2009)
34. Chesi, G., Henrion, D.: Guest editorial: Special issue on positive polynomials in control. IEEE Trans. on Automatic Control 54(5), 935–936 (2009)
35. Chesi, G., Hung, Y.S.: Analysis and synthesis of nonlinear systems with uncertain initial conditions. IEEE Trans. on Automatic Control 53(5), 1262–1267 (2008)
36. Chesi, G., Tesi, A., Vicino, A., Genesio, R.: On convexification of some minimum distance problems. In: European Control Conf., Karlsruhe, Germany (1999)
37. Chiang, H.-D., Hirsch, M.W., Wu, F.F.: Stability regions of nonlinear autonomous dynamical systems. IEEE Trans. on Automatic Control 33(1), 16–27 (1988)
38. Chiang, H.-D., Thorp, J.S.: Stability regions of nonlinear dynamical systems: a constructive methodology. IEEE Trans. on Automatic Control 34(12), 1229–1241 (1989)

39. Chiang, H.-D., Wu, F.F., Varaya, P.P.: Foundations of direct methods for power system transient stability analysis. IEEE Trans. on Circuits and Systems 34(2) (1987)
40. Choi, M., Lam, T., Reznick, B.: Sums of squares of real polynomials. In: Symposia in Pure Mathematics, pp. 103–126 (1995)
41. Chua, L.O., Desoer, C.A., Kuh, E.S.: Linear and Nonlinear Circuits. McGraw-Hill, New York (1987)
42. Coutinho, D.F., de Souza, C.E., Trofino, A.: Stability analysis of implicit polynomial systems. IEEE Trans. on Automatic Control 54(5), 1012–1018 (2009)
43. Cruck, E., Moitie, R., Suebe, N.: Estimation of basins of attraction for uncertain systems with affine and lipschitz dynamics. Dynamics Control 11, 211–227 (2001)
44. Curto, R.E., Fialkow, L.A.: The truncated complex K-moment problem. Transactions of the American Mathematical Society 352(6), 2825–2855 (2000)
45. Davison, E.J., Kurak, E.M.: A computational method for determining quadratic Lyapunov functions for nonlinear systems. Automatica 7, 627–636 (1971)
46. Din, M.S.E., Zhi, L.: Computing rational points in convex semi-algebraic sets and sums of squares decompositions. SIAM Journal on Optimization 20(6), 2876–2889 (2010)
47. El-Samad, H., Prajna, S., Papachristodoulou, A., Doyle, J., Khammash, M.: Advanced methods and algorithms for biological networks analysis. Proceedings of the IEEE 94(4), 832–853 (2006)
48. Genesio, R., Tartaglia, M., Vicino, A.: On the estimation of asymptotic stability regions: State of the art and new proposals. IEEE Trans. on Automatic Control 30, 747–755 (1985)
49. Geromel, J.C., Colaneri, P.: Robust stability of time varying polytopic systems. Systems and Control Letters 55(1), 81–85 (2006)
50. Hachicho, O., Tibken, B.: Estimating domains of attraction of a class of nonlinear dynamical systems with LMI methods based on the theory of moments. In: IEEE Conf. on Decision and Control, Las Vegas, Nevada, pp. 3150–3155 (2002)
51. Hardy, G., Littlewood, J.E., Pólya, G.: Inequalities, 2nd edn. Cambridge University Press, Cambridge (1988)
52. Hasenauer, J., Rumschinski, P., Waldherr, S., Borchers, S., Allgower, F., Findeisen, R.: Guaranteed steady-state bounds for uncertain chemical processes. In: Int. Symposium on Advanced Control of Chemical Processes (2010)
53. Henrion, D., Garulli, A. (eds.): Positive Polynomials in Control. Springer, Heidelberg (2005)
54. Henrion, D., Lasserre, J.-B.: GloptiPoly: Global optimization over polynomials with Matlab and SeDuMi. ACM Trans. on Mathematical Software 29(2), 165–194 (2003)
55. Henrion, D., Lasserre, J.-B.: Detecting global optimality and extracting solutions in GloptiPoly. In: Henrion, D., Garulli, A. (eds.) Positive Polynomials in Control, Springer, Heidelberg (2005)
56. Henrion, D., Lasserre, J.-B.: Convergent relaxations of polynomial matrix inequalities and static output feedback. IEEE Trans. on Automatic Control 51(2), 192–202 (2006)
57. Hol, C.W.J., Scherer, C.W.: Computing optimal fixed order $\mathscr{H}_\infty$-synthesis values by matrix sum of squares relaxations. In: IEEE Conf. on Decision and Control, Paradise Island, Bahamas, pp. 3147–3153 (2004)
58. Hol, C.W.J., Scherer, C.W.: A sum-of-squares approach to fixed order H_∞-synthesis. In: Henrion, D., Garulli, A. (eds.) Positive Polynomials in Control, pp. 45–71. Springer, Heidelberg (2005)
59. Hu, T., Lin, Z.: On enlarging the basin of attraction for linear systems under saturated linear feedback. Systems and Control Letters 40, 59–69 (2000)

60. Ichihara, H.: State feedback synthesis for polynomial systems with bounded disturbances. In: IEEE Conf. on Decision and Control, Cancun, Mexico, pp. 2520–2525 (2008)
61. Ichihara, H.: Optimal control for polynomial systems using matrix sum of squares relaxations. IEEE Trans. on Automatic Control 54(5), 1048–1053 (2009)
62. Jarvis-Wloszek, Z., Feeley, R., Tan, W., Sun, K., Packard, A.: Some control applications of sum of squares programming. In: IEEE Conf. on Decision and Control, Maui, Hawaii, pp. 4676–4681 (2003)
63. Jarvis-Wloszek, Z., Feeley, R., Tan, W., Sun, K., Packard, A.: Control applications of sum of squares programming. In: Henrion, D., Garulli, A. (eds.) Positive Polynomials in Control, pp. 3–22. Springer, Heidelberg (2005)
64. Kaltofen, E., Li, B., Yang, Z., Zhi, L.: Exact certification in global polynomial optimization via sums-of-squares of rational functions with rational coefficients. Journal of Symbolic Computation (2010) (to appear)
65. Khalil, H.K.: Nonlinear Systems. Prentice-Hall, Englewood Cliffs (2001)
66. Kojima, M.: Sums of squares relaxations of polynomial semidefinite programs. Technical report, Tokyo Institute of Technology (2003)
67. Kokotovic, P., Marino, R.: On vanishing stability regions in nonlinear systems with high-gain feedback. IEEE Trans. on Automatic Control 31(10), 967–970 (1986)
68. Lasserre, J.-B.: Global optimization with polynomials and the problem of moments. SIAM Journal of Optimization 11(3), 796–817 (2001)
69. Lasserre, J.-B., Netzer, T.: Sos approximations of nonnegative polynomials via simple high degree perturbations. Math. Zeitschrift 256, 99–112 (2006)
70. Laurent, M.: Sums of squares, moment matrices and optimization over polynomials. In: Putinar, M., Sullivant, S. (eds.) Emerging Applications of Algebraic Geometry. IMA Volumes in Mathematics and its Applications, vol. 149, pp. 157–270. Springer, Heidelberg (2009)
71. Levin, A.: An analytical method of estimating the domain of attraction for polynomial differential equations. IEEE Trans. on Automatic Control 39(12), 2471–2475 (1994)
72. Lofberg, J.: YALMIP: A toolbox for modeling and optimization in MATLAB. In: CACSD Conf., Taipei, Taiwan, pp. 284–289 (2004)
73. Loparo, K.A., Blankenship, G.L.: Estimating the domain of attraction of nonlinear feedback systems. IEEE Trans. on Automatic Control 23(4), 602–607 (1978)
74. Margolis, S.G., Vogt, W.G.: Control engineering applications of V. I. Zubov's construction procedure for Lyapunov functions. IEEE Trans. on Automatic Control 8, 104–113 (1963)
75. Nesterov, Y.: Squared functional systems and optimization problems. In: Frenk, H., Roos, K., Terlaky, T., Shang, S. (eds.) High Performance Optimization. Kluwer, Dordrecht (2000)
76. Nesterov, Y., Nemirovsky, A.: Interior-Point Polynomial Methods in Convex Programming. SIAM (1994)
77. Ohta, Y.: Piecewise linear estimate of attractive regions for linear systems with saturating control. In: IASTED Int. Conf. on Applied Modeling and Simulation, pp. 311–316 (2002)
78. Ohta, Y., Onishi, M.: Stability analysis by using piecewise linear lyapunov functions. In: IFAC World Congress, Beijing, China, pp. 515–520 (1999)
79. Oishi, Y.: A matrix-dilation approach to robust semidefinite programming and its error bound. In: American Control Conf., Minneapolis, MI, pp. 123–129 (2006)
80. Oishi, Y.: An asymptotically exact approach to robust semidefinite programming problems with function variables. IEEE Trans. on Automatic Control 54(5), 1000–1006 (2009)

81. Papachristodoulou, A., Prajna, S.: Analysis of non-polynomial systems using the sum of squares decomposition. In: Henrion, D., Garulli, A. (eds.) Positive Polynomials in Control. Springer, Heidelberg (2005)
82. Papachristodoulou, A., Prajna, S.: A tutorial on sum of squares techniques for system analysis. In: American Control Conf., Portland, Oregon, pp. 2686–2700 (2005)
83. Parrilo, P.A.: Structured semidefinite programs and semialgebraic geometry methods in robustness and optimization. PhD thesis, California Institute of Technology (2000)
84. Peaucelle, D., Sato, M.: LMI tests for positive definite polynomials: Slack variable approach. IEEE Trans. on Automatic Control 54(4), 886–891 (2009)
85. Peet, M., Papachristodoulou, A., Lall, S.: Positive forms and stability of linear time-delay systems. In: IEEE Conf. on Decision and Control. San Diego, CA, pp. 187–192 (2006)
86. Peterfreund, N., Baram, Y.: Convergence analysis of nonlinear dynamical systems by nested lyapunov functions. IEEE Trans. on Automatic Control 43(8), 1179–1184 (1998)
87. Powers, V., Wörmann, T.: An algorithm for sums of squares of real polynomials. Journal of Pure and Applied Linear Algebra 127, 99–104 (1998)
88. Prabhakara, F.S., El-Abiad, A.H., Koivo, A.J.: Application of generalized Zubov's method to power system stability. Int. Journal of Control 20, 203–212 (1974)
89. Prajna, S., Papachristodoulou, A.: Analysis of switched and hybrid systems - beyond piecewise quadratic methods. In: American Control Conf., Denver, Colorado (2003)
90. Prajna, S., Papachristodoulou, A., Parrilo, P.A.: SOSTOOLS: a general purpose sum of squares programming solver. In: IEEE Conf. on Decision and Control, Las Vegas, Nevada (2002)
91. Prajna, S., Papachristodoulou, A., Wu, F.: Nonlinear control synthesis by sum of squares optimization: a Lyapunov-based approach. In: Asian Control Conf. (2004)
92. Putinar, M.: Positive polynomials on compact semi-algebraic sets. Ind. Univ. Math. 42, 969–984 (1993)
93. Reznick, B.: Extremal psd forms with few terms. Duke Mathematical Journal 45(2), 363–374 (1978)
94. Reznick, B.: Some concrete aspects of Hilbert's 17th problem. Contemporary Mathematics 253, 251–272 (2000)
95. Reznick, B.: On hilbert's construction of positive polynomials (2007), submitted for publication, `http://www.math.uiuc.edu/~reznick`
96. Rodden, J.J.: Numerical applications of Lyapunov stability theory. In: Joint Automatic Control Conference, Stanford, California, pp. 261–268 (1967)
97. Rozgonyi, S., Hangos, K.M., Szederkenyi, G.: Determining the domain of attraction of hybrid non-linear systems using maximal lyapunov functions. Kybernetika 46, 19–37 (2010)
98. La Salle, J.P.: Some extensions of Lyapunov's second method. IRE Trans. on Circuits Theory 7, 520–527 (1960)
99. Sato, M., Peaucelle, D.: Robust stability/performance analysis for linear time-invariant polynomially parameter dependent systems using polynomially parameter-dependent Lyapunov functions. In: IEEE Conf. on Decision and Control, San Diego, CA, pp. 5807–5813 (2006)
100. Scheiderer, C.: Positivity and sums of squares: A guide to some recent results. In: Putinar, M., Sullivant, S. (eds.) Emerging Applications of Algebraic Geometry, IMA Volumes in Mathematics and its Applications, vol. 149, pp. 271–324. Springer, Heidelberg (2009)
101. Scherer, C.W.: LMI relaxations in robust control. European Journal of Control 12(1), 3–29 (2006)

102. Scherer, C.W., Hol, C.W.J.: Matrix sum-of-squares relaxations for robust semi-definite programs. Math. Programming Ser. B 107(1-2), 189–211 (2006)
103. Schmuedgen, K.: The K-moment problem for compact semi-algebraic sets. Mathematische Annalen 289, 203–206 (1991)
104. Shim, H., Seo, J.H.: Non-linear output feedback stabilization on a bounded region of attraction. Int. Journal of Control 73(5), 416–426 (2000)
105. Shor, N.Z.: Class of global minimum bounds of polynomial functions (in russian). Cybernetics 23, 731–734 (1987)
106. Shor, N.Z.: Nondifferentiable optimization and polynomial problems. Kluwer Academic Publishers, Dordrecht (1998)
107. De Souza, C.E., Coutinho, D.F., Fu, M.: Stability analysis of finite-level quantized linear control systems. In: European Control Conference, Budapest, Hungary, pp. 79–84 (2009)
108. Stengle, G.: A nullstellensatz and a positivstellensatz in semialgebraic geometry. Math. Ann. 207, 87–97 (1974)
109. Sturm, J.F.: Using SeDuMi 1.02, a MATLAB toolbox for optimization over symmetric cones. Optimization Methods and Software 11-12, 625–653 (1999)
110. Tan, W., Packard, A.: Stability region analysis using polynomial and composite polynomial Lyapunov functions and sum-of-squares programming. IEEE Trans. on Automatic Control 53(2), 565–571 (2008)
111. Tesi, A., Villoresi, F., Genesio, R.: On the stability domain estimation via a quadratic Lyapunov function: convexity and optimality properties for polynomial systems. IEEE Trans. on Automatic Control 41(11), 1650–1657 (1996)
112. Tibken, B.: Estimation of the domain of attraction for polynomial systems via LMI's. In: IEEE Conf. on Decision and Control, Sydney, Australia, pp. 3860–3864 (2000)
113. Tibken, B., Hachicho, O.: Estimation of the domain of attraction for polynomial systems using multidimensional grids. In: IEEE Conf. on Decision and Control, Sydney, Australia, pp. 3870–3874 (2000)
114. Topcu, U., Packard, A.: Local stability analysis for uncertain nonlinear systems. IEEE Trans. on Automatic Control 54(5), 1042–1047 (2009)
115. Topcu, U., Packard, A., Seiler, P.: Local stability analysis using simulations and sum-of-squares programming. Automatica 44(10), 2669–2675 (2008)
116. Topcu, U., Packard, A., Seiler, P., Balas, G.: Stability region estimation for systems with unmodeled dynamics. In: European Control Conference, Budapest, Hungary, pp. 85–90 (2009)
117. Trofino, A.: Robust stability and domain of attraction of uncertain nonlinear systems. In: American Control Conf., Chicago, Illinois, pp. 3707–3711 (2000)
118. Vandenberghe, L., Boyd, S.: Semidefinite programming. SIAM Review 38, 49–95 (1996)
119. Vannelli, A., Vidyasagar, M.: Maximal Lyapunov functions and domains of attraction for autonomous nonlinear systems. Automatica 21(1), 69–80 (1985)
120. Zubov, V.I.: The methods of Lyapunov and their applications. Leningrad University Press (1957)

Author Biography

Graziano Chesi received the Laurea in Information Engineering from the University of Florence, in 1997, and the Ph.D. in Systems Engineering from the University of Bologna, in 2001. From 2000 to 2006 he was with the Dipartimento di Ingegneria dell'Informazione of the University of Siena, and in 2006 he joined the Department of Electrical and Electronic Engineering of the University of Hong Kong. He was a visiting scientist at the Department of Engineering of the University of Cambridge from 1999 to 2000, and at the Department of Information Physics and Computing of the University of Tokyo from 2001 to 2004. He was the recipient of the Best Student Award of the Faculty of Engineering of the University of Florence in 1997.

Dr. Chesi has served as Associate Editor of *Automatica*, *BMC Research Notes*, the *European Journal of Control*, the *IEEE Transactions on Automatic Control*, and *Systems and Control Letters*. Also, he has served as Guest Editor of the Special Issue on *Positive Polynomials in Control* of the *IEEE Transactions on Automatic Control*, *Systems Biology* of the *International Journal of Robust and Nonlinear Control*, and *Visual Servoing* of *Mechatronics*. He is the Founder and Chair of the Technical Committee on Systems with Uncertainty of the IEEE Control Systems Society.

He is author of the book "Homogeneous Polynomial Forms for Robustness Analysis of Uncertain Systems" (Springer, 2009), editor of the book "Visual Servoing via Advanced Numerical Methods" (Springer, 2010), and author of the book "Domain of Attraction: Analysis and Control via SOS Programming" (Springer, 2011). He is the first author of more than 100 technical publications. His research interests include computer vision, nonlinear systems, optimization, robotics, robust control, and systems biology.

Dr. Graziano Chesi
Department of Electrical and Electronic Engineering
University of Hong Kong
Pokfulam Road
Hong Kong
Phone: +852-22194362, Fax: +852-25598738, Email: chesi@eee.hku.hk
Homepage: http://www.eee.hku.hk/~chesi

Author Biography

Index

Lecture Notes in Control and Information Sciences

Edited by M. Thoma, F. Allgöwer, M. Morari

Further volumes of this series can be found on our homepage:
springer.com

Vol. 415: Chesi, G.:
Domain of Attraction
283 p. 2011 [978-0-85729-958-1]

Vol. 414: Imine, H.; Fridman, L.; Shraim, H.; Djemai, M.:
Sliding Mode Based Analysis and Identification of Vehicle Dynamics
127 p. 2011 [978-3-642-22223-8]

Vol. 413: Eleftheriou, E.; Reza Moheimani, S.O.:
Control Technologies for Emerging Micro and Nanoscale Systems
289 p. 2011 [978-3-642-22172-9]

Vol. 412: Fridman, L.; Moreno, J.; Iriarte, R.:
Sliding Modes after the first Decade of the 21st Century
XXX p. 2011 [978-3-642-22163-7]

Vol. 411: Kaczorek, T.;
Selected Problems of Fractional Systems Theory
344 p. 2011 [978-3-642-20501-9]

Vol. 410: Bourlès, H., Marinescu, B.:
Linear Time-Varying Systems
637 p. 2011 [978-3-642-19726-0]

Vol. 409: Xia, Y., Fu, M., Liu, G.-P.:
Analysis and Synthesis of Networked Control Systems
198 p. 2011 [978-3-642-17924-2]

Vol. 408: Richter, J.H.;
Reconfigurable Control of Nonlinear Dynamical Systems
291 p. 2011 [978-3-642-17627-2]

Vol. 407: Lévine, J., Müllhaupt, P.:
Advances in the Theory of Control, Signals and Systems with Physical Modeling
380 p. 2010 [978-3-642-16134-6]

Vol. 406: Bemporad, A., Heemels, M., Johansson, M.:
Networked Control Systems
appro. 371 p. 2010 [978-0-85729-032-8]

Vol. 405: Stefanovic, M., Safonov, M.G.:
Safe Adaptive Control
appro. 153 p. 2010 [978-1-84996-452-4]

Vol. 404: Giri, F.; Bai, E.-W. (Eds.):
Block-oriented Nonlinear System Identification
425 p. 2010 [978-1-84996-512-5]

Vol. 403: Tóth, R.;
Modeling and Identification of Linear Parameter-Varying Systems
319 p. 2010 [978-3-642-13811-9]

Vol. 402: del Re, L.; Allgöwer, F.; Glielmo, L.; Guardiola, C.; Kolmanovsky, I. (Eds.):
Automotive Model Predictive Control
284 p. 2010 [978-1-84996-070-0]

Vol. 401: Chesi, G.; Hashimoto, K. (Eds.):
Visual Servoing via Advanced Numerical Methods
393 p. 2010 [978-1-84996-088-5]

Vol. 400: Tomás-Rodríguez, M.; Banks, S.P.:
Linear, Time-varying Approximations to Nonlinear Dynamical Systems
298 p. 2010 [978-1-84996-100-4]

Vol. 399: Edwards, C.; Lombaerts, T.; Smaili, H. (Eds.):
Fault Tolerant Flight Control
appro. 350 p. 2010 [978-3-642-11689-6]

Vol. 398: Hara, S.; Ohta, Y.; Willems, J.C.; Hisaya, F. (Eds.):
Perspectives in Mathematical System Theory, Control, and Signal Processing
appro. 370 p. 2010 [978-3-540-93917-7]

Vol. 397: Yang, H.; Jiang, B.; Cocquempot, V.:
Fault Tolerant Control Design for Hybrid Systems
191 p. 2010 [978-3-642-10680-4]

Vol. 396: Kozlowski, K. (Ed.):
Robot Motion and Control 2009
475 p. 2009 [978-1-84882-984-8]

Vol. 395: Talebi, H.A.; Abdollahi, F.; Patel, R.V.; Khorasani, K.:
Neural Network-Based State Estimation of Nonlinear Systems
appro. 175 p. 2010 [978-1-4419-1437-8]

Vol. 394: Pipeleers, G.; Demeulenaere, B.; Swevers, J.:
Optimal Linear Controller Design for Periodic Inputs
177 p. 2009 [978-1-84882-974-9]

Vol. 393: Ghosh, B.K.; Martin, C.F.; Zhou, Y.:
Emergent Problems in Nonlinear Systems and Control
285 p. 2009 [978-3-642-03626-2]

Vol. 392: Bandyopadhyay, B.; Deepak, F.; Kim, K.-S.:
Sliding Mode Control Using Novel Sliding Surfaces
137 p. 2009 [978-3-642-03447-3]

Vol. 391: Khaki-Sedigh, A.; Moaveni, B.:
Control Configuration Selection for Multivariable Plants
232 p. 2009 [978-3-642-03192-2]

Vol. 390: Chesi, G.; Garulli, A.; Tesi, A.; Vicino, A.:
Homogeneous Polynomial Forms for Robustness Analysis of Uncertain Systems
197 p. 2009 [978-1-84882-780-6]

Vol. 389: Bru, R.; Romero-Vivó, S. (Eds.):
Positive Systems
398 p. 2009 [978-3-642-02893-9]

Vol. 388: Jacques Loiseau, J.; Michiels, W.; Niculescu, S-I.; Sipahi, R. (Eds.):
Topics in Time Delay Systems
418 p. 2009 [978-3-642-02896-0]

Vol. 387: Xia, Y.; Fu, M.; Shi, P.:
Analysis and Synthesis of Dynamical Systems with Time-Delays
283 p. 2009 [978-3-642-02695-9]

Vol. 386: Huang, D.; Nguang, S.K.:
Robust Control for Uncertain Networked Control Systems with Random Delays
159 p. 2009 [978-1-84882-677-9]

Vol. 385: Jungers, R.:
The Joint Spectral Radius
144 p. 2009 [978-3-540-95979-3]

Vol. 384: Magni, L.; Raimondo, D.M.; Allgöwer, F. (Eds.):
Nonlinear Model Predictive Control
572 p. 2009 [978-3-642-01093-4]

Vol. 383: Sobhani-Tehrani E.; Khorasani K.;
Fault Diagnosis of Nonlinear Systems Using a Hybrid Approach
360 p. 2009 [978-0-387-92906-4]

Vol. 382: Bartoszewicz A.; Nowacka-Leverton A.;
Time-Varying Sliding Modes for Second and Third Order Systems
192 p. 2009 [978-3-540-92216-2]

Vol. 381: Hirsch M.J.; Commander C.W.; Pardalos P.M.; Murphey R. (Eds.)
Optimization and Cooperative Control Strategies: Proceedings of the 8th International Conference on Cooperative Control and Optimization
459 p. 2009 [978-3-540-88062-2]

Vol. 380: Basin M.
New Trends in Optimal Filtering and Control for Polynomial and Time-Delay Systems
206 p. 2008 [978-3-540-70802-5]

Vol. 379: Mellodge P.; Kachroo P.;
Model Abstraction in Dynamical Systems: Application to Mobile Robot Control
116 p. 2008 [978-3-540-70792-9]

Vol. 378: Femat R.; Solis-Perales G.;
Robust Synchronization of Chaotic Systems Via Feedback
199 p. 2008 [978-3-540-69306-2]

Vol. 377: Patan K.
Artificial Neural Networks for the Modelling and Fault Diagnosis of Technical Processes
206 p. 2008 [978-3-540-79871-2]

Vol. 376: Hasegawa Y.
Approximate and Noisy Realization of Discrete-Time Dynamical Systems
245 p. 2008 [978-3-540-79433-2]

Vol. 375: Bartolini G.; Fridman L.; Pisano A.; Usai E. (Eds.)
Modern Sliding Mode Control Theory
465 p. 2008 [978-3-540-79015-0]

Vol. 374: Huang B.; Kadali R.
Dynamic Modeling, Predictive Control and Performance Monitoring
240 p. 2008 [978-1-84800-232-6]

Vol. 373: Wang Q.-G.; Ye Z.; Cai W.-J.; Hang C.-C.
PID Control for Multivariable Processes
264 p. 2008 [978-3-540-78481-4]

Vol. 372: Zhou J.; Wen C.
Adaptive Backstepping Control of Uncertain Systems
241 p. 2008 [978-3-540-77806-6]

Vol. 371: Blondel V.D.; Boyd S.P.; Kimura H. (Eds.)
Recent Advances in Learning and Control
279 p. 2008 [978-1-84800-154-1]

Batch number: 10620143

Printed by Printforce, the Netherlands